Werner Blum
Peter L. Galbraith
Hans-Wolfgang Henn
Mogens Niss
(Editors)

Modelling and Applications in Mathematics Education

The 14th ICMI Study

Springer

Werner Blum
University of Kassel/ Germany

Hans-Wolfgang Henn
University of Dortmund/ Germany

Peter L. Galbraith
The University of Queensland/ Australia

Mogens Niss
Roskilde University/ Denmark

Additional material to this book can be downloaded from http://extras.springer.com

ISBN -13: 978-1-4899-9249-9
ISBN -10: 1-4899-9249-9

ISBN-13: 978-0-387-29822-1 (eBook)
ISBN-10: 0-387-29822-3 (eBook)

Printed on acid-free paper.

9 8 7 6 5 4 3 2 1

springer.com

Modelling and Applications in Mathematics Education

New ICMI Study Series

VOLUME 10

Published under the auspices of the International Commission on Mathematical Instruction under the general editorship of

Michèle Artigue, President   Bernard R. Hodgson, Secretary-General

The titles published in this series are listed at the end of this volume.

TABLE OF CONTENTS

PREFACE

Among the themes that have been central to mathematics education during the last 30 years are those of mathematical *modelling* and *applications* of mathematics to extra-mathematical fields. More generally we refer to these as *relations between mathematics and the extra-mathematical world* (sometimes also called the "real world") or preferably, according to Henry Pollak, the "rest of the world". That applications and modelling have been important themes in mathematics education can be inferred from the wealth of literature on these topics, including material generated from a multitude of national and international *conferences*. In particular let us mention firstly the ICMEs (the International Congresses on Mathematical Education), with their regular working or topic groups and lectures on applications and modelling; and secondly the series of ICTMAs (the International Conferences on the Teaching of Mathematical Modelling and Applications) which have been held biennially since 1983. Their Proceedings and Survey Lectures, have addressed the state-of-the-art at the relevant time, and contain many examples, studies, conceptual contributions and resources involving relations between the real world and mathematics, for all levels of the educational system. In *curricula and textbooks* we find today many more references to real world phenomena and problems than, say, twenty years ago. Yet while applications and modelling play more important roles in many countries' classrooms than in the past, there still exists a substantial gap between the ideals expressed in educational debate and innovative curricula on the one hand, and everyday teaching practice on the other. In particular, genuine modelling activities are still rather rare in mathematics classrooms.

Altogether, during the last few decades there has been considerable work in mathematics education that has centred on applications and modelling. Many activities have had a primary focus on practice, e.g. construction and trial of mathematical modelling examples for teaching and examination purposes, writing of application-oriented textbooks, implementation of applications and modelling in existing curricula, or development of innovative, modelling-oriented curricula. Several of these activities also contain research components such as: clarification of relevant concepts; investigation of competencies and identification of difficulties and strategies activated by students when dealing with application problems; observation and analysis of teaching; study of learning and communication processes in modelling-oriented lessons; and evaluation of alternative approaches used to assess performance in applications and modelling. In particular during the last ten years the number of genuine research contributions has increased considerably.

That applications and modelling have been, and continue to be, central themes in mathematics education is not at all surprising. Nearly all questions and problems in mathematics education, that is questions and problems concerning human learning and the teaching of mathematics, influence and are influenced by relations between mathematics and some aspects of the real world. For instance, one essential answer (albeit not the only one) to the question as to why all persons ought to learn mathematics is that it provides a means for understanding the world around us, for coping with everyday problems, or for preparing for future professions. When addressing the question of how individuals acquire mathematical knowledge, we cannot avoid the role of its relationship to reality, especially the relevance of situated learning (including the problem of the dependence of learning on specific contexts). General questions as to what "mathematics" is, as a part of our culture and as a social phenomenon, of how mathematics has emerged and developed, involve also "applications" of mathematics in other disciplines, in nature and society. Today mathematical models and modelling have penetrated a great variety of disciplines, leaving only a few fields (if any) where mathematical models do not play some role. This increasing involvement has been substantially supported and accelerated by the availability of powerful electronic tools, such as calculators and computers, with their enormous communication capabilities.

Relations between the real world and mathematics are particularly relevant within the current OECD (Organisation for Economic Co-operation and Development) PISA project. What is being tested in PISA (Programme for International Student Assessment), is mathematical literacy, that is, according to the PISA framework, "an individual's capacity to identify and understand the role that mathematics plays in the world, to make well-founded judgements and to use and engage in mathematics, in ways that meet the needs of that individual's life as a constructive, concerned, and reflective citizen." That means the emphasis in PISA is on the use of mathematical knowledge in a multitude of situations and contexts. In several countries, this project has initiated an intense discussion about aims and design of mathematics instruction in schools, and especially about the role of mathematical modelling, applications of mathematics and relations to the real world. Such deliberations are also occurring in countries outside the OECD.

This book is the Study Volume of ICMI Study 14 on "Applications and Modelling in Mathematics Education", which began effectively in 2002 with the development of the Discussion Document by the Programme Committee (published in Educational Studies in Mathematics 51(2002)1/2, pp 149-171). In mounting this Study, ICMI has taken into account the reasons mentioned above for the importance of relationships between mathematics and the real world, as well as the contemporary state of the educational debate, and of research and development in this field. This does not, of course, mean that we claim to know satisfactory answers to the essential questions in this area, and that the role of the Study is simply to provide a forum for putting these together. Rather, an important aim of the Study and this Volume has been to identify shortcomings, as well as to stimulate further research and development activities, in addition to reporting on existing research and practice.

Documenting the state-of-the-art in a field and identifying deficiencies and needed research requires a structuring framework. This is particularly important in an area which is as complex and difficult to survey as the teaching and learning of mathematical modelling and applications. As we have seen, this topic not only deals with most of the essential aspects of the teaching and learning of mathematics at large, it also touches upon a wide variety of versions of the real world outside mathematics that one seeks to model. Perceived in this way, the topic of applications and modelling may appear to encompass all of mathematics education plus much more. It is evident, therefore, that we need a way of conceptualising the field so as to reduce complexity to a meaningful and tractable level. That is why this Volume commences with an introductory Part I where we clarify some of the basic concepts and notions of the field, and offer a conceptualisation that helps to structure it and to identify important challenges and questions. This introductory part, at the same time, provides a concise access to the field for the uninitiated reader together with a brief sketch of its history.

Following from this introductory part, the Volume contains plenary papers given at the Study Conference (Dortmund, February 2004) and various papers that address important issues in the field. It is stressed, however, that this Study Volume is not simply the Proceedings of the Study Conference – rather, the production of this Volume has involved an independent process. Of course, the papers presented at the Study Conference provided a rich source for this Volume, and the majority of papers here were derived in some way from those Conference papers. However, many of the papers in this Volume have been produced independently of the Study Conference, in particular to fill gaps that became obvious during the Conference.

We would like to express our sincere thanks to the members of the Programme Committee for this Study who have contributed in various ways to producing this Volume. In particular, several members have acted as editors of Sections in this Volume. Without their work and devotion, this extensive Volume could not have been completed. Our thanks go equally to all the authors who have contributed to this Volume and thus helped to make it – so we hope – a rich source of information and inspiration for readers. We also thank ICMI very much for having given priority to this Study, and in particular its Secretary, Bernard Hodgson, for his sensitive way of channelling ICMI views and proposals into this Study while, at the same time, leaving the organisers and editors with all the freedom they wanted and needed to undertake this task. Eventually, we would like to thank the Publisher, Springer, also for their patience when the completion of this Volume was on their agenda.

Let us finish this Preface by expressing our hope that this ICMI Study 14 Volume will be of value both for mathematics educators, mathematics teachers and mathematicians as well as for interested professionals in other disciplines in which mathematics plays an essential role, and that it will contribute to a strengthening and further development of the field of applications and modelling in mathematics education, and to an intensification of various kinds of research and practice activities in the field.

Werner Blum, Kassel (Germany)
Peter Galbraith, Brisbane (Australia)
Hans-Wolfgang Henn, Dortmund (Germany)
Mogens Niss, Roskilde (Denmark)

Part 1

INTRODUCTION

Part 1

INTRODUCTION

Mogens Niss, Werner Blum and Peter Galbraith
Roskilde University, Denmark, Email: mn@mmf.ruc.dk /
University of Kassel, Germany, Email: blum@mathematik.uni-kassel.de/
University of Queensland, Australia, Email: p.galbraith@uq.edu.au

Abstract: In this part of the volume, we shall give an introduction both to the field of applications and modelling in mathematics education and to the present volume. In section 1, we present the field of applications and modelling to the mathematics educator who is not a specialist in the field. In Section 2, we explain the basic terms, notions and distinctions in applications and modelling. On this basis, we provide, in Section 3, the conceptualisation of the field adopted in this ICMI Study. This conceptualisation is centred on a number of issues which will be the subject of Section 4. In Section 5, we briefly outline the historical development of applications and modelling in mathematics education. Finally, in Section 6, the structure and organisation of the present book will be described and explained.

1. INTRODUCTION FOR THE UN-INITIATED READER

Our endeavour in this section is to briefly present the field of applications and mathematical modelling in mathematics education to interested mathematics educators who are not specialists in the field.

For the remainder of this section, we need a first terse definition of the basic concepts involved. An *application* of mathematics occurs every time mathematics is *applied*, for some purpose, to deal with some domain of the *extra-mathematical world*, for instance in order to understand it better, to investigate issues, to explain phenomena, to solve problems, to pave the way for decisions, etc. The extra-mathematical world can be another subject or discipline, an area of practice, a sphere of private or social life, etc. The term "real world" is often used to describe the world outside mathematics, even though, say, quantum physics or orbitals in chemistry may appear less than real to some. The extra-mathematical world is then a helpful way of indicating that part of the wider "real world" that is relevant to a particular issue or problem. In any application of mathematics a mathematical *model* is involved, explicitly or implicitly. A mathematical model consists of the extra-mathematical domain, D, of interest, some mathematical domain M, and a mapping from the extra-mathematical to the mathematical domain (see Fig. 1-1). Objects, relations, phenomena, assumptions, questions, etc. in D are identified and selected as relevant for the purpose and situation and are then mapped – translated – into objects, relations, phenomena, assumptions, questions, etc. pertaining to M. Within M, mathematical deliberations, manipulations and inferences are made, the outcomes of which are then translated back to D and interpreted as conclusions concerning that domain. This so-called *modelling cycle* may be iterated several times, on the basis of validation and evaluation of the model in relation to the domain, until the resulting conclusions concerning D are satisfactory in relation to the purpose of the model construction. The term *modelling* refers to the entire process, and everything involved in it – from structuring D, to deciding upon a suitable mathematical domain M and a suitable mapping from D to M, to working mathematically within M, to interpreting and evaluating conclusions with regard to D, and to repeating the cycle several times if needed or desirable.

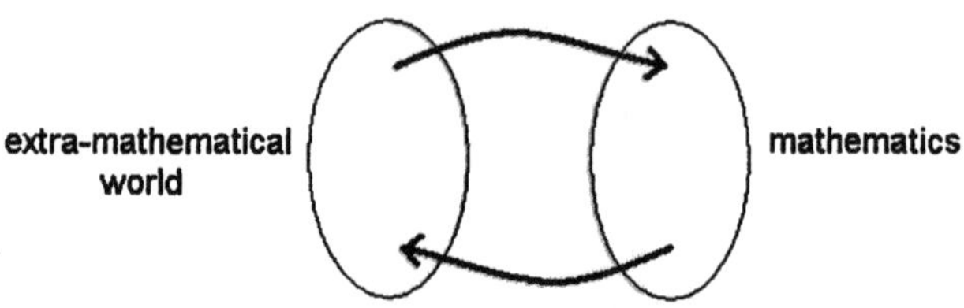

Figure 1-1. Mathematics and the rest of the world

At various times in the history of mathematics teaching and learning, it has been debated whether some forms of mathematical applications (or, in more recent times, modelling) should have a place in different sorts of mathematics curricula, or whether the extra-mathematical utilisation of mathematics should be the responsibility of those subjects that utilise the applications and the modelling. Sometimes – or in some places – curricula have focused on pure mathematics while leaving applications and modelling (if relevant) to other subjects. At other times – or in other places – curricula have made explicit room for applications and modelling. The very fact that there are, from time to time, such debates about the possible place and role of applications and modelling in the teaching and learning of mathematics suggests that there are issues to consider and think about. If we suppose that it has been agreed that applications and modelling should play some part in the mathematical education of a given category of students, hosts of further questions arise that are to do with "why?", "under what circumstances and conditions?", "what?", "when?", "how?", "taught by whom?", and so on. All of these questions need to be dealt with by research as well as by practice.

When it comes to the question of "why?" there exists a fundamental *duality* (not to be mistaken for a dichotomy) between the categories of possible answers. The first category focuses on "applications and modelling *for the learning of mathematics*", i.e. on the actual or potential ways in which applications and modelling may be a vehicle for facilitation and support of students' learning of mathematics as a subject. The other category focuses on "learning mathematics so as to *develop competency in applying mathematics and building mathematical models*" for areas and purposes that are basically extra-mathematical. We are dealing with a duality because the relationship between mathematical learning, and applications and modelling, has two different orientations, depending on which is the goal and which is the means. This duality plays out very differently at different educational levels, and for different types of curricula.

At the primary and lower secondary levels the duality is only seldom made explicit, as it is quite customary at these levels to insist on both orientations simultaneously, recognising that they are intrinsically intertwined. A major reason why we teach mathematics to typical students at primary or lower secondary level, is that they should become able to use mathematics in a variety of contexts and situations outside the classroom. This implies that applications and modelling should be on the agenda of teaching and learning, without necessarily using these terms. On the other hand, for most primary or lower secondary students it is difficult to motivate or learn mathematical concepts, methods, techniques, terminology, and results and to engage in mathematical activity, unless clear reference is being established to the use and relevance of mathematics to extra-mathematical contexts and situations, which are often also responsible for creating meaning and sense-making with regard to the mathematical entities at issue.

Also at further educational levels where mathematics is being taught for vocational or professional purposes, that are closely related to other areas or subjects, for instance in, say, carpentry, plumbing, banking, economics or engineering, the duality is sometimes deliberately kept implicit so as to blur what is the goal and what is the means of mathematics teaching. However, even though implicit, the duality is there nevertheless, provided of course, there is such a thing as "mathematics" explicitly mentioned in the curriculum.

In contrast, at upper secondary or tertiary level the duality between "applications and modelling for the learning of mathematics" and "learning mathematics for applications and modelling" is indeed, often, a significant one, worth further exploration.

Let us take a closer look at the two poles of this duality. Firstly, the "applications and modelling for the learning of mathematics" pole is to do with (a) demonstrating to students that mathematics is actually being used by people outside the mathematics classroom for a variety of reasons and purposes, thus helping to generate a richer image of the nature and role of mathematics; (b) helping to provide meaning and interpretation to mathematical entities and activities; and (c) – partly as a consequence of (a) and (b) – providing motivation for students to engage in the study of mathematics by helping to shape their beliefs and attitudes towards it. The second pole, "learning mathematics for applications and modelling", focuses (a) on one goal of the teaching and learning of mathematics, namely to equip students with the capability to bring mathematics to bear outside itself; and (b) on the fact that the extraneous use of mathematics is always brought about through mathematical models and modelling.

In principle one might think that even if an ultimate goal is to foster applications and modelling capabilities with students, this would not necessarily require applications and modelling to be dealt with in the classroom. In fact, from time to time, there has been a tendency amongst mathematics educators, including teachers, to assume that once someone has learnt theoretical mathematics in a proper and efficient way and to a satisfactory extent, that individual will be able to apply mathematics in other areas and contexts without further teaching. Hence there is, according to this view, no reason to spend precious time in the study of mathematics on dealing with applications and modelling. Moreover, the reasoning goes, that as the extraneous use of mathematics involves by definition, non-mathematical objects, phenomena, features, and facts, the mathematics classroom is not the right place to deal with such matters, particularly since the mathematics teacher more often than not would be an amateur in dealing with them, and hence by and large be unqualified for the task. Instead, to the extent that mathematics is relevant in other areas and subjects, the teachers and professionals in those areas and subjects, will master the mathematics involved well enough for them to be in charge of the applications and modelling themselves.

However, there is, today, ample evidence from practice and research that there is *no automatic transfer* from having learnt purely theoretical mathematics to being able to use it in situations that have not already been fully mathematised. Moreover, even if mathematics is being activated within other areas or subjects there is evidence that those aspects of models and modelling that are to do with the relationships between the mathematical representations and the domain of application, including validation of model assumptions and results, are not taken seriously. This suggests that if we want students to develop applications and modelling competency as one outcome of their mathematical education, applications and modelling have to be explicitly put on the agenda of the teaching and learning of mathematics.

For this to be possible teachers must be able to orchestrate teaching environments, situations and activities that can foster the development of applications and modelling competency, in many educational contexts alongside the development of other mathematical competencies. This gives rise to hosts of issues and decisions concerning allocation of time, design and planning of the settings for teaching and learning, choice of activities and materials, design and implementation of assessment instruments, etc., in addition to striking a balance between applications and modelling work and other kinds of mathematical work, perhaps with other foci. In that context, although there are intimate relationships between applications and modelling and mathematics at large, including its theoretical aspects, mathematical competence and mastery does indeed contain many other competencies in addition to applications and modelling competency. Furthermore, the successful development of applications and modelling competency presupposes that other mathematical competencies are also present, if only for the simple reason that work within the mathematical domain of a model cannot take place without such competencies.

In the same way as students do not become able to apply mathematics and to analyse and construct mathematical models as an automatic result of having learnt purely theoretical mathematics, teachers do not become able to orchestrate environments, situations and activities for applications and modelling as an automatic result of having been trained as mathematicians or mathematics teachers in traditional ways that focus entirely on purely mathematical subject matter. And if we want teachers of mathematics to become able to place applications and modelling on the agenda of their teaching in efficient, successful, and reflective ways, they need opportunities to develop that capacity during their pre-service education and through regular in-service activities of professional development.

In the hope that this section has provided a helpful first orientation to the non-specialist reader, we now go on to describe how the following sections of this introductory part of the Study Volume are designed to present a systematic development of the field. In the following Section 2 basic terms, notions, and distinctions – in continuation of the terms defined in the first paragraph of this section – are introduced and elaborated. Then follows Section 3 which introduces the conceptualisation of the field adopted for this ICMI Study, reflected both in the initiating Discussion Document produced by the International Programme Committee for this Study (Blum et. al., 2002) and in the present Volume. In Section 4, special attention is being paid to selected issues resulting from the way in which the field has been conceptualised by the Programme Committee. A brief historical sketch of the field, applications and modelling in mathematics education, is given in Section 5. Finally, Part 1 is concluded in Section 6, by a description and explanation of the structure of this Volume.

2. APPLICATIONS AND MODELLING – NOTIONS

Mathematical modelling and applications in education, by definition encompass elements from both the mathematical and educational domains. As a result some common notions in the field have different shades of meaning from those with similar names found elsewhere in the education community. We believe it is important to clarify important distinctions.

2.1 Models and Modelling

Firstly we distinguish our use of the terms *model* and *modelling* from other usages found in general education. We read frequently of model teachers, of modelling good teaching practice, and of modelling student understanding or classroom interaction, but these interpretations are not what we have in mind in using the terms. Similarly, models as physical objects (e.g. plaster models of geometrical solids or surfaces), mental models as used in a variety of learning contexts, and models as instantiations (e.g. of axiomatic geometry systems) illustrate usages that lie outside our particular field of activity.

In continuation of our outline of a definition in the first paragraph of the previous section, let us look a little more closely at the purpose of constructing mathematical models. The generic purpose of building and making use of a model is to understand or tackle problems in some segment of the real world. Here we use the term *problem* in a broad sense, encompassing not only practical problems, but also problems of a more intellectual nature that aim at describing, explaining, understanding or even designing parts of the world, including issues and questions pertaining to scientific disciplines. Dealing with such problems requires individuals to build, test, and apply mathematical models designed to answer questions of importance in real world settings. The world we live in contains tangible objects, tools, artefacts, and structures, both natural and humanly built. It is also a place of intangibles: ideas, expectations, values and power relationships. By *real world* we mean everything that is to do with nature, society or culture, including everyday life, as well as school and university subjects or scientific and scholarly disciplines different from mathematics. The extra-mathematical domain of Section 1, relevant to a particular modelling enterprise, will involve a subset of this real world.

The modelling perspective begins with the conceptualisation of some problem situation. Simplifying, structuring, and making this situation more precise – according to the problem solver's knowledge, goals, and interests – leads to the specification of a problem in terms of the language and concepts of the situation. Some of the problems we address through mathematical modelling tend to be of a practical nature: How do we optimise a particular design? What is the best route for a new freeway? Which borrowing options are the cheapest for a given purpose? Such problems also involve intangibles. What does best route mean? Cheapest? Most direct? Least disruptive to communities? Other problems are of a scientific nature, like: Are we able to identify mechanisms that may be responsible for observed variations in predator-prey populations?

If appropriate, real data are collected to provide more information on the situation of interest. These data frequently suggest the type of mathematical model that is appropriate to address the specified real-world problem. Through a process of mathematisation, the relevant objects, data, relations, conditions, and assumptions from the extra-mathematical domain are then translated into mathematics, resulting in a mathematical model through which to address the identified problem.

Now mathematical methods are used to derive mathematical results, relevant to questions arising from the translation of the real world problem. Such methods include logical deduction from mathematical assumptions, utilisation of theoretical results within mathematical topics, solving equations, performing symbolic manipulation or numerical computations, estimating parameters, performing statistical testing, simulation etc.

The ensuing mathematical results must then be translated back into the extra-mathematical domain within which the original problem was located – that is *interpreted* in relation to the original real world problem context. The problem solver then *validates* the model by checking whether interpreted mathematical outcomes are reasonable and compatible in terms of the information given in the original problem. At the same time the model is *evaluated* by checking whether the solution is appropriate and useful for its purposes. When one or both of these 'tests' is deemed unsatisfactory, the whole process needs to be repeated using a modified or a totally different model. Finally (if achieved) the solution of the original real world problem is stated, and where relevant, communicated to others.

While the sub-process leading from a real world problem situation to a mathematical model is sometimes called *mathematical modelling,* it has become customary (as indicated in Section 1) to use that notion also for the entire process consisting of structuring, generating real world facts and data, mathematising, working mathematically and interpreting/validating (perhaps several times round the loop) as just described. This latter position has been adopted for the present Volume.

We hold that the distinction between the *modelling process* and *models* is a particularly important one. In the course of the modelling process, as depicted above, one or more *mathematical models* may be produced and are thus integral parts of the greater whole. Sometimes a model will be an idiosyncratic, ad hoc construction, but often it will be a variation of a standard type (e.g. inverse proportionality, linear, exponential or logistic growth, the harmonic oscillator, a Poisson probability process, etc.). It follows that the study of *standard models* is important in providing a wide-ranging toolkit to enhance the options available to solvers, but more often than not a simple application of such models is not capable of capturing all the significant features of the problem to be modelled.

Different kinds of models (e.g. deterministic or stochastic models) can at times be developed using different formulations for the same problem and displaying different properties and qualities. Finally models can differ in their level of sophistication, and yet in different ways enhance understanding of a problem to a certain degree. For example simple modelling with arithmetic can provide useful insights at one level, for a problem whose complete solution may require sophisticated algebra or calculus. This feature is a major reason why modelling can begin with integrity in the elementary school.

2.2 Applications versus Modelling

Using mathematics to solve real world problems, in the broad sense adopted here, is often called *applying* mathematics, and a real world problem which has been addressed by means of mathematics is called an *application* of mathematics. Sometimes, though, the notions of "applying" or "application" are used for any kind of linking of the real world and mathematics.

During the last one or two decades the term "*applications and modelling*" has been increasingly used to denote all kinds of relationships whatsoever between the real world and mathematics. The term "modelling", on the one hand, tends to focus on the direction "reality → mathematics" and, on the other hand and more generally, emphasises the *processes* involved. Simply put, with *modelling* we are standing outside mathematics looking in: "Where can I find some mathematics to help me with this problem?" In contrast, the term "application", on the one hand, tends to focus on the opposite direction "mathematics → reality" and, more generally, emphasises the *objects* involved – in particular those parts of the real world which are (made) accessible to a mathematical treatment and to which corresponding mathematical models already exist. Again simply put, with *applications* we are standing inside mathematics looking out: "Where can I use this particular piece of mathematical knowledge?" It is in the comprehensive sense outlined here that we understand the term "applications and modelling" as used in the title of this Study.

2.3 Applied Mathematics

Applied Mathematics is a time-honoured descriptor for undergraduate or upper secondary courses (mainly in the Anglo-Saxon countries) that focus on applications of mathematics in fields such as solid and fluid mechanics. Typically they involve advanced topics (e.g. rigid dynamics and partial differential equations) and theory is developed and applied within the context of well-posed problems. Where they differ from the major emphasis in this volume is in starting from the point where a problem has already been formulated within a theoretical framework from physics. So the emphasis is in the application of (often) advanced designated mathematical techniques, rather than engaging in the complete modelling cycle, wherein the mathematics of relevance often needs to be first identified.

2.4 Applied Problem Solving

Applied Problem Solving is a term that admits a variety of interpretations and emphases. Sometimes it is used to denote the processes that are involved when a real world problem has to be solved. In this sense, it is only another term for modelling, where its use emphasises the *strategic* elements that are necessarily involved in the solution process. However, often the term is used for problem solving activities with any kind of extra-mathematical context whatsoever, including artificial or play-like contexts with only inessential references to extra-mathematical objects or phenomena. When we use this term, we always mean it in the narrower sense first mentioned.

2.5 Modelling and Application Problems

When it comes to curriculum materials, a variety of problem types appear under such headings that are variable in the degree to which they meet or attempt to genuinely meet real-world criteria. It is worth canvassing a few of the most common types.

Word Problems: These have been with us for centuries, and because they are couched in verbal terms, are often presented as applications of mathematics. Word problems are nothing more than a "dressing up" of a purely mathematical problem in words referring to a segment of the real world. In this case mathematising means merely "undressing" the problem, and the solving process then only consists of this undressing, the use of mathematics, and a straightforward interpretation. At their best, word problems allow for interesting and worthwhile activities located within what is effectively the solution and interpretation stages of the modelling cycle – translation takes place between the worlds of mathematics and words. At their worst, they promote mathematical tasks in an unrealistic disguise and recipe approaches to their solution.

Standard applications: Typified by problems like finding the largest cylindrical parcel that can be shipped according to certain postal requirements, standard applications are characterised by the fact that the appropriate model is immediately at hand. Such problems can be solved without further regard to the nature of the given real world context. In our example, this context can be stripped away easily to expose a purely mathematical question about maximizing volumes of cylinders under prescribed constraints. So, the translation processes involved in solving standard applications are straightforward, that is, again, only a limited subset of the modelling cycle is needed.

Modelling problems: A typical example of a modelling problem is the following: "Decide the best location for speed bumps to calm traffic along a road within the college campus." Here a particular question must first be specified, then a mathematical model must be formulated, solved and interpreted. Finally the proposed solution must be evaluated, both mathematically and in context, followed by recommendations argued in terms of the modelling effort. In problems like that, the complete modelling cycle is involved.

It is not suggested that these problems typify some hierarchy of desirability independent of circumstances. They serve different purposes, and the challenge is to select and use appropriate examples of each genre. The point here is to acknowledge that while many problems are presented as examples of applications and modelling, only a few survive if the full modelling process is used as the criterion.

2.6 Modelling Competency

By a "*competency*" we mean the ability of an individual to perform certain appropriate actions in problem situations where these actions are required or desirable. So *mathematical modelling competency* means the ability to identify relevant questions, variables, relations or assumptions in a given real world situation, to translate these into mathematics and to interpret and validate the solution of the resulting mathematical problem in relation to the given situation, as well as the ability to analyse or compare given models by investigating the assumptions being made, checking properties and scope of a given model etc. In short: modelling competency in our sense denotes the ability to perform the processes that are involved in the construction and investigation of mathematical models.

It is clear that modelling competency is not sufficient for solving real world tasks and problems. Typically, *other competencies* such as representing the mathematical objects involved in an appropriate way, arguing and justifying what is being done when applying mathematics or “simply” performing mathematical algorithms and procedures are necessary as well. Additionally, when problems are solved in a group (typical for learning environments where modelling is required), *social competencies* not specific to mathematics are needed for effective cooperative teamwork and for the mutual construction and testing of knowledge generated during modelling activity. The use of the term “competencies” in the context of modelling may involve several of these domains, typically in combination. Of course, these competencies may support and advance one another. If it is, for instance, the main objective of a modelling activity that mathematical conceptual structures emerge during the solution of a modelling problem, then modelling competency contributes to the enhancement of mathematical competence. The reverse situation applies when individuals develop the ability to choose mathematical approaches to model formulation, based on a network of mathematical knowledge and competencies.

3. CONCEPTUALISATION OF THE FIELD

What is “the reality” of applications and modelling in mathematics education, that is, what is the societal and systemic framework in which activities that have to do with the topic of this Volume take place? We consider this reality constituted essentially by two dimensions: The significant “*domains*” within which mathematical applications and modelling are manifested on the one hand, and the *educational levels* within which mathematical applications and modelling are taught and learnt, on the other.

More specifically, in the first dimension we discern three different domains, each forming a kind of continuum. The first domain consists of the very *notions of applications and modelling*, i.e. what is meant by an application of mathematics, and by mathematical modelling; their essential components in terms of concepts and processes; the epistemological characteristics of applications and modelling in terms of mathematics as a discipline and other disciplines and areas of practice. Here we also consider who uses mathematics; for what purposes; with what types of outcomes; what defines modelling competency, and so on. The second domain is that of the *classroom,* used here as a broad indicator of the location of teaching and learning activities pertaining to applications and modelling. While this includes the classroom in a literal sense, it also includes the student doing his or her homework, individual or group activity, the teacher's planning of learning activities, looking at students' products, written or other, and so forth. The third and final domain is the *system* domain, where the word system, refers to the whole institutional, political, structural, organisational, administrative, financial, social, and physical environment that exerts an influence on the teaching and learning of applications and modelling. While we have chosen not to consider individuals, in particular students and teachers, as constituting separate domains, this does not imply that individuals have been left out of our conceptualisation. The individual student is a member of the classroom, as defined above, when engaging in learning activities in applications and modelling. The individual teacher can also be regarded as a member of the applications and modelling classroom, namely when he or she is engaged in teaching, supervising, advising or assessing students. From another perspective, however, the teacher is also a member of the system domain. This happens when he or she speaks or acts on behalf of the 'system' (typically in the form of his or her institution) in matters concerning selection, placement, and examination of the individual student, or invokes rules, procedures or other boundary conditions in decisions on, say, curricular matters.

The second dimension is constituted by the *educational levels*. We have adopted a relatively crude division of levels, both in order to avoid excessive detail in the discussion, and to achieve a division which is compatible with the educational structure in most, maybe all, countries in the world. The levels adopted are the *primary*, the *secondary*, the *tertiary* levels, and the level of *teacher education*. Here we do not primarily refer to age levels but to intrinsic levels of the learners' knowledge and competencies. While a much more fine-grained division might have been adopted, the present one at least allows for the consideration of applications and modelling at all educational levels.

The *issues, problems and questions* that are dealt with in this Volume can be placed somewhere within this two-dimensional "reality space". They are constituted by certain objects, phenomena or situations, and drawn from combinations of applications and modelling contexts and educational levels, that these issues concern. To illustrate the point, let us give one *example* of such an issue.

Issue: Context dependence and transfer

One of the fundamental reasons for attributing a prominent position to applications and modelling in the teaching and learning of mathematics is the underlying assumption that students should be able to engage in applications and modelling activities outside the classroom, that involve areas and contexts that are new to them. In other words, it is assumed that applications and modelling competency developed in and for some types of areas and contexts can be transferred to other such types having different properties and characteristics. However, several research studies suggest that for many students this transfer is limited in scope and range, and that the knowledge and abilities of students relate considerably to the situation and the context in which they have been acquired. (See, e.g., Brown, Collins & Duguid, 1989; de Corte, Greer & Verschaffel, 1996; Niss, 1999). Hence we ask:

To what extent is applications and modelling competency transferable across areas and contexts? What teaching/learning environments are needed or suitable to foster such transferability?

This issue therefore concerns the classroom domain and (at least) the primary, secondary, and tertiary levels. Consequently the issue is situated within the "rectangle" constituted by the entire classroom domain and the first three educational levels. If the focus were to be limited to address, say, the secondary level, then the rectangle would be reduced accordingly.

In this Volume, various issues considered relevant to the field are addressed. Each of the issues constitutes its own segment of reality, but of course those segments may intersect. The reality space and the issues placed within it may be mapped as in Fig. 1-2:

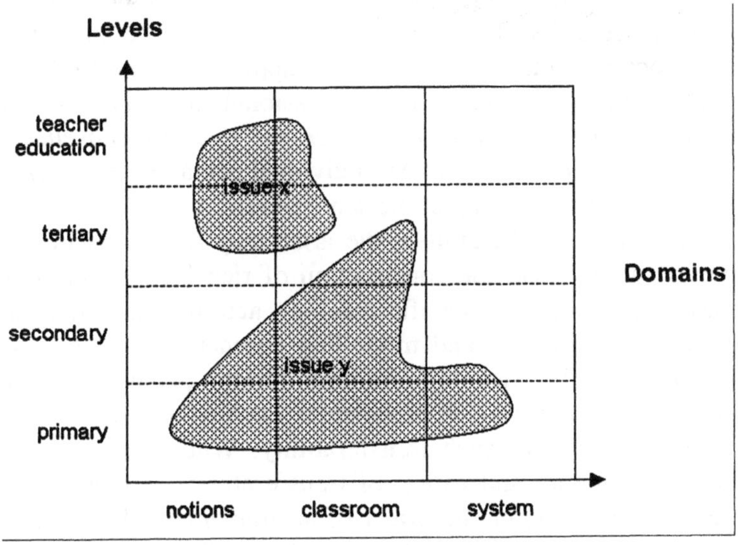

Figure 1-2. The "reality" of applications and modelling

What do we mean more precisely by an "*issue*"? The formulation of the issue given above as an example consists of two parts. First, there is a background part outlining a *challenge*, i.e. a dilemma or a problem, which may be of a political, practical, or intellectual nature. Second, there are particular *questions* that serve the purpose of pinpointing some crucial aspects of the challenge that deserve to be dealt with. We shall identify some particularly important issues in the next section of this introductory chapter.

From the point of view of this Study, an issue concerning applications and modelling in mathematics education may be viewed and approached from a variety of different *perspectives*, each indicating the category of answers sought. The most basic of these perspectives is that of *doing*, i.e. actual teaching and learning practice as enacted and carried out in the classroom (in the generalised sense outlined above). Here, the focus is on what does (or should) take place in everyday classrooms at given educational levels. Another perspective is the *development and design* of curricula, teaching and learning materials or activities, and so forth. Here, the focus is on establishing short or long term plans and conditions for future teaching and learning. A third perspective is that of *research*, which focuses on the generation of answers to research questions as yet unanswered. A fourth and final perspective is that of *policy* for which the focus is on the instruments, strategies and policies that are or ought to be adopted in order to place matters pertaining to applications and modelling on the agenda of practice or research in some desired way. Accordingly, a given issue may be addressed from one or some (perhaps all) of these four perspectives. To avoid a possible misinterpretation let us stress that the order in which we have presented these four perspectives does not imply a hierarchy. It appears that each of these perspectives can be linked to a particular professional role: The role of *teacher* or *student*, the role of *curriculum developer*, the role of *researcher*, and the role of *lobbyist* or *decision maker*. An individual can assume several or even all of these roles, but not usually at the same time.

In the above-mentioned example, the issue may be approached from the perspective of 'doing', e.g. the construction of rich learning environments, and the carrying through of specific teaching activities meant to underpin transferability of application and modelling competencies between certain areas and contexts. The 'development and design' perspective is adopted when finding or devising ways to orchestrate teaching and learning activities that are hoped to improve such transferability. When the emphasis is on understanding the nature and extent of transferability of competencies between areas and contexts, or the effect of an implemented design, then the 'research' perspective is invoked. Finally, the 'policy' perspective applies when the focus involves arguing or lobbying for, say, more room and time in the curriculum for rich first-hand experiences supporting a variety of applications and modelling activities drawn from different areas, contexts, and situations.

So, we can say in short that the present Volume essentially consists of the identification, presentation, and explication of important issues in the field, of reflections upon these from various perspectives, and of conclusions concerning the issues raised and dealt with by means of these perspectives.

4. ELABORATION OF SELECTED ISSUES

In this section, a number of issues that present challenges and associated questions are introduced. They have been grouped according to certain inherent emphases, and there is some natural overlap in the matters they address. The purpose here is to set the scene for the various sections in Part 3 of the present Volume where these issues are elaborated into more detail, and approached from different perspectives.

The specific connections with Part 3 of this Volume are as follows. The first two issues are closely connected with Section 3.1 on epistemology. The third issue relates to Section 3.2 on goals and authenticity, while Sections 3.3, 3.4, and 3.5, respectively, address the next three issues of modelling competencies and beliefs, of mathematical competencies, and of pedagogy. Issues 7 and 8 associated with implementation and practice form the topic areas of Section 3.6, while issue 9 concerning assessment and evaluation is dealt with in Section 3.7. Issue 10 (technology) is embedded in many (if not all) of the other issues. Consequently we decided not to include a special section on technology in this Volume, but to address it within other sections, in particular in Sections 3.4 and 3.5. The questions in this section associated with the issues, are drawn selectively from the various perspectives illustrated in the preceding section. They are representative, rather than exhaustive, and do not necessarily follow the order of the perspectives as introduced previously. Rather they have been chosen to reflect priority concerns within the respective issues, from which they follow seamlessly in the text.

4.1 Epistemology of applications and modelling

Essential characterisations of modelling and applications involve posing and solving problems located in the real-world, which for our purposes includes other discipline areas within which mathematics is applicable, activity within professions such as engineering or medicine, and general contexts of living as they impact on individuals, groups, and communities. The modelling enterprise involves identifying and addressing open-ended questions, creating, refining and validating models, and arguing the case for implementation of model informed outcomes. These share in common a linking of the field of mathematics with some aspects of the world, with the purpose of enhancing knowledge, but also ensuring or advancing the sustainability of health, education and environmental well-being, and the reduction of poverty and disadvantage.

We therefore face an imperative to articulate the relationship between applications and modelling, its values, methods and skills, and the world we live in. This overriding purpose directs us to identify specific questions that force a clarification of our assumptions and understandings such as:

What are the components of modelling as a process, and what is meant by or involved in each? How does knowledge of applications and modelling accumulate, evolve and change over time? What is the meaning and role of pure mathematical constructs such as abstraction, formalisation, generalisation, verification and proof in terms of applications and modelling activities? How are mathematical content knowledge, modelling know-how, and context-specific knowledge deployed for successful modelling outcomes? What opportunities for generalisability and transfer occur when working across contexts? What is necessary for an applications and modelling enterprise to be considered genuine?

4.2 Views of the modelling process

Various views of the modelling process co-exist within educational circles. These have to do both with perceptions of the modelling process, and constraints and opportunities perceived to exist within particular educational settings. Broadly speaking mathematical modelling activity can be viewed primarily as either a means or an end for educational purposes (see the explication of this duality in Section 1). Both approaches can exist at all levels of education, and share aspects in common, such as some version of the modelling process. And their goals are not necessarily mutually exclusive, since in seeking to solve genuine problems the need for new mathematical content may emerge, while real-world contexts can provide legitimate vehicles for the structured introduction of desired mathematics. Particular approaches, described by terms such as *emergent modelling*, and *model eliciting activities*, have been designed to focus attention on important educational challenges in the model building process. Always, for different purposes, and at different levels, the perspective of the modeller is central, and many specific questions continue to require action. For example:

Which versions of the modelling process are most appropriate for given purposes? To what extent are views and versions of the modelling process more or less applicable at different educational levels? What tensions exist between the 'modelling as means' and 'modelling as an end' approaches? How can essential modelling attributes (from the perspective of professional modellers) be introduced with integrity into modelling programmes at different educational levels?

4.3 Goals of applications and modelling and their curricular embedding

Applications and modelling can make fundamental contributions to the development of students' competencies, and consequently deserve a presence in mathematics curricula at all levels. In considering the balance to be achieved between applications and modelling activities, and other activities in mathematics classrooms, questions such as the following emerge:

To what extent is it possible (or desirable) to identify a core curriculum in applications and modelling within the general mathematical curriculum? Which applications, models and modelling processes should be included in such a curriculum? Is it beneficial to generate specific courses or programs on applications and modelling, or is it better to integrate applications and modelling into standard mathematical courses? What characterises the content of such courses at different educational levels from elementary to tertiary, including teacher education?

Since mathematics accounts for a large proportion of time in school, it needs to provide experiences and abilities that contribute to education for life after school, whether in further study, work, or in enhancing the quality of life. How and to what extent can applications and modelling provide supports for enriching a student's general education for these purposes? In this enterprise what is a suitable balance between creating one's own models of real situations and problems, and making judgements about models made by others?

With respect to the *authenticity* of application problems, there exists a wealth of applications and modelling problems and materials for use in mathematics classrooms at various educational levels, ranging from merely "dressed up" artificial problems, to those involving genuine problem situations. The concept of authenticity, and its contribution to the development of modelling ability is an important area for study; and not least is the task of reaching an agreed and reasonable meaning for the term 'authenticity' itself. This challenge has both teaching and research implications, directing attention to questions such as the following:

What are the requirements for problems to be regarded as authentic, as distinct from trivial or contrived? What authentic applications and modelling materials are available worldwide? Given the realities and pressures of classrooms, teachers' backgrounds and interests, students' experience, competencies, and motivations, how can teachers set up and implement authentic applications and modelling tasks? What effect does the authenticity of problems have on the ability of students to transfer acquired knowledge and competencies to other contexts and situations?

4.4 Modelling competencies and beliefs

One of the most important goals for students is the acquisition of *modelling competency*, and hence we need to ask how modelling competency can be characterised, and how it can be developed over time. This overall purpose implies the need to pursue specific questions such as the following:

Can specific sub-skills and sub-competencies of "modelling competency" be identified? How is modelling competency different from general problem solving ability? Are there identifiable stages in the development of modelling competency? What different competencies are needed for collaborative work in modelling and applications, compared with those needed when working alone? To what extent is the capacity to solve modelling and application problems impacted by the context in which a problem is set?

The successful development of applications and modelling competencies in learners is strongly dependent on the nurturing qualities of teachers, and this means that the inclusion of modelling in teacher pre-service and in-service education courses must be effectively promoted.

What is essential in a teacher education programme to equip prospective teachers to teach applications and modelling? Given the all-purpose curriculum needs of primary teachers, and their often-limited mathematical background and confidence, how can they be provided with real, non-trivial modelling situations?

Beliefs, attitudes and emotions play important roles in the development of critical and creative senses in all aspects of mathematics. Modelling aims, among other things, to provide students with a better apprehension of mathematical concepts, teaching them to formulate and to solve problems situated in specific contexts, awakening their critical and creative senses, and shaping their attitude towards mathematics and their picture of it.

So we ask to what extent can applications and modelling provide an environment to support both students and teachers in their development of appropriate beliefs about, and attitudes towards, mathematics? What are the implications of available research concerning beliefs, attitudes, and emotions for changing teaching practice and classroom cultures with respect to their contribution? In particular, can modelling effectively promote views of mathematics that extend beyond transmissive techniques, to advance its role as a tool for structuring other areas of knowledge? And what strategies are feasible for teacher education, both pre-service and in-service, to address the fear experienced by some teachers when faced with applications and modelling?

4.5 Mathematical competencies and their relation to applications and modelling

Effective mathematics learning involves a duality between the learning of strategies, concepts and skills, and competency in using this knowledge to address problems that are located both within mathematics, and in the rest of the world. Just as mathematical knowledge contributes to modelling competency, we consider also how modelling activities can contribute to the development of other mathematical competencies such as are listed in curricula worldwide. In particular, modelling problems provide contexts for coherent rather than piecemeal learning, providing a vehicle both for connecting individual pieces of mathematical knowledge and giving them purpose – a whole that is much more than the sum of the parts. Here we consider questions like the following:

How can modelling activities assist in the effective development of specific mathematical attributes, which are listed as desirable in the curriculum statements of various countries and programmes? Which particular mathematical competencies are applications and mathematical modelling approaches *most* suited to enhance? To what extent are motivational factors important in using modelling activities to develop mathematical competencies? In what ways can the use of modelling and application problems contribute to a balance of pure and applied mathematical competencies?

4.6 Modelling pedagogy

The pedagogy of applications and modelling intersects with the general pedagogy of mathematics instruction in many respects, but simultaneously involves a range of practices that are not part of the traditional mathematics classroom. While examples of successful applications and modelling initiatives have been documented in a variety of countries, and contexts, the extent of such programmes remains modest. Furthermore, approaches to teaching applications and modelling vary from the use of traditional methods and course structures, to those that involve a variety of innovative teaching practices, including an emphasis on group activity and the use of innovative assessment. When we seek to identify and articulate appropriate pedagogical principles and strategies for the development of applications and modelling courses and their teaching, we ask:

What research evidence is available to inform and support the pedagogical design and implementation of teaching strategies for courses with an applications and modelling focus? What are the areas of greatest need in supporting the design and implementation of such courses? What obstacles appear to inhibit necessary changes in classroom culture e.g. the introduction of interactive teamwork, or new forms of assessment in applications and modelling initiatives? What criteria are most helpful in selecting relevant elements from general theories of human development and/or learning? What criteria can be used to choose the most desirable option at a particular point within an applications and modelling teaching segment (e.g. between individual and group activity)? What documentations of successful group and other innovative learning practices exist?

4.7 Teaching/learning practice of applications and modelling

Applications and modelling activities vary in style and format across levels of education and across national contexts. Rarely (except in some specialised university programmes) do we find whole courses devoted to mathematical modelling. Applications when mentioned, are often restricted to illustrations purporting to show where some piece of recently learned mathematics is used in practice. In some places modelling requirements are included as a strand in a wider syllabus, in others curricular statements are sufficiently flexible to allow teachers to introduce a modelling approach on their own initiative. At senior secondary school level the influence of high stakes assessment has a definitive impact on whether and how modelling is included, and what kind of interpretation a modelling focus is likely to take. Questions of importance include the following:

What kinds of teaching programmes have been successful at different levels of education – elementary, secondary, tertiary? What are minimum requirements for a teaching programme to achieve a measure of success? How can modelling approaches be introduced successfully into a conservative school mathematics programme? What kind of challenges does a teacher face when 'going it alone', and what kind of support is most needed? What types of problems have been found to be most successful for student engagement and learning?

4.8 Implementation of applications and modelling in practice

Changing educational systems is a major challenge, as it impacts upon many different parties with sectional interests. With the increasing interest in and argument for mathematical modelling both inside and outside the mathematical community, there is a need to ensure that mathematical modelling is implemented in a sustained fashion at all levels of mathematics education. In spite of a variety of existing materials, innovative programmes, and sustained arguments for the inclusion of modelling in mathematics education, it is necessary to ask why its presence in everyday teaching practice remains limited in so many places. So we ask for example:

What are obstacles that traditionally impede the introduction of applications and mathematical modelling into curricula at different levels of education, and how can these be overcome? What documented evidence of success in overcoming impediments to the introduction of applications and modelling courses exist? What is required for developing a modelling environment within traditional courses at school or university, and how does one ensure that a mathematical modelling philosophy in curriculum documents is mirrored in classroom practice? What continuing education experiences such as support for teachers need to be provided to maintain initiatives in mathematical modelling?

4.9 Assessment and evaluation of applications and modelling

The teaching and learning of mathematics at all levels is closely related to the assessment of student achievement. To assess mathematical modelling performance is not easy, for the more complicated and open a problem, the more complicated it becomes to assess the quality of a solution. And when technology is available, assessment becomes even more complex. There are many indications that assessment modes traditionally used in mathematics education are not fully appropriate to assess modelling competency. Hence a need exists to investigate alternative assessment modes that are available to teachers, institutions and educational systems, that can capture the essential components of modelling competency – and to address obstacles to their implementation. Questions needing attention include the following:

What are the practical implications of assessing mathematical modelling as a process, rather than a product? If there is a change in the mathematics conception of students while experiencing and learning mathematical modelling, how can that change be assessed? When mathematical modelling is introduced into traditional courses at school or university, how should existing assessment procedures be adapted? When centralised testing of students is implemented, how do we ensure that mathematical modelling is assessed, and assessed validly? How does one reliably assess the contributions and achievement of individuals within group activities and projects?

The evaluation of the effectiveness of teaching programmes with application and modelling components is similarly challenging. In what ways do common evaluation procedures for mathematical programmes carry over to programmes that combine mathematics with applications and modelling? What counts as success when evaluating outcomes from a modelling programme? What do biologists, economists, industrial and financial planners, employers etc., look for in a student's or an employee's mathematical modelling abilities?

4.10 Technology with and for applications and modelling

Many technological devices are highly relevant for applications and modelling. They include calculators, computers, the Internet, and computational or graphical software as well as all kinds of instruments for measuring, for performing experiments etc. These devices provide not only increased computational power, but broaden the range of possibilities for approaches to teaching, learning and assessment. On the other hand, the use of calculators and computers may also bring associated problems and risks.

How should technology be used at different educational levels to effectively assist the development of students' modelling abilities, and to enrich their experience of open-ended mathematical situations in applications and modelling? What implications does technology have for the range of applications and modelling problems that can be introduced? How is the culture of the classroom influenced by the presence of technological devices? Will button pressing compromise the thinking and reflection necessary within modelling problems, or can these be enhanced by technology? What evidence of successful or failed practice in teaching and learning applications and modelling has been documented as a direct consequence of the introduction of technology? When does technology potentially kidnap learning possibilities, e.g. by rendering a task trivial, and when can it enrich them? Are there circumstances where modelling processes cannot be developed without technology? With particular regard to less affluent countries, can applications and modelling be undertaken successfully without any advanced technology? What are the implications of the availability of technology for the design of assessment items and practices, for use in contexts involving applications and modelling?

5. A BRIEF HISTORY OF THE FIELD

In this section we give a brief outline of the history of applications and modelling in mathematics education. Such a history has two aspects, one pertaining to applications and modelling in the practice of mathematics teaching and learning, and one pertaining to applications and modelling as a field of research and development within mathematics education. We shall deal with these two aspects in sequence.

5.1 Applications and modelling in mathematics instruction

As long as mathematics has been part of general schooling, applications and (less so) modelling have been integrated with the learning of mathematics as far as primary and lower secondary levels are concerned. Of course, this does not imply that every single element in the learning of mathematics is linked to objects, phenomena or problems in extra-mathematical domains, but rather that from time to time elementary mathematics (e.g. arithmetic and geometry) is confronted with some extra-mathematical reality. Applications and modelling concerning inventories and taxation, land mensuration, trade, currency, calendars, lunar cycles, solar eclipses, construction of temples, division of inheritance etc., were dealt with in Mesopotamian and Egyptian scribe schools, and in Indian, Chinese, and Arabian textbooks. The same was true of many mathematical (text) books published in Europe (including the UK) 1200 – 1800, where one could also find applications and modelling concerning architecture, art (perspective drawings), ballistics, construction of fortresses and bastions, optics, horoscopes, map making, navigation, gambling, insurance, heat propagation, population forecasts etc. Also ancient Greek mathematicians such as Archimedes and Eratosthenes were involved in applications to warfare, physics, geography etc.

Mathematics as a scientific discipline was always intimately intertwined with its neighbouring disciplines physics, astronomy and engineering, so much so that until the early 19th century mathematics was by and large perceived as a natural science that involved many applications and modelling activities. However the notion of applications and modelling in our sense was hardly expressible at that time as it was extremely difficult to disentangle mathematics from the fields it served, a disentanglement which is necessary if we are to discern between extra-mathematical domains lending themselves to mathematical representation, and mathematics as an independent edifice. So, while those who did receive some form of mathematical education were indeed exposed to what we, today, would term applications and modelling, this was in no way made explicit.

We know that things changed radically with the advent of abstract non-Euclidean geometry in the early 1800's, the development of abstract analysis, and abstract algebra, and the extraordinary development of pure mathematics during the second half of the 19th century and throughout the 20th century. This development, however, always went in parallel and sometimes even hand-in-hand with an equally strong development of advanced uses of mathematics brought about through sophisticated application and modelling activities, some of which rested on the creation of new mathematical topics (e.g. functional analysis, linear programming, coding theory, cryptography) designed to cope with issues and problems pertaining to the real world in the sense of this Volume.

Up until the mid-19th century it was extremely difficult to obtain teaching of pure mathematics – except for Euclid's "Elements" which was part of many secondary or tertiary level curricula. Only from the mid-19th century did the idea that mathematics education should (also) be provided for general formative reasons begin to gain momentum – for purposes of training the logical "thinking muscle" (Niss, 1996, p 23).

From the end of the 19th century, most post-elementary curricula began to contain both pure and applied components (but without explicit instances of modelling). Throughout the 20th century, the curricular balance between these components has been an issue of continuing debate amongst mathematics educators, as reflected in many curriculum documents. The choice made in different epochs regarding this balance was a reflection of societal trends as well as of intrinsic features of the teaching and learning of mathematics. Several authors (e.g. Galbraith, 1989; Niss, 1996) have identified "pendulum swings" in the emphases on pure versus applied aspects of mathematics in education. The pendulum swings can be seen, in part at least, as an instantiation of the duality discussed previously between "applications and modelling for the learning of mathematics" and "learning mathematics so as to develop competency in applying mathematics and building mathematical models". Moreover, to the extent that the teaching and learning of mathematics displays a balance between theoretical mathematics and the application of mathematics for extra-mathematical purposes, intrinsic pendulum swings are very likely to occur, caused by the simple fact that when one side of the balance receives too much emphasis to the disadvantage of the other, there is an inherent tendency to begin calling for more emphasis on the other side.

At intervals throughout the 20th century pragmatic and utility oriented movements characteristically gained momentum, to insist on a serious role for the applications of mathematics in curricula, geared to students who were not expected to become mathematics professionals, but rather users of mathematics at various levels of sophistication. By this means, established applications of mathematics – e.g. interest and annuities, geographical coordinates, technical drawings, consumer arithmetic and so on – were given a place in many curricula. However, utilitarian movements frequently dealt only with low-level mathematics, and generally speaking no attention was paid to the analysis of applications as models, let alone to modelling as a process. In the 1920's the pendulum swung, in many countries, back to a stronger emphasis on theoretical mathematics, which was seen as well suited to exercising a general formation and development of the individual, not with respect to logical thinking alone, but with respect also to mental dispositions and personal attitudes. The post-depression 1930's and the 1940's saw yet another pendulum swing back to an emphasis on utility and the application of mathematics, which was then succeeded by a renewed emphasis on purely theoretical mathematics in the late 1940's and the first half of the 1950's.

The late 1950's saw two new developments with very different orientations, which were, however, intertwined to some degree. In the UK, for example, voices were raised, in particular amongst industrialists (Cooper, 1985), that graduates from schools and universities were not able to use whatever mathematics they might have been taught to solve real world problems. At best they had learnt mathematics by rote and were able to solve routine problems with which they were familiar from the teaching they had received. However, when confronted with non-familiar situations they did not possess any tools to deal with them. This stimulated calls for mathematical instruction to take the application of mathematics seriously, to such an extent that students would become able to tackle open, unfamiliar problem situations themselves. The resulting development towards mathematical modelling as an educational enterprise (Pollak, 1968), began in engineering settings and the sciences, and spread to other fields during the next decades.

At the same time, the so-called New or Modern Mathematics movement emerged and gained momentum. While its aim was to restore a proper focus on theoretical mathematics in a renovated form as encountered in university mathematics – the "Bourbakist orientation" – the ultimate end was, in fact, to equip students with the prerequisites needed for dealing with mathematics in real world contexts, which according to the reformers of mathematics required a deep knowledge of and insight into theoretical mathematics. This is reflected in ICMI President Marshall Stone's chapter in the reform manifesto "New Thinking in School Mathematics" based on the famous Royaumont meeting in 1959. This insistence on the eventual utility of mathematics made the initial intertwining of the modern mathematics reform and practitioners' call for a focus on real world problem solving possible.

However the two movements soon diverged, so much so that they resulted in mutual opposition, and the divergence was amplified by the expansion of upper secondary and tertiary education from the late 1960's. This expansion meant that many more students, and new types of students, would receive some form of mathematics education, and for most the aim was not to prepare them to become mathematics professionals, but to teach them mathematics that could be used outside of mathematics itself. In turn this implied that the application of mathematics gradually became a priority issue in mathematics education. As lessons from the reform movements of the late 1950s and the 1960s strongly suggested that graduates from schools and universities could not be expected to (be able to) put mathematics to use just because they had been taught theoretical mathematics, applications and modelling had to be taught, and not just praised. This influence was acknowledged in curriculum development, through the production of instructional materials, and in teaching practice in a number of countries (e.g. the UK and Australia, Austria, Denmark, Germany, and the Netherlands) from the 1970s. Since then more and more countries have adopted a similar view, so that the active application of mathematics in open-ended situations by means of mathematical models and modelling has made its way into the teaching and learning of mathematics.

It is characteristic that the later phases of this development were stimulated and sustained by the gradual establishment of communities of mathematics educators working on issues of applications and modelling in mathematics education, as elaborated in the following section.

5.2 Applications and modelling as a field of research and development in mathematics education

It is easy to observe, throughout the history of mathematics education, that the application of mathematics as an educational endeavour has always had its more or less influential, and more or less singular, advocates in debates on mathematics teaching and learning. The more emphasis the *Zeitgeist* placed on theoretical mathematics, the more vocal these voices became. Alongside the trends sketched in the previous sub-section such advocates increased in number, gained more and more strength, and became (better) organised in the 1960's. The call for applications and modelling received strong patronage in the late 1960s (for a very outspoken example, see Hammersley, 1969), partly in reaction to a perceived neglect by enthusiastic adherents to 'modern mathematics' movements. Hans Freudenthal organised an instrumental conference in 1968 on the theme "How to Teach Mathematics so as to be Useful?" the papers from which were subsequently published in *Educational Studies in Mathematics* in 1968. Freudenthal's opening address (Freudenthal, 1968) was assigned the telling title "Why to Teach Mathematics so as to be Useful?"

Since then we may discern three phases of research and development concerning applications and modelling in mathematics education. In the first phase, which we might refer to as the *advocacy phase* (roughly in the decade 1965 – 1975), symbolically initiated by Freudenthal's conference, advocates of applications and modelling provided arguments in favour of the serious inclusion of such components in the teaching of mathematics.

This was continued in the second *development* phase, (1975 – 1990) which was mainly characterised by the development of actual curricula and materials at various levels so as to encompass applications and modelling components (e.g. Mathematics Applicable in the UK and COMAP/UMAP in the USA). The emphasis here included the design and conduct of – sometimes experimental – teaching units and particular modelling courses, the procuration and implementation of instructional materials, and the generation and cultivation of specific cases of applications and modelling for potential use in mathematics classrooms. During this phase attempts were made at systematising and analysing, at a theoretical level, the argument for applications and modelling in mathematics education, and of investigating theoretically (Blum & Niss, 1991) and historically (Kaiser-Messmer, 1986) the relationship between applications and modelling and other components of mathematics education. In other words a research perspective was emerging. The programmes of the *International Congresses on Mathematical Education* (the ICMEs) reflected these trends, in particular since ICME-3 in Karlsruhe, 1976. The development phase further saw the initiation of what later became an international community of mathematics educators developing and investigating applications and modelling at various educational levels. The most visible aspect of this initiation was the conference series *International Conferences on the Teaching of Mathematical Modelling and Applications* (the ICTMAs) and the resulting conference Proceedings. The first conference was held in Exeter (UK) in 1983 (Berry et al., 1984), and since then an ICTMA has been held every second year. It is no coincidence that the word "modelling" figures prominently in the title, and before the word "applications". This is in order to underline the priority given to the modelling process, a priority which was a focal point for the British polytechnic environment in which the ICTMAs were first conceived, and which remains so in the community.

With some caution, the current phase (since 1990) might be termed the *maturation phase*, in the sense that it is during the last one-and-a-half decades that empirical studies of teaching and learning of applications and modelling have been added to the theoretical emphases of the previous phases. This does not imply, however, that there is an abundance of such research at our disposal today, but the volume is growing, with studies increasingly undertaken by younger researchers entering the field. It is also during this phase that the community around the ICTMAs established itself as an organised community, the *International Community of Teachers of Mathematical Modelling and Applications*, also carrying ICTMA as its acronym, and which was adopted by the *International Commission on Mathematical Instruction* as an Affiliated Study Group in 2004. During the same period, work in North America (e.g. Lesh & Doerr, 2003) has involved systematic research and development in the modelling field, mainly at elementary and secondary school levels.

The present ICMI Study Volume might be said to formally mark the maturation of applications and modelling as a research discipline in the field of mathematics education.

6. ABOUT THE STRUCTURE OF THIS VOLUME

This Volume consists of six parts. Following this introductory Part 1, the subsequent Part 2 contains the eight plenary papers of the Study Conference in Dortmund, 2004. In Part 3, the core part of this Volume, several crucial issues in the field of applications and modelling in mathematics education are addressed, grouped into seven Sections; these sections have resulted from Working Groups at the Conference, organised by members of the Programme Committee, and are edited by these members, in some cases supported by co-editors from outside the Committee. The chapters in Part 4 discuss some questions on applications and modelling that are specific to the various educational levels; these chapters, too, have resulted from Working Groups at the Conference, organised mostly by members of the Programme Committee, and have been written by the organisers. In addition to the chapters in Parts 3 and 4, Part 5 presents four selected case studies from different parts of the world; in each of these, some particularly interesting development in the field of applications and modelling is reported. Finally, Part 6 contains a short bibliography. In the following, we will describe the content of each Part and its contribution to this Volume in some more detail.

The papers in **Part 2**, based on the plenary presentations at the Study Conference in Dortmund and ordered here alphabetically, cover a wide range of topics – in particular, all the issues listed in Section 4 are addressed by this collection of papers. In Chapter 2.1, Claudi Alsina pleads for the inclusion of real world objects and situations into everyday mathematics teaching and gives several concrete examples. Morten Blomhøj and Tomas Højgaard Jensen elaborate in Chapter 2.2 on mathematical competencies, based on the Danish KOM project, and on the role of modelling in this framework. Drawing on observations of students working in a technological setting, Jere Confrey and Alan Maloney analyse in Chapter 2.3 mathematical modelling from an epistemological point of view. In Chapter 2.4, Helen Doerr discusses the new challenge for pre-service and in-service teacher education programmes raised by the inclusion of applications and modelling in mathematics teaching. Mainly based on current work in assessment, Peter Galbraith in Chapter 2.5 emphasises the need for further related research in the field of applications and modelling. Brian Greer, Lieven Verschaffel and Swapna Mukhopadhyay make a plea in Chapter 2.6 for making mathematical modelling activities accessible for primary school children by way of appropriate real world problems, while in Chapter 2.7, Gabriele Kaiser and Katja Maaß report on empirical studies into students' and teachers' beliefs, and consider ways that modelling competencies can be developed in everyday teaching. The final Chapter 2.8 contains reflections by Henry Pollak on various issues in mathematics education, based on interviews with him prior to the conference, which he was unable to attend. These video interviews are included on the CD available with this book.

The seven sections in **Part 3** all have the same structure. The introductory chapter, written by the section editors, contains a framework for the issue under discussion as well as a description of how each paper included in this section contributes to shedding light on certain aspects of this issue. So all remaining chapters of a section can be regarded as samples of how certain questions, combined with the issue considered in the section, may be approached. Section 3.1, edited by Jere Confrey, deals with epistemology; here, certain aspects of the relationship between mathematics and the real world are analysed within an educational setting. Section 3.2, edited by Peter Galbraith, combines the challenge of goal setting within mathematical modelling activities, with questions of what role authenticity should play in the selection of tasks and problems for certain purposes. In Section 3.3, edited by Brian Greer and Lieven Verschaffel, several kinds and levels of modelling activities, together with various facets of modelling competency are presented and discussed. The extensive Section 3.4, edited by Eric Muller and Hugh Burkhardt, contains papers dealing with roles that applications and modelling activities can play in advancing mathematical skills and abilities, including reflections on the role of the teacher and of technology. Pedagogical aspects of modelling are the focus of Section 3.5, edited by Hans-Wolfgang Henn, again including technological aspects. Section 3.6, edited by Thomas Lingefjärd, addresses questions of how to implement applications and modelling into teacher education programmes. Finally, Section 3.7, also edited by Peter Galbraith, focuses on the theory and practice of assessment and evaluation related to applications and modelling.

The papers in **Part 4** deal with five educational levels: primary, lower secondary, upper secondary, tertiary, and teacher education. These papers are not intended to be comprehensive surveys of what is going on around the world within this theme of the Study. Rather, some specific examples have been chosen that reflect priorities identified by the corresponding group members during the Conference.

Part 5 presents four cases: A study of the modelling activities of prospective mathematics and science teachers in the USA, a study into the implementation of modelling in the secondary school mathematics curriculum in the province of Ontario/Canada, an investigation into the influences on the sustaining of application-oriented curriculum change for senior secondary mathematics in two Australian states, and a report on teaching experiments with modelling social issues in South African schools. All four cases indicate exemplary activities aimed towards a broader implementation of mathematical modelling in curricula and classrooms.

Part 6 contains a concise bibliography which includes some basic references (in English language) published on the topic of applications and modelling in mathematics education in recent years, most of them generated in the context of the ICME and ICTMA conference series. Many more publications can be found in the separate lists of references contained in each of the chapters in this Volume.

REFERENCES

Berry, J., Burghes, D.N., Huntley, I.D., James , & D.G.J, Moscardini, A.O. (Eds.) (1984). *Teaching and Applying Mathematical Modelling.* Chichester: Ellis Horwood.

Blum, W. et. al., (2002). ICMI Study 14: Applications and Modelling in Mathematics Education – Discussion Document. *Educational Studies in Mathematics, 51*(1/2), 149-171.

Blum, W., & Niss, M (1991). Applied Mathematical Problem Solving, Modelling, Applications, and Links to Other Subjects – State, Trends and Issues in Mathematics Instruction. *Educational Studies in Mathematics 22,* 37-68.

Brown, J.S., Collins, A., & Duguid, P. (1989). Situated cognition and the culture of learning. *Educational Researcher 18,* 32-42.

Cooper, B. (1985). *Renegotiating Secondary School Mathematics: A Study of Curriculum Change and Stability.* London and Philadelphia: The Falmer Press.

De Corte, E., Greer, B., & Verschaffel, L. (1996). Mathematics teaching and learning. In Berliner, D.C., & Calfee, R.C. (Eds.), *Handbook of Educational Psychology* (pp. 491-549). New York: Macmillan.

Freudenthal, H. (1968). Why to Teach Mathematics so as to Be Useful? *Educational Studies in Mathematics 1,* 3-8.

Galbraith, P.L. (1989). Mathematics Education and the Future: A Long Wave View of Change. *For the Learning of Mathematics.* 8(3), 27-33.

Hammersley, J.M. (1968). On the Enfeeblement of Mathematical Skills by "Modern Mathematics" and by Similar Soft Intellectual Trash in Schools and Universities. In *Bull. Inst. Math. Appl.,* 66-85.

Kaiser-Messmer, G. (1986). *Anwendungen im Mathenatikunterricht, 2 volumes.* Bad Salzdetfurth: Franzbecker.

Lesh, R.A., & Doerr, H. (Eds.) (2002). *Beyond Constructivism: A Models and Modelling Perspective on Teaching, Learning, and Problem Solving in Mathematics Education.* Mahwah: Lawrence Erlbaum.

Niss, M. (1999). Aspects of the nature and state of research in mathematics education. *Educational Studies in Mathematics 40,* 1-24

Niss, M. (1996). Goals of Mathematics Teaching. In Bishop, A., Clements, K., Keitel, C., Kilpatrick, & J., Laborde, C. (Eds.). *International Handbook of Mathematical Education.* (pp. 11-47). Dordrecht / Boston / London: Kluwer Academic Publishers.

Pollak, H. (1968). On some of the Problems of Teaching Applications of Mathematics. *Educational Studies in Mathematics 1,* 24-30.

Part 2

PLENARIES

Chapter 2.1

LESS CHALK, LESS WORDS, LESS SYMBOLS ... MORE OBJECTS, MORE CONTEXT, MORE ACTIONS

Claudi Alsina
E.T.S. Arquitectura de Barcelona. Technical University of Catalonia (UPC),
Email: claudio.alsina@upc.edu

Abstract: We will show how real objects, real places and real challenges may play an important role in the process of teaching mathematics by means of modelling and applications.

1. INTRODUCTION

Throughout this contribution we will defend the idea that realistic teaching is an appropriate method for quantitative literacy training (Steen, 1998, 2001).

An important consequence of teaching "via applications" is that the classical way of delivering lectures needs to be changed. Teaching with applications today means stopping the "talk & chalk" method; no longer using an old textbook and instead offering a very lively guiding program, based upon various information sources, which opens new windows to appreciate the context of the students and their creativity as individuals and as a group (Alsina, 1998b, 2003)

Following the discussion document of this ICMI Study: "by *real world* we mean everything that has to do with nature, society or culture, including everyday life as well as school and university subjects or scientific and scholarly disciplines different from mathematics."

So we will focus only on *objects* and *instruments*, on *everyday* situations, on *frequent* or *recent events*, on real challenges and showing how this realistic approach may play an important role in the process of teaching by means of modelling and applications (see, e.g., the educational projects "Modelling our world" (COMAP, 1998) and "Math in context" (Romberg & de Lange, 1998).

The ideas that we will present here come from our own experience and research in teaching mathematics as a service subject, training groups of teachers and doing workshops with high school students.

2. MATHEMATICAL TAKING OFF AND LANDING

The following diagram shows the classical way to deal, step by step, with the procedures for modelling-working-applying:

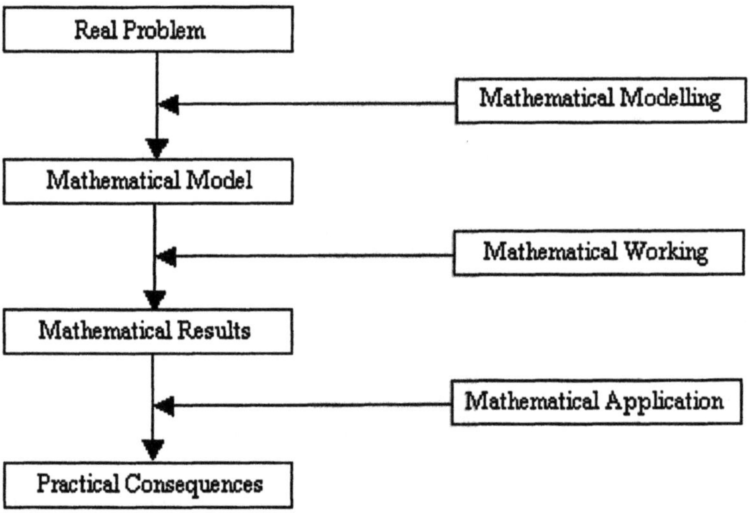

While most contributions in this field focus their attention on the central parts of this diagram, our aim here is to fix our views on the two boxes at either extreme: *the realities to be considered and skills for deriving practical consequences.* Too often, both ends become theoretical: word problems versus word solutions. If this is the case then we lose the possibility of motivating and providing students applied competencies (Niss, 1992, 2001).

3. LET US USE REAL OBJECTS

Our chief concern in this section is to pay attention to the wide range of daily life objects that may be used for teaching purposes, either for introducing new shapes and problems or for motivating a concrete visual approach to useful mathematics (Bolt, 1991; Steen, 1994; Alsina,1998a,b). All these objects constitute real applications of mathematics and by observing their characteristics and functional properties, students may appreciate the creativity behind them or may discover their limitations. As teachers may bring objects to the classroom and students may bring their own, this is a free material at our disposal.

Example: Modelling in the rain – with umbrellas

Today's umbrellas are sophisticated folding structures but they share with older ones a beautiful geometrical fact: the regular 8-gon determined by their extreme points. When you observe the moving octagonal pyramid of the structure you can discover how several articulated parallelograms change angles (and areas) but keep their perimeters. All these parallelograms determine a moving bipyramid. If you join two extreme points of the 8-gon you obtain a moving 7-gon which can be used to mark the 7-gon in a given circle (one of the impossible solutions with rules and compasses!).

It is interesting (Alsina, 2003) to collect umbrellas, fans, hats, etc, from all around the world (see also Gheverghese, 1996), i.e., daily life objects that exhibit flexibility.

Example: Polyhedra and polygons in context.

Nature exhibits a very restrictive collection of polyhedra. Only in some specific classes of minerals does one find basic shapes such as cubes or prisms. However, designers have produced a wide range of objects that have polyhedral forms. Packaging, logistics and beauty have motivated these designs.

Table 2.1-1. Polyhedra and daily life objects

CUBES	Dice, soup cubes, presents in boxes, hat boxes…
TETRAHEDRA	Tetra Pack ®, 3D puzzles, tripods, 4-faced dice...
OCTAHEDRA	Diamonds for cutting, table structures, kites, 8-faced dice...
ICOSAHEDRA	MAA logo, 20-faced dice, domes, sculptures…
DODECAHEDRA	Holders, 12-faced dice, no parking signs…
PRISMS	Toblerone packages, cookie boxes, pencils…
BOXES	TetraBrik ® packaging, cakes, Chanel nº 5 box, packages…
PYRAMIDS	Egyptian pyramids, the top of an obelisk, Sharkowski pieces…
BIPYRAMIDS	Whipping tops, jewels…
OTHER POLYHEDRA	Jewels, footballs, puzzles…

In our geometry education we anticipate the study of n-gons to the knowledge of polyhedra. This is, possibly, a mistake. Our visual experience goes, in general, from 3D to 2D.

Nevertheless, n-gons appear also by themselves in some planar objects or graphical designs.

Table 2.1-2. Polygons and daily life objects

TRIANGLES	Traffic signs, damage signals, musical instrument
QUADRILATERALS	Paper sheets, tile, cookies, cubes, brooches, snacks
PENTAGONS	Chrysler logo, napkin knot, tables
HEXAGONS	Tiles, plates, pencil sections, kite
OCTAGONS	Wind's directions, tables, trays, domes
n-GONS	Hours in a watch (12), cookies, commercial logos
STAR GONS	Sea star, star of David, tyre, clasps

Example: Curves in our life

Table 2.1-3. Curves in our life

Curve	Daily life examples
Straight line	Edge of a sheet of paper, string with a plumb-bob...
Circle	Plate, rim of a glass, coin, wheel, ring...
Ellipse	Profile of a hat, inclined liquid in a glass...
Parabola	Parabolic antenna, hand near ear...
Hyperbola	Profile of a bell, arcs in a hexagonal pencil...
Sinusoid	Snake's movement, sea waves, roofs...
Cycloid	Trajectory of a point in a wheel, pizza maker...
Catenary	Train wires, hanging chain...
Spirals	Classical discus, tape in a cassette, CDs...

Example: Transformations in daily life experiences

We may identify the basic geometrical transformations whose effects are seen in daily life movements: *translation* when walking down a street: *rotation* when the hands of a watch move or when we open a door; *symmetry* as a mirror effect: *similitude* when making reduced or enlarged Xerox copies; *affinities* when folding a box; *projectivities* when making shadows or photographs; *homeomorphisms* when folding a T-shirt, etc.

4. LET US EXPLORE REAL PLACES

Where are we teaching? Are we in a big city? Are we in a small village? Are we in a developed country...? We must be sensitive to our location. Some environments are rich in motivating contexts; others are not. We may take advantage of the location or, alternately, we may need to supply "additional motivation".

Do we have factories to visit? Are interesting measurements available in the area? Do we have notable buildings? How is public transportation organized? How is pollution measured...? If we get positive answers to these questions then we will have interesting places to generate mathematical activities at hand. Otherwise we can "bring" appropriate input to the classroom by means of Internet, books or pictures...

Wherever we are, in addition to geographical or architectural possibilities, there are social demands, social issues to be faced, cultural activities, etc.

We need to take these motivating situations into account as much as possible: working conditions, retirement plans, economic indices, inflation, theatre, television, book reading, local dances, music, musical instruments and cuisine—all social and cultural realities may have some mathematical interest.

On a compulsory level we prepare future citizens in a very specific social context. Our mathematics teaching may benefit from local characteristics and it is our goal to prepare students to be *critical citizens and good professionals in whatever their context.*

Examples of local applications	**Examples of global applications**
• Geometrical characteristics of the school • Distances from school; times • Geographical coordinates of the school • Cost of food at school • Ratios of ingredients in popular dishes • Geometry of particular buildings in town • Mathematics in folk dances • Different scales in local charts • Statistical study of minorities in local society	• Demographic issues: perspectives • Timing in travelling • Mathematics in democracy • Ecological problems: The Kyoto agreement • Mathematics and traffic (cars, roads, petrol) • Locations in the planet (GPS system) • Air traffic control and capacities • Digital images as messages • Implications of air conditioning in housing • Statistics on process: imports and exports

• History of calendars. Local holidays • Mathematics and sports. World records • Art exhibits in the town • Mathematics in newspapers and magazines • Mathematics in consumer issues. Indices • Numbers and classical tales • Numbers in popular sayings • Mathematics and music • Alcohol rates and driving: waiting times	• A visit to a car factory: sequential working • A visit to a food factory: quality control • Codes, phones, messages, Internet • Mathematics and genomics • Art: painting, sculptures, buildings • Fair division: geometry and equity • Mathematics and information: CDs, DVDs.

5. LET US FACE REAL CHALLENGES

In the previous sections we have been using objects and places to provide visual images and to make mathematics visible. Let us consider now the challenge of facing realistic problems and finding realistic solutions.

One may know a lot of things about cubes, observing minerals and houses, making cardboard models, enjoying interactive 3D-programs in the computer, etc., but occasionally it is useful to face the real problem of making a real cubic box, such as one that can be used to contain a present. Say, for example, that you want to open (and to close) just one face, and you want to transport the box – design problems may be very instructional.

Example: Function and design

Most shapes that we have around us are the result of a design process: houses, streets, cars, beds, bells, pencils… In this designed reality there is a strong mathematical component, from measurements to shapes. Most of these objects were designed to satisfy some desirable function. As part of the classical dialogue between form and function designers look for "optimal solutions". But "optimal" may hide different ideas: minimal quantity of element, low cost, ecological aims, transportability goals, etc.

It's interesting to know how designers work and how they find the best design solutions. Let us recount here, in some detail, the story of Jacob Rabinov.

Rabinov worked in New York, making 225 patents for all sorts of devices. When he retired, he wrote the acclaimed book *Inventing for Fun and Profit.* One of Rabinov's favourite topics was "screws and screwdrivers". He wanted to avoid the problem of so many screws being removed due to the fact that so many screwdrivers could be used for the same class of screw (even coins!). Thus, Rabinov took advantage of geometry and created a screw whose head made it impossible to manipulate with any conventional driver. Here we reproduce his description:

> *If you make a triangular depression with sides in the shape of three arcs, were each point of the triangle is the center of curvature of the opposite arc, you have a triangular hole that can be driven with a specially shaped screwdriver, but not by any flat screwdriver. If you insert a flat blade, the blade will pivot at each corner and slide over the opposite curved surface, hit the next corner and slide again, and so on. Such a screw should look very attractive and would be very difficult to open without the proper tool.*

These three arcs quoted above form a Reuleaux triangle, a convex figure of constant width, which is not a circle.

Example: Beware of the steps in a staircase!

This is an example to be studied with materials on a 1:1 scale, and which has a universal value: all humans need stairs which are easy to climb and almost all human beings use shoes. Stairs are important objects. Measure them! (e.g. using electronic measurements). The ideal steps have two important measurements: H (height) and D (depth), related by the affine equation, 2H + D = 63 cm. The inclination *tan* A = H:D is also interesting. What are the upper and lower bounds for H, D and A? When is it convenient to have a ramp and not a stair? In vertical ladders (such as in submarines) you face the steps to go down, but in normal stairs you come down the other way around: when is it better to face the steps?

In houses, stairs, streets, singular buildings, parks, mountains or plains – not far away from the classroom site – we find 1:1 models to provide a rich setting in which to practice actual measuring, drawing techniques, indirect measurements, finding of data, etc. The best lecture in the blackboard on the inclination of streets can't replace the real experience (at least once!) of effectively measuring the inclination in a real street. We want real challenges, not artificial questions.

6. IMPLEMENTATION IS THE ANSWER – WHAT ARE THE QUESTIONS?

One major challenge in mathematics education is to achieve the goal that realistic modelling and realistic application be indeed implemented in courses. This, however, is not so easy. We have clear evidence that there are many difficulties entailed in introducing this approach to our daily teaching (Breiteig, Huntley, & Kaiser-Meβmer, 1993). We would like to make some remarks:

We need teachers who are confident

The main objection is that pre-service education does not provide teachers enough knowledge and experiences to be confident in dealing with applications and modelling. Thus, many improvements on pre-service and in-service training need to be made.

The level of learners

Clearly, the level of learners will orientate us as to which choice to make concerning applications. While a tender, fictional tale on numeracy may be appropriate in kindergarten, there is no way to tell the same story in a high school. Each generation of students has topics that are relevant to them – and we want them to be interested.

Often in recreational mathematics, problems are presented in a fiction-real context which insinuates that the result to students' discovery will be a crucial issue in their lives. Crossing rivers, climbing castles, covering chess-boards with tetraminos – who does these things today? Useless mathematics cannot become useful even if it is presented in a fun way.

"Cooked" examples to illustrate some mathematical concepts or results are related to situations which are not interesting for the students, or even teachers. Let us recall the old problem "if 5 workers in a building will end the work in 3 weeks, when will the building be finished if 25 workers are assigned?"

Applications (and especially research activities related to them) are ideal for inducing cooperative work or teamwork. Good assessment, e.g., in setting up projects, needs to take into account individual and cooperative aspects. Preparation for cooperative work is a crucial goal in today's circumstances (Galbraith, 1998; deLange, 1996; Blum, Niss, & Huntley, 1981; Blum & Niss, 1991; Tanton, 2001).

The applied approach is time consuming

This is a key issue. Indeed, many teachers do not organize active learning visits, experiments, etc., because there is a lack of time. It seems there is no problem of time for chalkboard expositions. What is true is that our proposals imply a careful preparation of agendas.

Technology is not the solution

The growing power of technologies may induce some people to believe that these new devices are the essential tools for providing support for well-structured experiences; that simulations and images may be enough to eliminate completely the need for "real experiments" and hands-on materials. This is not possible. New software gives new mathematical insights but can't replace "learning by making". In our discussion here, technology serves to complement what we are presenting.

Hands-on materials are not always available

While textbooks and classroom materials are produced in big quantities, with a wide range of alternatives, very few kites can be purchased in the market. Of course, many materials can be easily made and real objects are everywhere. New commercial initiatives, however, would truly be welcome. Fortunately, many "free" opportunities exist around us.

Realistic activities need to be properly integrated

There is the risk that realistic experiences can become isolated, like islands in the sea of formal instruction. Such experiences are not useful if they are not combined with the usual activities of everyday teaching practice. These actions serve to assist students to learn important concepts, and of giving them new opportunities to develop new skills.

REFERENCES

Alsina, C. (1998a). *Contar bien para vivir mejor.* Barcelona: Editorial Rubes.

Alsina, C. (1998b). Neither a microscope nor a telescope, just a mathscope. In P. Galbraith et al. (Eds.), *Mathematical Modelling, Teaching and Assessment in a Technology Rich World,*(pp. 3-10). Chichester: Ellis Horwood.

Alsina, C. (2002). Too much is not enough. Teaching maths through useful applications with local and global perspectives. *Educational Studies in Mathematics 50*, 239-250.

Blum, W., Niss, M., & Huntley, I. (Eds.). (1989). *Modelling, Applications and Applied Problem Solving – Teaching Mathematics in a Real Context.* Chichester: Ellis Horwood.

Blum, W., & Niss, M. (1991). Applied Mathematical Problem Solving, Modelling, Applications, and Links to Other Subjects - State, Trends and Issues in Mathematics Instruction, *Educational Studies in Mathematics 22*(1), 37-68.

Bolt, B. (1991). *Mathematics meets technology.* Cambridge: University Press.

Breiteig, T., Huntley, I., & Kaiser-Meßmer, G. (Eds.). (1993). *Teaching and Learning Mathematics in Context.* Chichester: Ellis Horwood.

COMAP (1998). *Mathematics: Modeling Our World.* Cincinnati: South-Western Ed. Pub.

De Lange, J. (1996). Using and Applying Mathematics in Education. In Bishop, A. et al. (Eds.), *International Handbook of Mathematics Education V.1*, (pp. 49-97). Dordrecht: Kluwer Acad. Pub.

Galbraith, P. et al. (Eds.). (1998). *Mathematical Modelling – Teaching and Assessment in a Technology-Rich World.* Chichester: Ellis Horwood.

Gheverghese J. G. (1996). *La Cresta del Pavo Real. Las Matemáticas y sus raíces no europeas.* Madrid: Pirámide.

Matos, J.F. et al. (Eds.). (2001). *Modelling and Mathematics Education.* Chichester: Ellis Horwood.

Niss, M. (1992). Applications and Modelling in School Mathematics - Directions for Future Development. In Wirszup, I., & Streit, R. (Eds.), *Development in School Mathematics Education Around the World V.3*, (pp. 346-361). Reston: NCTM.

Niss, M. (2001). Issues and Problems of Research on the Teaching and Learning of Applications and Modelling. In: Matos et al., loc. cit, (pp. 72-88).

Pollak, H. O. (1997). Solving Problems in the Real World. In Steen, L.A. (Ed.), *Why Numbers Count: Quantitative Literacy for Tomorrow's America*, (pp. 91-105). New York: The College Board.

Romberg, T.A., & de Lange, J. (Eds.). (1998). *Mathematics in Context.* Chicago: EBEC.

Steen, L.A. (1994). *For all practical purposes.* COMAP, Lexington: W.H. Freeman Co. New York. Versión española: *Matemáticas en la vida cotidiana.* Madrid: Addison-Wesley, 1999.

Steen, L.A. (1998). Numeracy: The New Literacy for a Data-Drenched Society. *Educational Leadership, 57/2* (October), 8-13.

Steen, L.A. (Ed.). (2001). *Mathematics and Democracy. The Case for Quantitative Literacy.* Princeton: National Council on Education and the Disciplines.

Tanton, J. (2001). *Solve this. Math activities for students and clubs.* Washington: MAA.

Chapter 2.2

WHAT'S ALL THE FUSS ABOUT COMPETENCIES?

Experiences with using a competence perspective on mathematics education to develop the teaching of mathematical modelling

Morten Blomhøj and Tomas Højgaard Jensen
Roskilde University, Denmark, Email: blomhoej@ruc.dk, Denmark,
The Danish University of Education, Email: thje@dpu.dk

Abstract: This paper deals with applying a description of a set of mathematical competencies with the aim of developing mathematics education in general and in particular the work with mathematical modelling. Hence it offers a presentation of the general idea of working with mathematical competencies as well as an analysis of some potentials of putting this idea into educational practice. Three challenges form the basis of the analysis: The fight against syllabusitis, the dilemma of teaching directed autonomy and the description of progress in mathematical modelling competency.

1. INTRODUCTION

Mathematics education is full of buzzwords. These are words that add flavour to an analysis, a discussion or the planning of a teaching practice just by being mentioned. "Metacognition", "project work" and "responsibility for one's own learning" are good examples.

An underlying agenda for the structuring of this article is, that there are good arguments against using such buzzwords, the danger of replacing words for thoughtfulness being one. Consequently, one should always take a critical stance and ask the question: For what kind of challenges is this a potentially useful concept, and how should we understand and use the concept in the light of this?

Within recent years "competence" has been added to the list of buzzwords, at least in the northwestern part of Europe. In what follows, we shall analyse the cognate concept "mathematical competence" by attempting to answer the critical question posed above.

Three potential uses of the concept are analyzed. In each case the analysis is spanned by a general problematique pertinent to working with mathematical modelling in mathematics education and one or more developmental projects attempting to use the competence perspective to deal with this problematique.

2. FIGHTING SYLLABUSITIS IN MATHEMATICS EDUCATION

What constitutes mathematics as a subject? Many things, of course, but we feel convinced that everyone will agree, that mathematics has to do with certain objects, concepts and procedures that we (tautologically) consider as mathematical. Many people use this relation to subject matter to characterize the subject. "Mathematics is the subject dealing with numbers, geometry, functions, calculations etc." is not a rare type of answer to the question of what constitutes mathematics.

What, then, does it mean to master mathematics? With reference to the above it is tempting to identify mastering mathematics with proficiency in mathematical subject matter. However, this belief if transformed into educational practice, is severely damaging. Damaging to the effect that is has been given the name of a disease, namely *syllabusitis* (Jensen, 1995). It is a disease because it fails to acknowledge a lot of important aspects: Problem solving, reasoning and proving and – in the context of this paper not least – modelling, just to mention some. Combined with the hardly ever challenged viewpoint that the aim of mathematics education is to make people better at mathematics, a curriculum infected by syllabusitis therefore fails to set an appropriate level of ambition and makes the educational struggle unfocused. Hence, it is important to address the following problematique:

Problematique 1: How can we describe what it means to master mathematics in a way that supports the fight against syllabusitis in mathematics education?

3. THE KOM PROJECT

This problematique was a main ingredient in a proposal by Mogens Niss for applying a set of mathematical competencies as a tool for developing mathematics education (Niss, 1999). The so-called KOM project (Niss & Jensen, to appear), running from 2000 – 2002 and chaired by Mogens Niss with Tomas Højgaard Jensen as the academic secretary, thoroughly introduced, developed and exemplified this general idea at all educational levels from primary school to university (cf. Niss (2003) for an actual presentation of the project).

The definition of the term "competence" in the KOM project (Niss & Jensen, to appear, ch. 4) is semantically identical to the one we use: *Competence* is someone's insightful readiness to act in response to the challenges of a given situation (cf. Blomhøj & Jensen, 2003). A consequence of this definition is that it makes competence headed for action, based on but identical to neither knowledge nor skills. Secondly, the situatedness should be noticed, since this defines competence development as a continuous process and highlights the absurdity of labelling anyone either incompetent or completely competent (Ibid.).

In our opinion these are good reasons for applying competence as an analytical concept in mathematics education, but in order to transform it into a developmental tool we need to be more specific. The straightforward approach is to talk about *mathematical competence* when the challenges in the definition of competence are mathematical, but this is no more useful and no less tautological than the above-mentioned definition of mathematics as the subject dealing with mathematical subject matter. The important move is to focus on *a mathematical competency* defined as someone's insightful readiness to act in response to *a certain kind of mathematical challenge* of a given situation, and then identify, explicitly formulate and exemplify a set of mathematical competencies that can be agreed upon as independent dimensions in the spanning of mathematical competence. The core of the KOM project was to carry out such an analysis, of which the result is visualized in condensed form in Fig. 2.2-1.

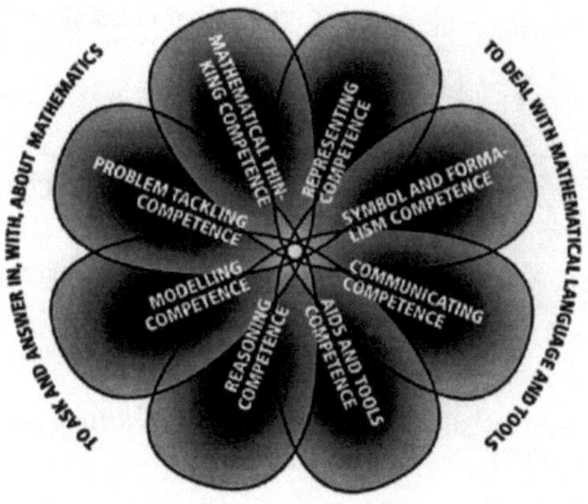

Figure 2.2-1. A visual representation – the "KOM flower" – of the eight mathematical competencies presented and exemplified in the KOM report (Niss & Jensen, to appear, ch. 4).

This set of mathematical competencies has the potential of replacing the syllabus as the focus of attention when working with the development of mathematics education, simply because it offers a vocabulary for a focused discussion of what it means to master mathematics. Often when a syllabus attracts all the attention in a developmental process, it is because the traditional specificity of the syllabus makes us feel comfortable in the discussion.

4. THE DILEMMA OF TEACHING DIRECTED AUTONOMY

Where does the discussion of the role of mathematical modelling in mathematics education appear in all this? The KOM project does not specifically focus on this matter, but on a more general level the suggested competence framework assigns a central role to mathematical modelling, namely as a natural constituent in the developing of *mathematical modelling competency*. In short this competency is defined as someone's insightful readiness to carry through all parts of a mathematical modelling process in a certain context (Blomhøj & Jensen, 2003). Fig. 2.2-2 shows our model of this process, inspired by and quite similar to many other models of this process found in the literature.

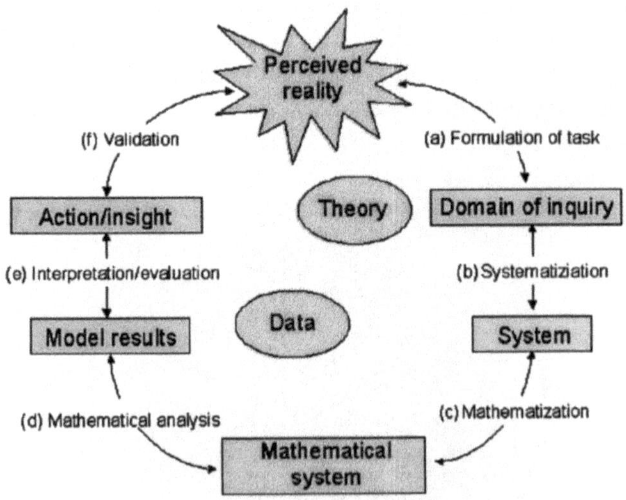

Figure 2.2-2. A visual representation of the mathematical modelling process (adapted from Blomhøj & Jensen, 2003).

Seeing the role of mathematical modelling as a natural constituent of the development of mathematical modelling competency derives from the assumption, that such development requires at least part of the teaching to be based on the holistic approach (Ibid.), i.e. students are challenged to work with full-scale mathematical modelling and have responsibility for directing the entire process.

By virtue of the "underdetermined" nature of the initial parts of the mathematical modelling process, the key characteristic of this challenge is to learn to cope with a feeling of "perplexity due to too many roads to take and no compass given" (Ibid.). But what do we as authorities consider as qualified choices in this situation? Those maintaining the educational focus – in casu developing mathematical modelling competency, which confronts us with *the dilemma of teaching directed autonomy*: The simultaneous need for student directed working processes and for maintaining educational focus (Jensen, to appear, ch. 9). The students need to be responsible for most of the decisions, but the decisions they make also need to be "the right ones"! This brings us to our second highlighted problematique:

Problematique 2: How can the dilemma of teaching directed autonomy be overcome when attempting to develop mathematical modelling competency?

5. THE ALLERØD EXPERIMENT

The problematique was one important aspect of a longitudinal developmental research project (Jensen, to appear) named after the upper secondary school where the teaching took place from 2000 – 2002. It involved a class of 25 students, their mathematics teacher and Tomas Højgaard Jensen as the researcher who initiated the experiment and took part in the planning and evaluation of the teaching.

The aim of the experiment was making development of mathematical modelling competency the hub of the general mathematics education. The instructional focus was to use student directed project work initiated by invitations to mathematical modelling such as:

- How far ahead must the road be clear in order to make a safe overtaking?
- What are the maximum sizes of a board if one is to turn a corner?
- Which means of transport is the best?

In the curriculum a set of mathematical competencies and a set of subject areas spanned the mathematical content in a matrix structure (Jensen, 2000, to appear), cf. Fig. 2.2-3. As a direct consequence of the aim of the project, the set of competencies were made the hub of the curriculum by creating a (often missing) link between the overall goals of the teaching and the syllabus. The intention was to use some of the competencies – not least mathematical modelling competency – as "guiding stars" for the students' attempt to structure their own project work, simply by pointing out the development of one of these competencies as the goal of each project work.

Subject area **Competency**	Numbers	Arithmetic	Algebra	Geometry	....	Area N
Math. thinking						
Problem tackling						
Modelling						
Reasoning						
Representing						
Symbol-form.						
Communication						
Aids and tools						

Figure 2.2-3. A matrix structure for describing the mathematical content of a piece of mathematics education (Niss & Jensen, to appear, ch. 8, and Jensen, to appear, ch. 9).

This turned out to be a very promising approach in the struggle to resolve the dilemma of teaching directed autonomy: The set of competencies as an independent dimension in the forming of the matrix structure made it possible to set up a very clear "contract" for each project work. Once having understood the nature and core elements of the competency in focus, the students could decide (after discussions with each other and with the teacher) which choices would be in accordance with the educational focus and with their personal interest.

The main pedagogical challenge of using this approach was to develop methods to help the students understand the nature and core elements of the different competencies in focus. It will take us too far to discuss the methods developed and used in the Allerød experiment here (cf. Jensen, to appear), but it is safe to say that it is a challenge calling for more research and developmental work.

6. PROGRESS IN MATHEMATICAL MODELLING COMPETENCY

Once having identified mathematical modelling competency as a central element in general education a third problem become apparent, namely:

Problematique 3: How can progress in mathematical modelling competency be described in ways that support the development of good and coherent teaching practices at different educational levels?

7. THE SAME MODELLING TASK AT THREE DIFFERENT LEVELS

Drawing on analyses from developmental projects at the lower secondary level (Blomhøj, 1993; Blomhøj & Skånstrøm, 2002), in teacher education (Blomhøj, 2000, 2003), and at first year university level (Blomhøj et al., 2001; Blomhøj & Jensen, 2003), we shall illustrate how progress in mathematical modelling competency need to be described along more than one dimension.

Analytically one can distinguish (at least) three different dimensions in mathematical modelling competency: A dimension describing the degree of coverage, meaning which parts of the modelling process the students are working with and at what level of reflection, a dimension that has to do with the technical level of the students activities involved in the modelling process, meaning what kind of mathematics they use and how flexible they do it, and a dimension that has to do with variation in the types of situations and contexts in which the students can actually activate their mathematical modelling competency, in short called the radius of action. In the KOM project these dimensions are proposed as a general approach to describing progression in the possession of a given mathematical competency (Niss & Jensen, to appear, ch. 9).

In the following we illustrate how progress in mathematical modelling competency in relation to a specific situation can be described as interplay between progress in the degree of coverage and progress in the technical level. For this purpose we use a modelling task, which we have used in developmental projects on mathematical modelling at all levels from lower secondary to first year university teaching and in teacher education. The task has typically been given as a group task, with 6 – 8 lessons distributed over two weeks to write up a report. The task has its point of departure in the following authentic text from a Danish traffic safety campaign (our translation):

A car driving 60 km/h passes a car driving at a speed of 50 km/h. When the cars are right beside each other a girl appears some meters ahead. The drivers react in the same way and the cars have brakes of equal quality. The car with 50 km/h stops right in front of the girl, while the other car, with the initial speed of 60 km/h, hits the girl with 44 km/h. Seven out of ten die in such an accident.

Given this text the students are simply asked: Can this be true?

At all educational levels, the first challenge for the students are to recognize that the claim in the campaign must be based on some kind of mathematical model, and that it therefore makes sense to try to model the traffic situation described in the campaign text and evaluate the claim against such a model.

The challenge is to use a holistic approach to the mathematical modelling process (cf. Fig. 2.2-2) and to see the potentials of a mathematical model connecting the specific description of the traffic situation and the claim of the campaign. Although this is certainly relevant in order to develop mathematical modelling competency, in this situation the students are not challenged to formulate a relevant problem themselves (process (a) in the modelling process) in order better to understand the phenomena in hand. If the students take the campaign text as a linguistic description of the system that they have to mathematize (and most students do), also process (b) has been taken care of in the task formulation. So, the task context takes care of the initial part of the modelling process.

In some of the projects, at lover secondary level and in teacher education, the students were introduced to modelling dynamical phenomena with difference equations and spreadsheet, while at the university course the students were expected to be able to use calculus to model the situation and hereby to be able to produce analytical results from analyzing their models. In relation to the degree of coverage the important thing is that in both cases the students are working with mathematization (process (c)) of a non-mathematical system. The way this is done at different educational levels can be seen as an example of progress in the technical level of the students' modelling activities.

However, at all levels the students typically try at some stage to model the situation without taking the time of reaction into account, meaning that they assume that the two cars start braking at the same point. Such a model produces the result that the car with the initial speed of 60 km/h hits the girl with 33 km/h and not the 44 km/h claimed in the campaign. Moreover this result is not depending on the braking effect of the cars.

Facing this result, students normally – especially if supported by a dialogue with the teacher – feel challenged to modify the model so that it may support the claim in the campaign. In this process the students experience, in a very concrete form, the cyclic nature of the modelling process. If embedded in the students' perception of mathematical modelling this constitutes a progress in the degree of coverage in their mathematical modelling competency.

After having included the time of reaction in the model, the need for realistic parameter values for the braking effect and the time of reaction gradually becomes apparent for the students. Finding such values, from the literature, contacting the authorities behind the campaign or from experimenting with the model, also constitutes a progress in the degree of coverage in the students' mathematical modelling competency (e.g. relations between the

"Mathematical system" and "Data" and/or "Theory" in Fig. 2.2-2). How this is actually done belongs to the technical dimension.

At this stage of the modelling process the students are able to produce new model results. Fig. 2.2-4 shows an example of model results produced by a group of 9th graders. However, as can be seen from the dialogue with the teacher (our translation), having set up a model and produced some results using e.g. a spreadsheet does not necessarily imply that the students on their own are able to interpret or evaluate the results in relation to the situation modelled (process (e)).

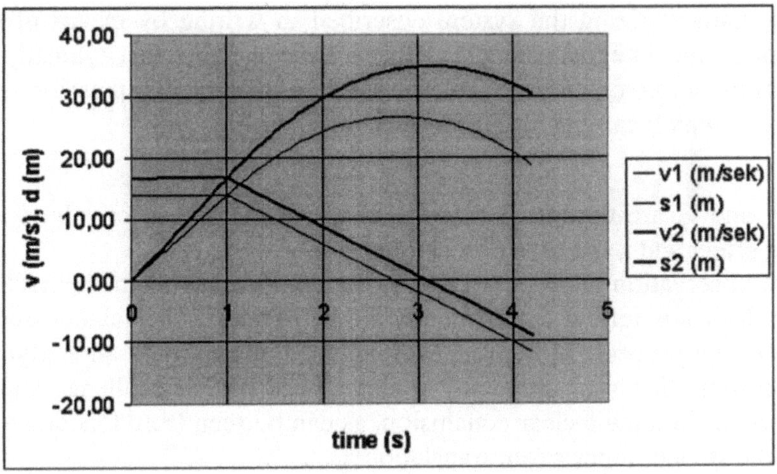

Figure 2.2-4. Graphs showing the velocity and position of car 1 and car 2 (the graphs for car 2 with the initial speed of 60 km/h are in bold).

T: Where does the girl stand?
S1: There! [Points at the point where the velocity graph for car 1 is zero.]
T: In 2,7 sec.?
S2: No, she is standing here! [Points at the top point of the position graph of car 1.]
T: Where? How many meters from the spot where the drivers first saw the girl?
S2: 26 meters. [Point it out on the second axis.]
T: So what about car 2? [The teacher leaves the place.]

....

T: What did you find out?
S1: Car 2 has passed that spot before car 1, the girl is dead before car 1 even stops. [Laughter.]
S2: Car 2 hits the girl with 11 m/sec.

Following this dialogue the students question the meaning of the decreasing position of car 2 after the time where its velocity becomes zero, and eventually this was included in the students' reflections on the validity of the model in their report (process (f)).

The dialogue illustrates the necessity for the teacher to challenge the students in order for them to interpret and reflect upon the outcome of their modelling activities. The teacher deliberately challenges the students' degree of coverage with respect to process (e) and (f) in the modelling process.

At the university course all the students possessed the technical prerequisite for mathematizing the system described in writing by means of using differential and integral calculus. However despite this fact typically only few groups (approx. 10%) are able to yield an analytical expression for the velocity of which car 2 is hitting the girl:

$$v_{2hit}^2 = v_2^2 - v_1^2 + 2bt_r(v_2 - v_1)$$

Here v_1 and v_2 are the initial velocity of car 1 and 2 respectively, b is the braking effect and t_r the time of reaction.

This observation shows clearly that the competency to mathematizing a system does not follow automatically from mastering the mathematics involved in the process. Moreover, even after having reached an analytic expression most first year university students need further challenge and support in order to draw a clear conclusion, as can be seen from this quote from one of the student reports (our translation):

> *According to our model the claim is only true when $b \cdot t_r = 11{,}61$ m/s. Inserting $g = 9{,}82$ m/s^2 as the maximal brake effect yields $t_r = 1{,}18$ s as the minimal time of reaction. This is a slow reaction for drivers, who are not under influence of alcohol or other drugs. We therefore conclude that the claim "10=44" is slightly exaggerated.*

Nearly the same degree of coverage in terms of the level of reflection can be reached by 9th graders (the first quote below) and teacher students (the second quote below) based on spreadsheet analyses of a difference equation model (our translation):

> *Experimenting with the model we find that the speed with which the second car hits the girl increases as we increase the time of reaction or the braking effect. But when changing these figures the position of the girl is also changed. Of course it is good to have good brakes.*

> *The Council for the Improvement of Traffic Safety has used a time of reaction of 1.5 sec. and a braking effect of 8 m/s^2. In this case the car with an initial speed of 60 km/h hits the girl with 43 km/h. This is only realistic for drivers, who have been drinking!*

8. SUMMING UP AND CONCLUSIONS

This paper has been framed by the following three problematiques:

Problematique 1: How can we describe what it means to master mathematics in a way that supports the fight against syllabusitis in mathematics education?

Problematique 2: How can the dilemma of teaching directed autonomy be overcome when attempting to develop mathematical modelling competency?

Problematique 3: How can progress in mathematical modelling competency be described in a way that support the development of good and coherent teaching practices at different educational levels?

Having presented the general idea of working with a set of mathematical competencies as laid out in the KOM project, the attempt to use a holistic approach to the teaching of mathematical modelling in the Allerød experiment and the description of progress in mathematical modelling competency when working with the same task at different educational levels, we are now in a position to sum up our conclusions as follows:

Point 1: A competence description of mathematical mastery makes it easier to discuss and tackle syllabusitis: By using a syllabus as the hub of mathematics education we fail to set the appropriate level of ambition.

Point 2: A matrix structured competence based curriculum can be a way to deal with a fundamental challenge when attempting to develop someone's mathematical modelling competency: The dilemma of teaching directed autonomy.

Point 3: In order to describe and support progress in students' mathematical modelling competency we need three dimensions:

- Degree of coverage, according to which part of the modelling process the students work with and the level of their reflections.
- Technical level, according to which kind of mathematics the students use and how flexible they are in their use of mathematics.
- Radius of action, according to the domain of situations in which the students are able to perform modelling activities.

We have illustrated the need for and interplay between the first two dimensions when analysing progress in mathematical modelling competency in relation to a specific situation. The limited space prevents us from illustrating the necessity of operating also with the third dimension, radius of action, when describing progress in mathematical modelling competency in general.

REFERENCES

Blomhøj, M. (1993). Modellerings betydning for tilegnelsen af matematiske begreber. *Nordisk MatematikkDidaktikk 1*, 16-37.

Blomhøj, M. (2000). Fuld fart frem og bremsen i – modellering i matematik. In *Matematik og Undervisning*. Norden: De nordiske matematiklærerforeninger, 117-126.

Blomhøj, M. (2003). Modellering som undervisningsform. In Skovmose, O., & Blomhøj, M. (Eds.), *Kan det virkelig passe?* (pp. 51-71). Copenhagen, Denmark: L&R Uddannelse.

Blomhøj, M. (2004). Mathematical modelling – a theory for practice. In Clarke, B. et al. (Eds.), *International perspectives on learning and teaching mathematics*, (pp.145-160). Göteborg University, Sweden: National center for mathematics education.

Blomhøj, M., Jensen, T.H., Kjeldsen, T.H., & Ottesen, J. (2001). Matematisk modellering ved den naturvidenskabelige basisuddannelse – udvikling af et kursus. *Tekster fra IMFUFA* 402, Roskilde University, Denmark. To be ordered from imfufa@ruc.dk.

Blomhøj, M., & Skånstrøm, M. (2003). Matematikmorgener – et udviklingsarbejde. In Holden, I. (Ed.), *Utvikling av matematikkundervisning i samspill mellom praksis og forskning – nye arbejdsformer i matematikkundervisningen*. Skriftserie for Nasjonalt Senter for Matematikk i Opplæringen 1, (pp. 61-72). Trondheim, Norway: NTNU.

Blomhøj, M., & Jensen, T.H. (2003). Developing mathematical modelling competence: Conceptual clarification and educational planning. *Teaching Mathematics and its Applications 22*(3), 123-139.

Blomhøj, M., Kjeldsen, T.H., & Ottesen. J. (2004). *BASE – et grundkursus i matematisk modellering*, 5. ed. Notes for a course on mathematical modelling at the Bachelor Science Study Programme at Roskilde University, Denmark. To be ordered from imfufa@ruc.dk.

Jensen, J.H. (1995). Faglighed og pensumitis. *Uddannelse* 9. Copenhagen, Denmark: The Ministry of Education, 464-468.

Jensen, T.H. (2000). Forsøgsprojektet "Problemløsning og modellering i den gymnasiale matematikundervisning". *LMFK-bladet* 7, 37-52.

Jensen, T.H. (to appear). *Udvikling af matematisk modelleringskompetence som matematikundervisningens omdrejningspunkt – hvorfor ikke?* Doctoral dissertation, Roskilde University, Denmark. Under preparation for publication in the series *Tekster fra IMFUFA*. To be ordered from imfufa@ruc.dk.

Niss, M. (1999). Kompetencer og uddannelsesbeskrivelse. *Uddannelse* 9. Copenhagen, Denmark: The Ministry of Education, 21-29.

Niss, M. (2003). Mathematical competencies and the learning of mathematics: The danish KOM project. In Gagatsis, A., & Papastavridis, S. (Eds.), *3rd Mediterranean Conference on Mathematical Education*, (pp. 115-124). Athens, Greece: Hellenic Mathematical Society and Cyprus Mathematical Society.

Niss, M., & Jensen, T.H. (Eds.). (2002). Kompetencer og matematiklæring – Idéer og inspiration til udvikling af matematikundervisning i Danmark. *Uddannelsesstyrelsens temahæfteserie* 18. Copenhagen, Denmark: The Ministry of Education.

Niss, M., & Jensen, T.H. (Eds.). (to appear). Competencies and Mathematical Learning – Ideas and inspiration for the development of mathematics teaching and learning in Denmark. English translation of part I-VI of Niss & Jensen (2002). Under preparation for publication in the series *Tekster fra IMFUFA*, Roskilde University, Denmark. To be ordered from imfufa@ruc.dk.

Chapter 2.3

A THEORY OF MATHEMATICAL MODELLING IN TECHNOLOGICAL SETTINGS

Jere Confrey and Alan Maloney
Washington University in St. Louis, USA, Email: jconfrey@wustl.edu / amaloney@wustl.edu

Abstract: A theory of mathematical modelling in education is offered, based on Dewey's description of inquiry. One aim is that a model provide a mapping between two stages of experience, rather than necessarily a mapping to a particular version of reality; a second aim is that it prepare students for further inquiry and reasoning experience. Two clinical interviews of students engaged in modelling provide examples of progress from an indeterminate to a determinate situation, and of modelling's potential in differentiated instruction.

1. INTRODUCTION

One can identify four distinct but related approaches to technology in mathematics instruction:
1. teach concepts and skills without computers, and provide these technological tools as resources after mastery;
2. introduce technology to make patterns visible more readily, and to support mathematical concepts;
3. teach new content necessitated by a technologically enhanced environment (estimation, checking, iterative methods);
4. focus on applications, problem solving, and modelling, and use the technology as a tool for their solution.

Selecting and prioritizing these approaches should engender a reconsideration of what mathematical knowledge is and why learning it should hold a central place in education. We take the position that mathematical modelling should be a central goal of mathematics instruction. We provide a theory of mathematical modelling that permits one to navigate among these varied approaches to technology, and then illustrate our theory with summaries of two investigations by experienced teachers.

Drawing upon an evolutionary view of knowledge, we believe that mathematical knowledge changes and evolves. We view mathematics not as a reservoir of ultimate knowledge, but rather as a tool that permits people to make sense of experience, gain predictive judgment, and offer explanation. This view is espoused by the pragmatists (Pierce, James, Dewey), with the perspective that knowledge should be subjected to criteria of "functional fitness" akin to the constructivist concept of viability, and to Piagetian accommodation and assimilation. We concur with socio-cultural theorists who identify cultural heritage and language as particular influences on development of knowledge as a human endeavor (Bornstein & Bruner, 1989; Vygotsky, 1930), and draw on the experience of education and technology researchers who recognize that uses of technology influence the evolution of both knowledge and pedagogy (Linn & Hsi, 2000). Mathematical knowledge, in our view, should contribute to our ability to identify, address and solve problems presented by our cultural and environmental surrounds.

Accordingly, we assert a primary role for the process of inquiry: it can lead us to means both to address outstanding problems and to secure results that can be preserved to inform future problem solving efforts. However, classroom use of "inquiry" often refers primarily to pedagogical processes, and can signal engagement without product, accomplishment, or proficiency. To avoid a tendency toward a narrowly process-oriented use of "inquiry," we draw on Dewey's definition of inquiry:

> ...the controlled or directed transformation of an indeterminate situation into one that is so determinate in its constituent distinctions and relations as to convert the elements of the original situation into a unified whole (Dewey, 1938, in McDermott, p. 226)

This profound definition of inquiry captures many of the key modelling ideas in the Discussion Document (DD) for the present volume, and related literature, in which it is suggested that mathematical modelling:

- Is the "process leading from a problem situation to a mathematical model" (DD),
- Includes the activities of "structuring, mathematising, working mathematically and interpreting/validating" (DD),
- Is for the "purposes of predicting, describing, explaining, understanding or even designing part of the world," (DD), and
- "involves the posing of genuine, non-rhetorical questions to which clear and specific answers are to be sought" (Niss, 2001, p. 3).

For Dewey, inquiry is a set of processes and outcomes, including:

- recognition of the indeterminate situation, which "becomes problematic in the very process of being subjected to inquiry" (Dewey, 1973, p. 229);
- anticipation, possibility, or predictions to be examined for their capacity for resolving the situation;
- reasoning, which draws on an established body of knowledge as a means to convert the indeterminate into the determinate, examining the original idea, providing evidence, and developing refined meanings more relevant to the problem than was the initial idea or prediction; and
- an identifiable determinate outcome.

Dewey's definition of inquiry offers important considerations that can help to unify evolving knowledge, pedagogy, and technology, and improve current conceptions of classroom inquiry, within a theory of modelling. First, "reasoning," a key underpinning of his theory of inquiry, is essential to mathematical modelling, in our view: problematic situations and ideas are necessarily considered in relation to previous experience and bodies of knowledge. Systems and structures of knowledge – neither inert nor irrelevant to modelling-oriented approaches – provide means to relate local constituents of problems to distal ones in the transformation of the indeterminate situation. Deweyian inquiry can progress through iterative rounds of problem definition and resolution – a generative learning process.

Secondly, Deweyian inquiry requires an outcome, a determinate situation as a unified whole. He wrote,

> ...inquiry is competent in any given case in the degree in which the operations involved in it actually do terminate in the establishment of an objectively unified existential situation. (Dewey, 1973, p. 227)

We see Dewey's definition as the core of a definition of modelling, and therefore repurpose it as *modelling*, as it signals the production of a terminus, the model. We identify two subparts of that activity: inquiry and reasoning. We place these in dialectic relation to one another. Inquiry is viewed a means to gain purchase on the indeterminate situation. Reasoning is the drawing on bodies of knowledge into the service of transforming the indeterminate situation into a determinate outcome.

Finally, most approaches to modelling describe a model as a map of reality, assuming the need to evaluate the fit between the model and the "real world." We focus instead on the concept of "fitness", referring to whether the model evolves to transform an indeterminate into a determinate situation. The mapping, then, is between two stages of experience (itself a key component of Deweyian educational philosophy). Fitness is pragmatically defined as the degree to which the mapping assists one in preparing oneself for future experience. This has led us to propose the following theory of modelling:

Mathematical modelling is the process of encountering an indeterminate situation, problematizing it, and bringing inquiry, reasoning, and mathematical structures to bear to transform the situation. The modelling produces an outcome – a model – which is a description or a representation of the situation, drawn from the mathematical disciplines, in relation to the person's experience, which itself has changed through the modelling process.

The activity of modelling does not depend on mapping to a particular definition of "reality;" in theory or for assessment. Rather it demands coordinating the justified results with the method of inquiry to provide a means to address outstanding problems. What is produced, represented, and recorded in inquiry into an indeterminate situation is a set of representations that are themselves key artifacts of the modelling process. These include observations, responses, measurements, interactions, indicators, and descriptions. Together these may be described as data, coding systems, methods of sampling and data collection. They are typically mediated by various technologies, and are close to the observed phenomena, but are not the phenomena themselves. It is through the coordination of these key artifacts, together with means of relating them through inquiry, reasoning, and experiment, that the indeterminate situation is converted into the determinate situation, the unified whole that we call a model.

2. THE CASES

We provide excerpts and analysis of student interviews to illustrate how our theoretical analysis of modelling can be applied to student work. The students in these cases were graduate students in education who had enrolled in a class on technology in mathematics and science education led by one of the authors. The class worked with computers and motion detectors for approximately three weeks, exploring a variety of curve shapes in relation to position-time, velocity-time, acceleration-time, falling body problems, and quadratic equations. In working with falling bodies, the students had used software tools (Function Probe) for reflecting, translating, and stretching graphs to fit curves to particular datasets.

As an assessment, each student investigated the motion of a vertically suspended spring (a "Slinky") with motion detector and computer. Their investigations were videotaped and extensively analyzed. Each student was initially asked to predict the appearance of position-time and velocity-time graphs of the spring's motion. Thus, each student made extensive use of direct observation (of the spring's motion), qualitative graphing, and technology (motion detector and computer), coordinated among these, and refined their reasoning about the spring's motion.

Student 1. A former second-grade teacher, M had not taken mathematics or science courses since early in her undergraduate career, and was insecure about her knowledge of those disciplines. In modelling the spring motion, her primary challenges turned out to be 1) predicting graphs of position-time and velocity-time and coordinating them with the spring's motion, 2) understanding the relationship between velocity as slope on a position-time graph and velocity as a point on a velocity-time graph, for constantly changing motion that reversed direction, and 3) interpreting negative velocity.

She initially predicted that the graph of position-time would be represented by a damped periodic function (our language for her picture; Fig. 2.3-1.a). Her first drawing shows a constant and diminishing velocity for each drop and rise of the spring, and diminishing period and amplitude.

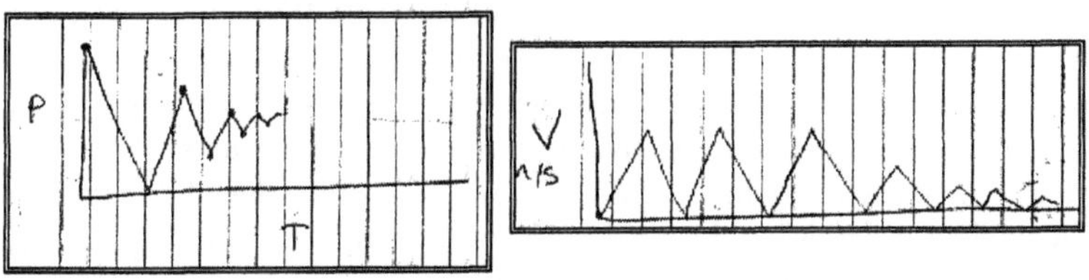

Figure 2.3-1. a. Initial prediction of position-time graph (M).
b. Initial prediction of velocity-time graph (M).

Her initial conjecture about the velocity of the spring was that velocity would be constant, and the same, for the falling (extension) and rising (compression) of the spring, including the same sign. She believed (incorrectly) that her velocity-time graph represented this (Fig. 2.3-1.b). M realized correctly that when the spring is fully extended and reverses direction, there is a momentary stop, with velocity going to zero, and showed the graph anchored to the x-axis (though she had not recognized at this point that the spring also stops momentarily at the top of its travel). Her graph achieved the continuity she expected from the smooth spring motion she observed. However, she conflated the representations of position-time and velocity-time, believing in both cases that the saw-tooth shape represented constant, equal speed during both dropping and rising.

To understand velocity-time graphs better, M tried to disentangle two related conceptual obstacles. The first was viewing velocity as a variable in itself, as a point. She understood velocity on a position-time graph as a derived quantity portrayed by slope, i.e. position change during a time interval. However, in her first try at a new simple velocity-time graph (Fig. 2.3-2.a), she stated that the steeper graph had greater velocity.

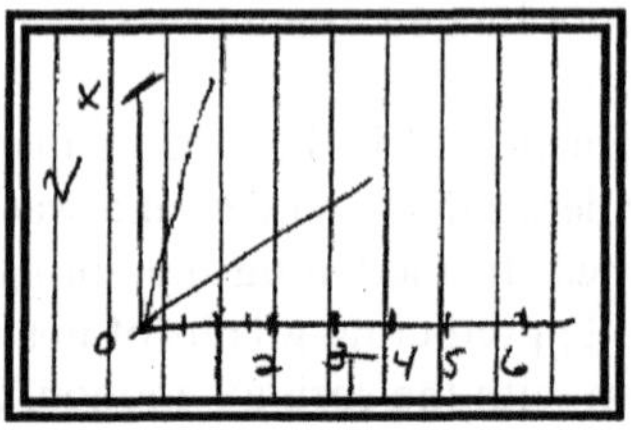

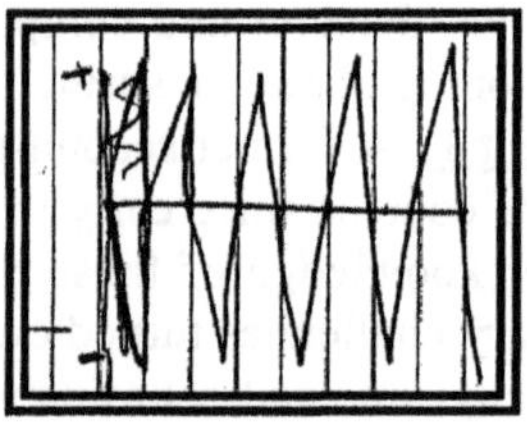

Figure 2.3-2. a. Velocity-time graph (M) *b.* Second velocity-time graph (M)

Her second conceptual obstacle was interpreting negative velocity, initially viewed as negative slope of a line segment located in quadrant 1 of position-time graph, not as a coordinate point (t,v) with $v<0$. An interview question led her to begin to coordinate these ideas:

M: "I don't understand the velocity being negative.... I understand that the negativity of the velocity is really an indication of direction. Right, when it negative... when it goes from 0 to –5 is it going faster or is it going slower?

I: ...if the velocity is zero, the object is not moving; we both agree on that... But what if the object is going negative five? And what are the units?

M: Meters per second;... that's... 5 meters per second going the other direction."

The exchange led M to define the meaning of the coordinate point (1, –5) on a velocity-time graph, but she then proposed (incorrectly) that the graph of a constant negative velocity for the spring should be represented with a straight slanted line reaching to (1, –5). Subsequently she drew vertical line segments to represent that the spring stopped instantaneously at that point (Fig. 2.3-2.b). During the interview, she did not resolve whether the change from zero to –5 m/s means that the spring speeds up or slows down. She perceived that the slanted line was consistent with graphs she had drawn previously for a falling ball, forgetting that for a falling body, velocity steadily increases in a single direction.

However, she observed the spring motion several times, attempting to coordinate her graph with her observations, each informing the other. She shortened the spring and moved it up and down slowly. She conjectured that the velocity increases as the spring drops, stating:

M: "It increases until it gets to the stopping point. Definitely. Slower and slower and faster and faster....." [explaining further] "... the weight of this is going to make it increase as it gets farther away from the starting point. The velocity is getting faster and then it stops. OK. So this is negative... stop, positive, stop. So, it's never positive always negative. Because it starts at zero [at the top] and this is a negative velocity [spring extending] and then it stops and it does the same thing coming up and then stops, so it's never going to be positive. Oh yeah because here is zero. So I think I was right."

At this point, M was satisfied with the velocity-time graph in Fig. 2.3-2.b. M then used the motion detector to produce the graphs in the Fig. 2.3-3.a and 2.3-3.b (time-synchronized p/t and v/t graphs), resulting in a crucial insight. Examining each graph separately, she explained to herself why the lines are curved. Then, examining the velocity curve, she said

M: "...it's going farther and farther away from the motion detector. It is going faster and faster and then slower and slower. So within one trip, it's going to go…O-O-OH!... within one trip on the way... right 'cause it has to slow down in order to stop...Half a trip is faster and half is slowing down... [but]... that does *not* make sense."

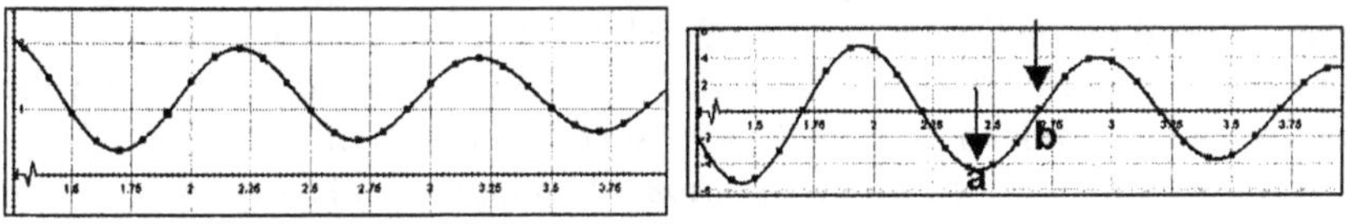

Figure 2.3-3. a. M's p/t graph, from motion detector
b. M's v/t graph, from motion detector

This (precarious) insight helps her obtain the consistency she expects in her velocity-time graph when the spring changes direction. She again examined the actual movement of the spring to consider why her initial conjecture – that the spring behaves like a falling body – did not make sense. She began to perceive in the spring's motion the change in velocity observed in her graphs, and tried to stabilize her prediction by looking at the section in Fig. 2.3-2.b between locations a and b:

M: "So it goes from 2.5 to 3.2 [sec], then it's going to here and its moving 2 m per sec, I am halfway, and then it gets to the top, why is it decreasing, oh because it has to stop. It's going faster [then] slower."

She was now convinced that velocity became zero when the spring reached its maximum *or* minimum height, and recognized the graphical representation (v=0) of velocity crossing the horizontal axis once each for a trip up and a trip down. Her next challenge was to understand that the spring passes through the highest speed halfway through both its compression and extension, as she tentatively asserted previously, rather than like a falling body (constantly increasing speed). In the final sections of the interview, going beyond the logical conclusion that the spring must slow down in order to stop, she began to inquire about forces, gravity, and what she called the spring's "resistance."

<u>Student 2.</u> D majored in mathematics, and later taught mathematics in middle school. He had studied physics in college, and was far more experienced with both graphical representation of periodic motion and modelling using formal equations. He quickly established a predictive mathematical

model based on his prior knowledge, $\sin(x)\cdot e^{-x}$. Considering his more advanced knowledge, one might wonder if the spring task is at an appropriate level for him, but three themes emerge from this case which provide some insight into the nature of modelling, generalize beyond the particular example, and demonstrate the flexibility of modelling tasks. These are:

1. assuring collection of high quality, reproducible data
2. extending the analysis of the initial experiment, and
3. varying the experiment to transform the basic model.

Reproducibility. D carefully set up the experiment to provide consistent reproducible data. He ensured identical time-and-distance starting points for each replicate by configuring the motion detector to begin recording data when the end of the Slinky passed a fixed distance from the detector. He discussed the importance of a "clean run", possible effects of outliers on the graphs, and methods to average data for specific periods of different trials.

Extension of analysis and prediction. D sought to identify and explain sources of variation or deviation from his original predictions. In particular, he noted the presence of apparent harmonic oscillations in the spring data. He recognized the need to modify his original model. In one interaction, D noted that when the spring got to the top of its travel, a reproducible oscillation resulted in a "chopping off" of the overall graph. Rescaling the graph, D commented:

> D: You can really see different sine waves here. There's this major force here, but then it's going quickly here and... there's a first harmonic going on right there. ...it has a positive derivative here, so it's accelerating along with the major frequency. And then here that first harmonic is decelerating, so we can see it kind of looks like it goes straight up and then turns, and doesn't accelerate quite as quickly but I bet if we did a Fourier transform we'd see a real strong line at... at um, about two and a half...a frequency of two fifths Hertz I guess that is. And then we'd see one at half that frequency, er... at double that frequency, that's ah 5 Hertz, um, that's, I think we'd see two bright points there… it was reproducible, too: all three trials did that.

D extended his exploration of the original model and its harmonics further, conjecturing that systematic variance or distortion could be amplified by modifying the setup. He bunched 20 coils together at the end of the spring, then released them when he released the entire spring:

> D: I'm going to try some sort of chaotic system, here, and I think probably the best would be just bunch up a certain number of these, and I want to make it reproducible, bunch up 20 [coils], and then let it go. And see, so that'll be you know this much of an effect, [manually demonstrates just the lowest 20 coils oscillating], along with the lar-

ger [spring]... I would hypothesize that this will cause a lot more harmonics than both of those frequencies. And there will be some relation between the whole length and the length [of coils at the bottom] I choose. It's not just going to be two oscillations, two frequencies anymore, but it'll be a good way to start that.

In the following, we note his careful attention to data quality, adjusting the sampling frequency to ensure the capture of the phenomena of interest. It also illustrates how D extended his model, using the coiled start to stimulate investigation of harmonics:

D: I set the sampling frequency to 30 Hertz. It was originally at 10 Hertz [which] would have lost a lot of these oscillations...

D: ...there's a major oscillation..., and then you can see these minor oscillations within that... actually sometimes have a greater velocity than the major oscillation, than the movement from the major oscillation. So for example, here, while the general direction was going down, there was an upward harmonic that had enough velocity to actually make the [spring] go up, even though [overall] it was coming down.

Transforming the basic model. The third theme emerged as D explored variations on the spring setup in order to understand their effects on the basic model. This was stimulated in part by the question "how would your model vary if you were to shorten the spring or add some weight to it?" In one experiment he shortened the spring. An interesting tactical move was to shorten the spring by a multiple (½) of its number of coils, rather than a fixed number (additive).

D was confident that halving the spring length would transform the position-time curve: halve the amplitude, double the frequency (halve the period) and vertically translate the position-time curve (to account for the different distance between the detector and the spring). A subsequent exchange illustrated D's strong and flexible background knowledge of transformations of the trigonometric curve. However, he was less certain whether there would be a change in the e^{-x} (damping) component of the mathematical model. He conjectured that there would be a horizontal shrink, and that this would be the same as a vertical stretch (a misconception). Visually, it was difficult to see that a horizontal translation, not a horizontal stretch, of an exponential equates to a vertical stretch. This differs from quadratic or linear functions, for which horizontal stretches can be equated to a vertical stretches.

3. CASE SUMMARIES

We suggest that for M, the computer-generated representation itself became a determinate model for the previously indeterminate situation of the spring's motion. M used a number of landmarks to accomplish this, including reasoning she had previously acquired about position-time graphs, and then derived inquiry into velocity-time graphs from her observations of the spring and previous experience with motion detectors. Comparing hand-drawn graphs to the technology-mediated representations served as feedback to her conjectures.

M's determinate situation evolved to be the coordination of the technology-mediated graphs with the motion of the spring. She illustrated how modelling can proceed from rudimentary knowledge of the constituents of position, time, and speed, and a limited but solid experience of those ideas applied to rolling and falling objects. It is compelling how she repeatedly applied these constituents until she formed a consistent model that coordinates among her observations and two representations mediated through the technology, and changed her own perception of the spring's motion. Within our theory's dialectic of inquiry and reasoning she demonstrated stronger emphasis on inquiry.

In D's case, we see how, in the same modelling situation, a more knowledgeable student can draw upon mathematical reasoning about trig functions, exponential decay, and transformations and harmonics. We also witness a more sophisticated experience with data sampling, averaging, and the means of ensuring data quality.

D's determinate situation was to generalize on the movement of springs as variations on the equation "sin times *e* raised to the negative time, with some constants." More advanced in his mathematical preparation, he approached the indeterminate situation with a basic equation in mind as a model, proceeded to examine it in varied circumstances and extended it to include finer distinctions and variations. D demonstrated more structured mathematical reasoning, based on his more developed prior knowledge, in his expression of the inquiry-reasoning dialectic. Both students demonstrated alternative representations, sampling, observation, and data use as key elements of the modelling process.

4. DISCUSSION AND CONCLUSIONS

Both students encounter an indeterminate situation, bring mathematical structures to bear, and transform the situation to a determinate form, resulting in a model. Their models are not actually models "of reality," but rather descriptions or representations drawn from mathematical discipline in relation to their experience of the spring's motion. They identify different forms of evidence and bring different levels of prior knowledge to bear. Their own experience changes as a result of the modelling exercise.

For both students, inquiry and reasoning led to finer and more focused experience of the spring's motion. Considering their different mathematical backgrounds and their progress in this very brief investigation, we suggest that modelling may be a vehicle for instruction differentiated across levels of experience within a single classroom setting.

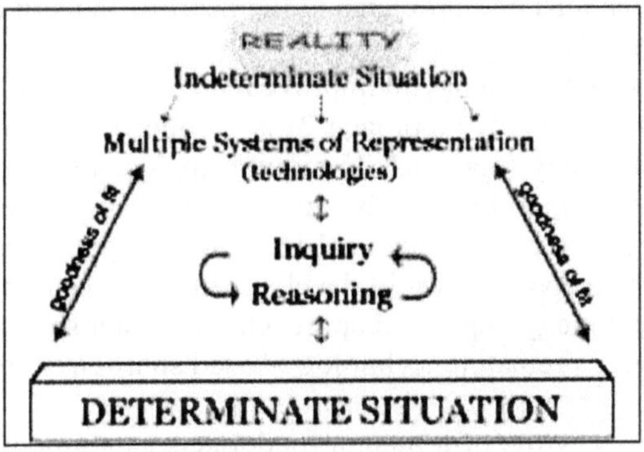

Figure 2.3-4. A model of modelling

Fig. 2.3-4 presents a graphic of our theory of modelling. Modelling can be an extraordinarily powerful organizer for mathematical instruction. It can anchor mathematical learning in the prior and current experience of a student, and build further experience through iterative rounds of inquiry and reasoning that incorporate mathematical problematizing. The entire modelling process is embedded with mutually coordinated observations, representations, and mathematization. The outcome of a modelling episode would be a condensed statement and summary of the students' mathematization efforts, which can be used both by students and instructors as the starting point for subsequent investigation. It is rich territory for metacognitive reflection.

Early in this chapter, we identified four approaches to the use of technology in mathematics instruction. We now return to these and comment on the role and priority of each approach, assuming a focus on modelling as a central organizer of the study of mathematics.

What are the roles of technology in a modelling approach to mathematical instruction? Whether electronic or mechanical, technology can incorporate and generate representations which are themselves the subject of or tools for identifying and transforming an indeterminate to a determinate situation.

Technology plays a central role in coordinating the inquiry, reasoning, and systematizing that lead to a determinate situation.

A modelling approach to mathematical instruction accommodates – even requires – all four of the approaches to technology listed at the beginning of this chapter. Technology permeates the everyday environment of students; it is only logical that it provide content for multiple disciplines, both explicitly mathematical and those, for reasons pragmatic, pedagogical, or theoretical, predisposed to mathematical treatment (approach 3). A growing ensemble of technologies serves as qualitative and quantitative tools for student investigation of applications and problems at the heart of modern mathematics curricula (approach 4).

Mediation with technology to improve recognition of patterns of behavior of mathematical concepts (approach 2) and concept mastery prior to introduction of technology (approach 1) require implementation of effective pedagogical strategies, and should be nuanced, not rigidly traditional. Understanding the genesis of mathematical ideas should guide decisions on technology use in conceptual and skill development. It is incumbent on educators to integrate technology-supported approaches to foster students' concept mastery, but to avoid premature technology-based short-circuiting of conceptual development.

Just as Dewey wrote of a map, a model acts as a guide: "that which we call a science or study puts the net product of our past experience in the form which makes it most available for the future. It economizes the workings of the mind in every way." (Dewey, 1973) It is this view of modelling in mathematics which we believe will move us into the next phase of mathematics teaching and learning.

REFERENCES

Blum, W. et al. (2002). ICMI Study 14: Applications and Modelling in Mathematics Education – Discussion Document. *Educational Studies in Mathematics*, *51*, 149-171

Bornstein, M. H., & Bruner, J. (eds.) (1989). *Interaction in Human Development*. Hillsdale, NJ: Lawrence Earlbaum Associates.

Niss, M. (2001). Issues and Problems of Research on the Teaching and Learning of Applications and Modelling. In J. F. Matos et al. (Eds) (2001), *Modelling and Mathematics Education: ICTMA-9,* (pp. 72-88). Chichester: Ellis Horwood,

Dewey, J. (1973). *The Philosophy of John Dewey.* J. J. McDermott (Ed.), University of Chicago Press.

Linn, M. C., & Hsi, S. (2000). *Computers, Teachers, Peers: Science Learning Partners.* Mahwah, NJ: Lawrence Earlbaum Associates.

Vygotsky, L. V. (1930). *Mind and Society.* Cambridge: Harvard University Press.

Chapter 2.4

WHAT KNOWLEDGE DO TEACHERS NEED FOR TEACHING MATHEMATICS THROUGH APPLICATIONS AND MODELLING?

Helen M. Doerr[1]
Syracuse University, Syracuse, NY, USA, Email: hmdoerr@syr.edu

Abstract: This paper begins by describing teachers' knowledge as the creation and development of increasingly sophisticated models or ways of interpreting the tasks of teaching. One study illuminates several ways that pre-service teachers perceive the processes of modelling and the limits of their experiences with stochastic models. Results from a second study indicate that teachers need to have a broad and deep understanding of the diversity of approaches that students might take with modeling tasks. The second study also suggests a reversal in the usual roles of teachers and students by engaging students as evaluators of models.

1. INTRODUCTION

The call for contributions to the ICMI *Study on Applications and Modelling in Mathematics Education* observes that only rarely do mathematics teacher education programs include an orientation to mathematical modelling or the use of modelling in prospective teachers' mathematics courses. This suggests that one reason for the limited use of applications and modelling at the primary and secondary levels of schooling is the lack of knowledge by those who are expected to teach mathematics through applications and modelling. However, the research base on the knowledge needed for teaching, at least in the United States, has established that subject matter knowledge alone, while necessary, is insufficient for quality teaching. This raises the issue, then, as to the scope of the knowledge that teachers need in order to be effective in using applications and modelling in their practice. As

I will argue below, the pedagogical knowledge for teaching modelling would appear to differ in some significant ways from traditional and reform-based methods for teaching mathematics.

In this paper, I will frame a discussion of the issues that applications and modelling raise for the mathematics education community when we focus our attention on teachers and teaching. First, I will describe a theoretical perspective on the nature and the development of teachers' knowledge. Second, I will provide results from my research on several aspects of the subject matter knowledge of pre-service teachers within the context of an undergraduate course in mathematical modelling. Third, I will provide an analysis of an example from a research project on teachers' pedagogical knowledge when teaching mathematics through modelling tasks. This example, drawn from the practice of an experienced secondary school teacher, illuminates the kinds of pedagogical knowledge that seem to be necessary when teaching from a modelling perspective. I will conclude with some comments about the challenges that this research raises about the knowledge that teachers need when teaching mathematics through applications and modelling.

2. THE NATURE OF TEACHERS' KNOWLEDGE

The starting point for conceptualizing the nature and development of the knowledge needed to teach mathematics through modelling and applications is that teaching is primarily about the creation and refinement of sophisticated models or ways of interpreting the tasks of teaching. These tasks include choosing appropriate modelling applications for students, knowing how students' models might develop over the course of several lessons or several applications, selecting activities and curricular materials that might further that development, and devising strategies for engaging students in the critical assessment of their models.

A modelling perspective on teachers' knowledge foregrounds the notion that teachers have models for teaching (Doerr & Lesh, 2003). These models are the systems of interpretation that teachers use to see students' ways of thinking, to respond to students' ideas, to differentiate the nuances of contexts in their practice, to see generalized understandings that cut across contexts, and to revise their own thinking in light of their experiences. In examining teachers' knowledge, we focus on how the teacher thinks about the context, what alternatives she considers, what purposes she has in mind, what elements of the situation she attends to and what meanings and relationships those elements have for her.

A central question for research on teacher knowledge is the examination of how teachers' models for teaching mathematics develop. It is clear that teachers come to their pre-service teacher programs with models of teaching already in place, based on years of apprenticeship as observers of practice. Furthermore, teachers' models of practice (or systems for interpreting practice) are significantly broader in scope and more complex than the kinds of models students develop. The results from two research projects that examined the subject matter knowledge of pre-service teachers and the complexities of the pedagogical knowledge of an experienced teacher illuminate some of the central characteristics of teachers' models for interpreting practice and provide some insight into the challenges inherent in the development of such models.

3. SUBJECT MATTER KNOWLEDGE IN PRE-SERVICE TEACHER EDUCATION

Few studies have directly addressed the knowledge of mathematical modelling that pre-service and in-service teachers hold and how that knowledge is acquired (e.g., Dugdale, 1994; Lingefjard, 2002; Zbiek, 1998). To examine the modelling knowledge of those preparing to teach, I designed an undergraduate course in mathematical modelling. The primary goals of the course were to introduce pre-service teachers (N=8) to some basic ideas and techniques in mathematical modelling by engaging them in the process of building mathematical models. The course content drew on problem situations from physics, biology, and mathematics itself. The course began with several empirical models and then moved to an analysis of discrete dynamical systems and stochastic models. We finished the course with some examples of continuous models. The technological tools included graphing calculators and calculator probes for data collection, Maple, spreadsheets, and a simple dynamic systems simulation language. The students worked in small groups and completed five modelling projects over the course of the semester. Several classes devoted time for students to work collaboratively on the projects and to present their findings.

The students' class work, their final projects, class discussions, and written assignments were the data corpus for this research study. The research questions focused on examining the nature of pre-service teachers' knowledge and perceptions about mathematical modelling. The analysis of the data yielded three significant findings. The first finding related to the mathematical knowledge of the pre-service teachers with respect to probabilistic situations. A serious misconception about binomial distributions and the probabilities of independent events occurred among the pre-service teachers in the same ways that I have found among secondary school students (Doerr, 2000). In particular, when creating a simulation for stochastic exponential growth using a random number generator, several pre-service teachers erroneously used a random number from a uniform distribution as an appropriate number in a context that called for a binomial distribution. A subsequent project involved creating a simulation for a stochastic logistic growth situation. In this context, the need for a random number from a binomial distribution was even less obvious; nearly all of the students made the error of choosing a random number from a uniform distribution. This finding confirmed results in the literature that would suggest that formal instruction in probability has limited impact on learners' abilities to reason probabilistically. However, it was also the case that all of the students were able to adjust their incorrect conceptions to mathematically correct ones through a process of explaining and justifying their models to each other. This suggests that mathematical modelling is a potentially powerful context for the mathematics learning of pre-service teachers.

The second finding directly addresses the perceptions and beliefs held by the pre-service teachers as to the nature of the modelling process. As part of the course, the students completed several readings that discussed modelling at a meta-level (Bassanezi, 1994; Weigand & Weller, 1998). Weigand and Weller (1998) present a description of modelling that involves a six-step process: analyzing (A), simulating (S), modelling with equations (M), working experimentally (W), interpreting (I), and explaining (E). Throughout the course, the pre-service teachers were asked to describe their own specific modelling processes in terms of these steps. Initially, the pre-service teachers saw Weigand and Weller's steps in the modelling process as an unproblematic description of how modelling was really done. They saw the steps as occurring in sequence. Early in the course, when asked to map the processes they had used to create a model, most students created maps similar to that in Fig. 2.4-1.

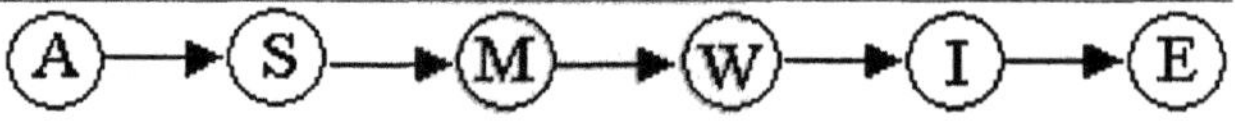

Figure 2.4-1. A sequential view of the modelling process

But later in the course, the pre-service teachers made a striking shift from seeing modelling as a fairly linear, sequential activity to seeing modelling as a non-linear, cyclic activity. A typical student map of the modelling process now looked more like the one shown in Fig. 2.4-2. In this case, the students saw their process as beginning with the simulation step (S) and then moving to interpreting (I), then modelling with equations (M) and so on.

The pre-service teachers engaged in extended discussions about the meaning of the terms that were used to identify each of these modelling steps. They began to give more nuanced meanings to the steps, describing their activities as "thinking about what is going on in the situation", "working and re-working the math equations to get them right", and asking questions such as "does everything we're doing make sense?" and "why do our ideas work?" This shift in the perceptions of the pre-service teachers came about as they reflected upon their experiences in developing models.

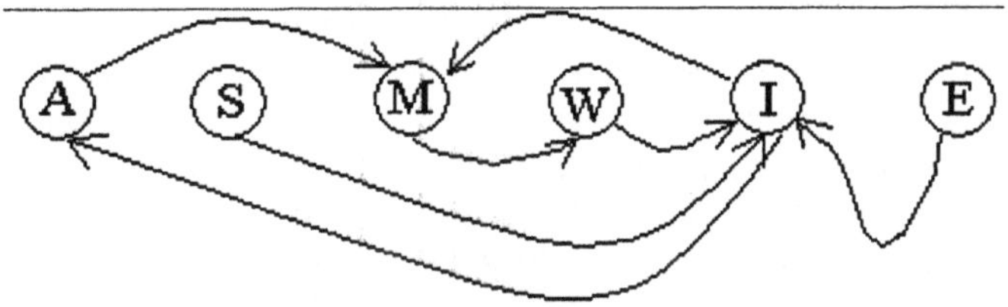

Figure 2.4-2. A non-linear, cyclic view of the modelling process

The third finding from the analysis contradicts the findings of Zbiek (1998), who in her study of pre-service secondary teachers found that many of the pre-service teachers tended to use regression analyses when available and that they often used curve-fitting uncritically in their approach to problem situations. We found no evidence to confirm these tendencies. Even though curve-fitting software was readily available, students rarely used it and when curve fitting was done, students always attended to the meaning of the resulting equations and coefficients in the problem tasks. This result does not suggest that these pre-service teachers were more sophisticated than those in Zbiek's study, but rather it argues that the nature of the modelling tasks, the range of tools available, the norms for argumentation, and the standards for quality of a solution were significant in influencing the types of modelling behavior that occurred in each setting.

Collectively, these findings suggest that pre-service teachers are likely to encounter greater difficulties in developing stochastic models than in developing models of deterministic phenomena. This result can in part be accounted for by the well-known misconceptions from the research on probabilistic reasoning and by the dominance in mathematics courses of continuous functions and their applications in physics. It leaves unanswered, however, questions about how to best approach the development of both kinds of models. The findings also suggest that by reflecting on their own modelling activity, pre-service teachers can come to understand the cyclic nature of the modelling process and appreciate the interconnectedness of the cognitive activities involved in the process. Finally, these results suggest that the use of regression models by pre-service teachers seems to be dependent on the kinds of modelling activities that they experience. This implies that pre-service teachers need to be exposed to a range of modelling activities that provide multiple opportunities for explanations and justifications of the modelling decisions that were made.

4. PEDAGOGICAL KNOWLEDGE IN ACTION

This set of results is drawn from the analysis of a teaching episode with an experienced secondary mathematics teacher who was using a sequence of modelling tasks related to exponential growth and decay. My intention in the analysis of the classroom data is to illustrate some of the pedagogical demands that are made on the teacher when a modelling approach is taken to the teaching and learning of mathematics. In this particular lesson with 16 – 17 year old students in a pre-calculus class, the students had been working on a task to model the doubling that occurs in bacteria growth. Finding appropriate graphs, equations, and tables to represent that growth was relatively straightforward for this group of students. One portion of the task focused the students' attention on the problematic issue of quantifying the growth rate and what units might be used to measure that rate:

> A biologist knows that the population of a bacteria culture doubles every 15 minutes. After 1 hour and 15 minutes, her assistant found that 80,000 bacteria were present.
>
> Examine the rate at which the bacteria culture is growing. How fast is the culture growing at 1 hour? At 1.5 hours? At 2 hours? How are you making these estimates? What are the units for this rate? Do your estimates make sense in terms of your graph?

The teacher, who had used these modelling tasks the previous year, knew that examining the rate of change would be problematic and challenging for her students. She recognized that the notion of rate of change was an important idea throughout the pre-calculus course. Understanding the changes in a model and ways of representing that change is a fundamental mathematical idea and one that foreshadows the development of important concepts in calculus. The teacher had chosen to focus on this particular aspect of the bacteria growth model because of the richness of the rate context.

Several groups of students presented their work on the board, including tables, graphs, and equations. The discussion of these solutions started out slowly with some comments on the tables and the different units for the rates and some comments on the functions, which were different as well. But the most interesting discussion occurred as the students talked about the rate of change. The teacher was able to pull in many student-to-student arguments as well as many elaborated student explanations. She was careful in listening and seeing how the students elaborated their rate concepts as the discussion evolved. The students had four different ways of presenting rates:

> *(1) Sara's method:* Sara found the bacteria present at 1 hour. Her equation was $y = 2500 \cdot 2^{4x}$ where x was in hours. Her explanation of the "4" was that it took four quarters to double and hence four of these quarters

("4x") would give you one doubling. She then found the bacteria present at 1.00001 hours and divided the increase in the bacteria by .00001 and called this quantity "bacteria per hour". She then calculated the rate at 1.5 hours and 2.0 hours and insightfully observed that the rate itself is also increasing by a doubling factor! In her arguments in class, Sara pointed out that rate could be thought of as the slope of "the little line segments between the points" of the graph.

(2) Bryan's method: Bryan took the amount of bacteria present at 60 minutes and divided it by 60, yielding 40,000/60 bacteria per minute. His equation was $y = 2500 \cdot 2^{x/15}$. Bryan was adamant that his equation and his estimate for the rate were correct! Bryan said that he still didn't see what was wrong with his approach. This seemed to be both a need for resolution of multiple methods and a need to reconcile his view with the other competing views in class. The teacher made the decision to continue with the discussion. This brought Jack (another student) to Bryan's side, and he led other students to try to see Bryan's point of view. Sara and others appeared to appreciate what Bryan was saying but weren't convinced that it was correct, however they had difficulty in explaining a flaw in the reasoning. The graphic representation of Bryan's estimate could be seen as the slope of the line joining the point (60, 40000) and the origin. The teacher drew this segment on the graph as the discussion evolved.

(3) Peter's method: Peter found the bacteria present at 1 hour and then said since it is 40,000, that you should divide the 40,000, since that is also the amount that it will increase in the next interval, by 15 to get the rate of increase. The teacher was initially unclear about how Peter was finding the rate. It appeared that the 40,000, which was the amount of bacteria present at 1 hour, was being divided by the time interval. But it was clear that the student was thinking that the 40,000 was both the amount and the increase in the amount, and hence you could divide it by the time interval and get the rate. The teacher re-cast Peter's description into the language of the change in the amount of bacteria divided by the change in the time interval and wrote (80,000 – 40,000)/15. Later the teacher commented that last year, several students had taken this approach and she had had trouble grading their papers because the students had not made clear how they were thinking about the quantity.

(4) Mark's method: Mark had written $y = 2500 \cdot 2^{2x}$ as his equation. The discussion of his solution focused on the rate at 1.5 hours, or the 6th time interval. Mark had used x to represent the number of 15 minute time intervals, rather than the actual time in hours or minutes (as had been done by the other students). Mark changed the table interval on his calculator

to .001 and found 160,111 bacteria at 6.001. Mark had written the rate as (160,111 – 160,000)/.001 and then described the rate "as per 15 minute interval." The teacher asked, "How are we getting the 15 minutes?" Mark replied that he was using time as 15 minute intervals.

During the discussion of Mark's method, Peter commented that this was just finding the slope between two points. Later, Mark argued that if we thought of the graph of the bacteria population as a position graph, then what we are trying to find is its velocity graph. The teacher quickly picked up on this as the connection to early work that the students had done with a simulation environment (Kaput & Roschelle, 1997) in exploring the relationship between a velocity and a position graph. She then asked, "How do we find a velocity graph from a position graph?" and the students answered, "by finding the slope." Bryan however stayed strong in his position by arguing that he was finding the slope at a point and asked, "what does that mean?" and "why can't I do it that way?" As class ended, Bryan and Mark continued to argue this point. After class, the teacher indicated that she wanted to have the students "commit" to their ideas and to think about the concept of rate, before pursuing it further in class. In this way, the teacher saw how a central concept such as rate of change is not understood "all at once" but is revisited through a sequence of modelling tasks.

This teaching episode suggests two major implications for the pedagogical knowledge of the teacher when teaching with modelling tasks. First, the teacher needs to have a broad and deep understanding of the diversity of approaches that students might take. Trying to quickly grasp the mathematics presented in the four approaches described above, while simultaneously devising appropriate responses, is not an easy task for the teacher. The difficulties in doing this should not be underestimated. To acquire such understanding, the teacher must engage in listening to the students as they interpret and explain their models. In the case above, the teacher recognized the ambiguity in how one student (Peter) was finding the rate, since the value of the function at the particular point in time was also equal to the increase over the next time interval. The teacher cast the student's representation into the language of rate of change so as to clarify the underlying mathematics.

The teacher also needed to carefully listen to another student's description of the rate as being "per 15 minutes", an approach that the teacher had not expected. In this instance, the teacher probed the student's thinking and attempted to understand the mathematics being expressed by that student. Later, the teacher supported the development of the students' ideas by elaborating on the connections that the students made to earlier representations that they had used. Modelling tasks provide the opportunity for students to develop a diversity of approaches to expressing their interpretations of a given situation. While this created a rich source of mathematical discussion for the students, it also placed substantial pedagogical knowledge demands on the teacher. This case illustrates four characteristics of the teachers' knowledge: (1) to be able to listen for anticipated ambiguities, (2) to offer useful representations of student ideas, (3) to hear unexpected approaches, and (4) to support students in making connections to other representations. How teachers acquire this knowledge, both in their preparation programs and in practice, remains an open question for researchers.

The second implication for the pedagogical knowledge of the teacher is illuminated in the shift that occurs in giving explanations and justifications. Rather than the teacher giving explanations and justifications to the students, the discussion of the models created a learning context in which the students were giving explanations and justifications to each other and to the teacher. This shift signals an important aspect of learning that takes place when using applications and modelling: the task for the teacher becomes one of putting the students in situations where they can interpret, explain, justify and evaluate the "goodness" of their models. In the case of the competing models for finding the rate of growth, the teacher encouraged the students to share their thinking and make sense of the explanations that were given by others. At the end of the discussion, however, she chose to give the students time to "commit" to their own ideas, perhaps re-evaluate them, before continuing with class discussion. In this way, the teacher gave the students the task of refining and revising their models, rather than proceeding to evaluate them herself. This change in pedagogical strategy is a major shift from more traditional instruction in mathematics where a primary role of the teacher is to evaluate students' work.

5. CONCLUDING REMARKS

The brief synopses of research that I have presented here are intended to suggest some of the challenges for teacher education programs that are raised by the use of modelling and applications for the teaching and learning of mathematics. Teacher education programs need to address both the subject matter knowledge of teachers and the development of new kinds of pedagogical knowledge. In particular, pre-service teachers need to gain experiences in their preparation programs with stochastic models; such a change would imply a shift away from the current dominance of deterministic models in the mathematical preparation of teachers. The difficulties that all learners have with probabilistic concepts make such a shift especially challenging. Pre-service teachers need to encounter modelling experiences that provide for a range of contexts and tools and that engage them in meta-level analyses of their modelling activity.

Teaching mathematics through modelling provides substantial challenges to our current ideas about pedagogy. When engaged in such teaching, teachers are likely to encounter substantial diversity in student thinking. This places new demands on teachers for listening to students, responding with useful representations, hearing unexpected approaches, and making connections to other mathematical ideas. A modelling approach to teaching mathematics calls for a major reversal in the usual roles of teachers and students. Students need to do more evaluating of their own ideas and teachers need to create opportunities where this evaluation can productively occur. Current research in the preparation and development of teachers in taking on these new roles is limited. International research in this area could provide the needed coherence for the development of a knowledge base of effective pedagogies when teaching mathematics through applications and modelling.

REFERENCES

Bassanezi, R. (1994). Modelling as a teaching-learning strategy. *For the Learning of Mathematics, 14*(2), 31-35.

Doerr, H. M. (2000). How can I find a pattern in this random data? The convergence of multiplicative and probabilistic reasoning. *Journal of Mathematical Behavior, 18*(4), 431-454.

Doerr, H. M., & Lesh, R. A. (2003). A modeling perspective on teacher development. In R. A. Lesh & H. M. Doerr (Eds.), *Beyond constructivism: Models and modeling perspectives on mathematics problem solving, learning and teaching* (pp. 125-139). Mahwah, NJ: Lawrence Erlbaum Associates.

Dugdale, S. (1994). K-12 teachers' use of a spreadsheet for mathematical modeling and problem solving. *Journal of Computers in Mathematics and Science Teaching, 13*(1), 43-68.

Kaput, J., & Roschelle, J. (1997). Deepening the impact of technology beyond assistance with traditional formalism in order to democratize access to ideas underlying calculus. In E. Pehkonen (Ed.), *Proceedings of the 21st Conference of the International Group for the Psychology of Mathematics Education* (pp. 105-112). Lahti, Finland: University of Helsinki.

Lingefjard, T. (2002). Mathematical modeling for preservice teachers: A problem from anesthesiology. *International Journal of Computers for Mathematical Learning, 7*, 117-143.

Weigand, H. & Weller, H. (1998). Modelling real-life problems involving periodic processes with computer algebra. *The International Journal of Computer Algebra in Mathematics Education, 5*(4), 251-267.

Zbiek, R. M. (1998). Prospective teachers' use of computing tools to develop and validate functions as mathematical models. *Journal for Research in Mathematics Education, 29*(2), 184-201.

[1] This material is based upon work supported by the National Science Foundation (NSF) under Grant No. 9722235. Any opinions, findings and conclusions or recommendations expressed in this material are those of the author and do not necessarily reflect the views of the NSF.

Chapter 2.5

BEYOND THE LOW HANGING FRUIT

Peter Galbraith
University of Queensland, Brisbane, Australia, Email: p.galbraith@mailbox.uq.edu.au

Abstract: This chapter argues that while there have been some notable achievements incorporating aspects of applications and modelling into educational programmes, this is no cause for complacency. In addition to making current knowledge and initiatives available to a wider spectrum of the educational community, future advancement requires that new questions be posed, existing conceptualisations deepened, and dilemmas identified and addressed. The chapter samples a selection of issues, some ongoing, some emerging that are illustrative of these challenges.

1. INTRODUCTION

Picking the *low hanging fruit* is used in the English-speaking world to mean collecting the fruit that is easy to reach, rather than stretching for rewards on higher branches. As a metaphor it refers to the temptation to repeat or re-invent activities, or move sideways to a new area of interest, rather than address tougher issues necessary for deepening progress in a field of study.

When the low hanging fruit is also sweet, the incentive to search the higher branches is diminished further. In the case of applications and modelling a shared excitement unites many who have enthused about early experiences in the field, for example when students unleash latent power that for whatever reasons had remained fettered in their previous mathematical life. However this very exhilaration can work against further progress, both individually, and particularly at a system level, by creating a sense of adequate achievement that obscures the reality that there is so much more to do. This Study Conference and its aftermath have a key part to play in disturbing complacency, and this chapter visits some areas where efforts have been expended, and some successes have been achieved, but where more is needed to move attainments beyond levels that might be too readily accepted. Given limited space, the areas chosen have mainly to do with authenticity in modelling activity, and with challenging issues associated with assessment.

2. AS IF FRIENDS

The expression 'as if friends' has been used to describe threats to the progress of an initiative that occur through the impact of apparently supportive and related concepts or activities, that can fundamentally undermine or retard the initiative. Two examples selected for illustration with respect to the progress of applications and modelling are *word problems*, and the *curve fitting* capacity of electronic calculating devices.

2.1 Word Problems versus Modelling Problems

Certainly it would be fair to say that *word problems* are widely construed as close relatives of *modelling problems,* for both word problems and modelling problems are couched in verbal clothes. In other ways the two may differ markedly, specifically with respect to meaningfulness and purpose, for modelling problems have real-world connections, which word problems often do not have. It was concern about the message conveyed by word problems that prompted Henry Pollak to write a short article (Pollak, 1969) that contributed seminally to the arrival of applications and modelling as a subject of importance in mainstream mathematics education. A contemporary example illustrating the same concerns expressed by Pollak is shown below.

> A take-away food shop sells hamburgers, sausages, and pizzas. On one day the number of hamburgers sold was three times the number of pizzas, and the number of sausages sold was five times the number of pizzas. The number of hamburgers and pizzas sold was in total 176. How many of each type of food was sold?

While this problem is couched in the language of the real world there is no sense in which it represents how a vendor would make decisions essential to her/his livelihood. It does not show how mathematics can be applied to enhance understanding of real problems. The point here is that this very issue, a major motivation behind the article by Pollak more than 30 years ago, remains an issue today. As was stated then, word problems have value in curricula and we can learn from advances made in the understanding of how students cope with them. However if they are to play an enhancing (rather than a subverting) role in helping students to apply their mathematics to real problems, it is essential to clarify those additional or different features that characterise examples that genuinely involve applications and modelling. Clearly this remains unfinished business.

2.2 Modelling as Curve Fitting

The view of modelling as a holistic process encompassing formulation, solution, interpretation, and evaluation as essential components is inherent in an approach within which links between the real and mathematical worlds are considered at every stage. It is salutary then to identify viewpoints in the mathematics education community that appear to differ explicitly, or by implication, from this view. It has become increasingly clear that some authors are using the term *mathematical modelling* to simply to describe procedures for fitting curves to sets of data points. Now it is beyond dispute that such skills are important for understanding and working with real world phenomena, but there is concern when such approaches ignore essential aspects of the real world context within which a problem is located. The purpose of one exercise was to fit a graph to national population data, using regression methods to search for the best fitting member among a range of 'possible' functions, that included, polynomials, exponentials, and rational functions – as driven by their availability on CAS technology. Missing entirely was any consideration of an underlying growth model responsible for the data generation in the first place – for example how a rate of change of population defined by births, deaths and migration could possibly lead to polynomial patterns of growth as explored in the paper.
Using curve fitting as a synonym for modelling creates an aberration of the modelling concept. One of the qualities we need to continue to emphasise is the holistic nature of the modelling process, versus the detail present within component phases, and that the latter, while important, must be validated in terms of the total context. Subversion of reality by choices available on the menus of calculators represents a substantial distortion of modelling practice, leading to bad modelling habits as well as inappropriate outcomes.

3. ASSESSMENT IN APPLICATIONS AND MODELLING

In addressing assessment it is recognised that a range of significantly different emphases exist within the field of applications and modelling, and that their perceived importance varies with context and educational level. The following have been selected as representative of contemporary activity.

- Assessment of tertiary modelling projects

- Diagnostic approaches to modelling competencies
- Issues in task design
- High stakes assessment at secondary level

3.1 Tertiary Modelling Projects

An early influence on the introduction of applications and modelling in the tertiary sector arose from statements by employers concerning perceived difficulties experienced by new graduates in working in teams, and solving real mathematical problems in the workplace (O'Carrol et.al., 1987). These aspects have been consistently addressed since that time, with emphases on formulation of problems, communication skills, various means of assessing quality in problem solutions, and in reporting outcomes of modelling projects. As a variation Hamson (2001) described how students coped when presented with formulated environmental models, for which they were required to examine and analyse outcomes. This approach suggests another way of assessing skills associated with the difficult *formulation* phase of modelling. It reverses the usual order by asking students to work 'backwards' – that is to articulate what a real problem could look like, given its formulation as a mathematical problem.

Several criterion-based approaches have been developed to evaluate aspects of the modelling process; use of videos of oral presentations rated on systematic criteria associated mainly with presentation skills (Le Masurier in Haines et.al, 1993); construction of rating scales for the assessment of modelling work reported in poster format (Houston, 1997); criteria for assessing extended mathematical modelling tasks in terms of communication, execution, and interpretation skills (Goldfinch and Goodall, 1995); criteria emerging from the use of self, peer, and tutor assessment of group oral presentations (Crouch and Haines in Izard, 1997). A central concern within the latter approach involves the consistency of raters. Crouch and Haines noted inconsistencies between peer and self-assessments, with those with good modelling performances often not recognising quality work carried out by peers. Houston found that peers were generally consistent with the lecturer but varied along the dimension of leniency-stringency, while Goldfinch & Goodall found that peers while generally consistent with the lecturer, again differed in stringency, and advocated specific training in the use of rating scales. All reported studies within this genre identified difficulties with inexperienced judges in identifying and rewarding quality work. Apart from course grades, the development of peer assessment capability bears directly on goals to enhance the ability of graduates to work in teams, in which the capacity to evaluate validly the models of others is essential to success. One suggestion to make raters more accountable in peer assessment is to 'assess the assessor' by including a judgement of the quality with which individuals identify significant aspects of others' models. Many excellent suggestions for assessing aspects of applications and modelling in higher education are also illustrated in publications from the HEFCE Project (1996).

3.2 Diagnostic approaches

Haines et al. (e.g. 2003) describe studies that probe student understanding within intermediate stages of the modelling cycle. Do students become better at these intermediate phases, e.g. identifying key assumptions, clarifying model purpose, formulating a precise problem etc...? Using multiple choice tools the authors highlighted difficulties that undergraduate students experience at early stages of the modelling process: (a) in identifying broad assumptions that influence a simple model (b) in posing clarifying assumptions and making a related mathematical formulation. They focused on *how* students reached decisions to select multiple-choice alternatives, when sets of alternatives were provided. It can be inferred that this approach has a diagnostic component in that student facility with respect to particular modelling skills is specifically targeted. A challenge remains to relate performance on such sub-skills to overall modelling ability, as a defined link if established, opens the way for the use of such multiple-choice testing also as a summative procedure in assessing modelling competence. Establishing this link unambiguously remains a challenge.

Houston and Neill (2003) expanded the above approach to include a wider group of students at the University of Ulster, and the subsequent analysis of student performance assisted materially in the preparation of a new modelling module. This outcome identifies a second *diagnostic* function for the approach, directed towards course evaluation and re-design, through the location of strengths and weaknesses in existing programs.

Other results of interest from these studies include the identification of intuitive modellers among those with no formal prior experience in mathematical modelling, and we are then left wondering at the extent of lost potential in terms of those with parallel latent capacities in secondary education.

3.3 Issues in task design

The design of modelling tasks is a necessary precursor to their inclusion in assessment programs. Initiatives such as the OECD Programme for International Student Assessment (PISA) has included items with applications and modelling content (Turner, 2004). Application skills sampled by the items, include the need to make assumptions, choose a mathematical approach, and interpret outcomes. The items do not provide for extended modelling work, but it is interesting that even so the omission rates were generally very high across countries, pointing to deficits in the confidence, as well as competence, with which students approach contextualised problems. Regarding the design of short items of this type, Izard et al. (2003) discuss the appropriateness of item-response theory for analysing the performance of students on questions with modelling content. This is a beautifully written analysis with potential to enhance what can be achieved at item level, noting also that the term *item* limits the scope to which the theory applies within the wider field of mathematical modelling. Particularly helpful within diagnostic assessment, and potentially within PISA type testing, it is not yet applicable, for example, to extended modelling problems involving several iterations around the modelling loop.

Turner noted that the complexity of mathematical modelling activities that 15-year-old students can cope with appears to be rather low, and this turns attention to the presently poorly understood relationship between task complexity and task difficulty. While this nexus has been addressed at the general level within test theory, again for short items, the term *complexity* takes on several layers of meaning when extended modelling tasks become the focus. Stillman (2002), and Stillman and Galbraith (2003) discuss an empirical approach to estimating the *complexity* of mathematical application tasks among senior secondary school students and teachers. Six types of complexity were theorised and identified, each with a series of sub-ranges: *conceptual; mathematical; linguistic, intellectual; representational; and contextual.* Data indicate that students and teachers appear to focus on only some of the possible complexity components, and individuals differ in what they attend to. Given the limits on human information processing capacity it seems likely that the search for a common construct for the complexity of extended application and modelling problems, and its impact on task difficulty is likely to be a long one.

3.4 Secondary school high stakes assessment

The extended nature of modelling problems creates issues concerning valid and reliable evidence when high stakes assessment is involved. Philosophical issues emerge with respect to legitimate measures of modelling competence, leading to different procedures being adopted by different authorities, within and across national boundaries. Can modelling ability be estimated, on the basis of performance on a single common problem, or is a pattern of performance across several problems needed to provide a reliable estimate? Both approaches have been used, and the implications for students, teachers, and for school organisation are profoundly different.

Eid (2001) analysed senior school examination questions used in two German states, to identify an almost complete lack of application content. He identified sample questions that *could* form modelling based assessment tasks, and raised the central dilemma of how much of a substantial modelling task can be included in a formal examination governed by entrenched system procedures. He therefore raises issues relevant to *potential* practice.

In terms of *actual* practices Australia has, for some time, mandated investigative activities, including applications and modelling, within formal assessment structures in some of its states. From the early nineties the state of Victoria prescribed an *investigative project*, where the investigations provided opportunity for open-ended creative thinking, but were structured so that the requirement, for example, to formulate a mathematical problem from a general real life situation was avoided. All students did the prescribed investigation over the same specified time period of four weeks, teachers were given instructions as to how the project should be supervised, and a list of criteria were provided for the report format. The modus operandi assumed that *problem solving* or *investigative* ability can be assessed by means of a single task, even though for some students this was the first time they were faced with such an experience in their school life. This assumption begs the question of how much modelling ability can be enhanced by experience, and appears in tension with the development of modelling skills through targeted teaching programs. By contrast the state of Queensland has mandated applications and modelling as a component of formal assessment, in which all assessment is school-based, with peer review processes charged with monitoring comparability of quality. Here teachers can construct their own approved projects within a system of assessment which is criterion based, and exit performance is assigned on the basis of cumulative assessment data, obtained through as many as six problems undertaken over a two-year period.

Such different practices lead us to consider the influence of *primary drivers* of high stakes assessment schemes, for in addressing the challenge presented by workable valid assessment of applications and modelling performance in the context of high-stakes assessment, we cannot avoid the direct and profound impact of system procedures. The risk is real that the integrity of applications and modelling will be compromised by proclaimed needs to conform to what are presented as non-negotiable attributes of a mandated, external, assessment system.

Of course this issue knows no national or state boundaries, and attention is thus directed to the philosophy of assessment, in whatever forms it is practised. In considering the practice of using a single common extended modelling task undertaken simultaneously by all students, we might ask what, in terms of the activity of mathematicians, does modelling resemble most? It can be argued that the open ended properties involved in modelling problems, the need for the solver to make and act on assumptions, to formulate a mathematical problem from a general non-mathematised context, to carry out validation activities, to revisit earlier stages of the modelling process, and to produce a critical report, is much more like research activity than other kinds of mathematical learning typically assessed by some form of common test. That is, philosophically the research thesis is a much closer fit for the type of activity within a modelling project than are questions on traditional timed examination papers. And in thesis work expertise is gained and tested over time, in a process within which reflection and action on feedback is an integral part of the learning process, and where ultimately the quality of the work is assessed by criterion-referenced judgments exercised by examiners, guided by canons of scholarship characteristic of the field. Furthermore the process is applied to a wide range of problems and examiners respond with qualitative judgments by comparing performance with respect to agreed scholarly indicators.

But surely this is Untrustworthy? Subjective? Surely in order to assess the relative merits of research candidates in a given area of mathematics, every student should be given the same thesis topic, enrolled at the same time, and given the same due date for submission! Ridiculous – of course it is – and yet essentially similar arguments are used to defend approaches to student assessment of mathematical modelling activity. The purpose here is to raise the issue of tension and compromise, to highlight the question of drivers of assessment programs, where power is located, and the extent to which its exercise threatens to compromise the integrity of the fundamental objectives that applications and modelling encompass.

In reflecting on such issues there is no suggestion that other, and arguably more valid procedures, are easy to apply. The alternative of profiling students using their longitudinal performance across several tasks also has difficulties, for this latter approach depends on the efficiency and efficacy of a moderation system. The thesis system works because examiners are scholars with a strong shared understanding of the canons of disciplinary scholarship underpinning the criteria that are severally applied to candidates' submissions. This is much more difficult to achieve within mass secondary education, when hundreds or thousands of individual teachers are involved. The task of attaining and maintaining comparability through peer review processes is a huge one, and remains a challenge of major proportions. But it also arguably has sufficient integrity to warrant a substantial investment in professional development and support.

4. FINAL REFLECTIONS

In considering applications and modelling within educational settings there is a danger that artefacts of education (e.g. assessment issues) can blind us to other potential goals, purposes and measures of achievement. This element has been there from the start when Henry Pollak asked, not what word problems could achieve in terms of student performance, but what they contributed to the capacity of students to apply their mathematics to problems outside the classroom. Early ICTMA conferences[1] were motivated by concerns expressed by employers that graduates were ill equipped to work in teams on non-standard problems in the field. Institutional or systemic assessment procedures do not necessarily contribute to enhancing purposes such as these, and formal assessment data alone stand to overlook important potential indicators of success. Applications and modelling is unusual within the mathematics curriculum, in that evidence of successful, and at times unforseen teaching and learning, may occur independently of formal or informal teaching or assessment measures. Examples of this occurred within a modelling program (Galbraith & Clatworthy, 1990), where a student within his hobby of hydroponics, spontaneously and idiosyncratically applied the modelling process to re-invigorate the growth environment of plants. Linjefjard and Holmquist (2001) describe how modelling courses for prospective teachers produced impacts that affected their views of the mathematical world, in addition to modelling specific attainments, while McNab et al. (2004) refer to the way in which primary school children were reconnected to their world and life-experiences through the medium of modelling. Such examples illustrate perhaps that authenticity remains the supreme challenge, and that other artefacts of education (including assessment) should remain its servant. Whatever our personal priorities, the road to better outcomes for applications and modelling leads ever on.

REFERENCES

Eid, W. (2001). Mathematical modelling by the pupils themselves – Possibilities and limitations in school-leaving examination papers. In J. Matos, S.K. Houston, W. Blum, & S. Carreira (Eds). *Modelling and Mathematics Education: Applications in Science and Technology* (pp. 354-365). Chichester: Horwood Publishing.

Galbraith, P.L. & Clatworthy, N.J. (1990). Beyond Standard Models: Meeting the Challenge of Modelling. *Educational Studies in Mathematics*. 21, 2, 137-163.

Goldfinch, J., & Goodall, G. (1995). A study of the reliability of a marking scheme for mathematical modelling assignments. In L. Burton & J. Izard (Eds.). *Using FACETS to Investigate Innovative Mathematics Assessments* (pp. 14-28). Birmingham: Assessment Research Group.

Hamson, M. (2001). Mathematical modelling with environmental students. In J. Matos, S.K.

Houston, W. Blum, & S. Carreira (Eds). *Modelling and Mathematics Education: Applications in Science and Technology* (pp. 280-287). Chichester: Horwood Publishing .

Haines, C., Izard, J., & Le Masurier, D. (1993). Modelling intentions realised: Assessing the full range of developed skills. In T. Breitag, I. Huntley, & G. Kaiser-Messmer (Eds.). *Teaching and Learning Mathematics in Context* (pp. 200-212). London: Ellis Horwood.

Haines, C., Crouch, R., & Fitzharris, A. (2003). Deconstructing mathematical modelling: approaches to problem solving. In Qi-Xiao Ye, Werner Blum, Ken Houston, & Qi-Yuan Jiang (Eds). *Mathematical Modelling in Education and Culture: ICTMA 10* (pp. 41-53). Chichester: Horwood Publishing.

HEFCE. (1996). *Mathematics learning and assessment: Sharing innovative practices.* London: Arnold.

Houston, S.K. (1997). Evaluating rating scales for the assessment of posters Some mathematical characteristics of students entering applied mathematics degree courses. In K. Houston, W. Blum, I. Huntley, & N. Neill (Eds.). *Teaching and Learning Mathematical Modelling-Innovation, Investigation, and Applications.* (pp.135-148). Chichester: Albion Publishing.

Houston, K., & Neill, N. (2003). Assessing modelling skills. In S. Lamon, W. Parker, & K. Houston (Eds.). *Mathematical Modelling: A Way of Life* (pp. 155-164). Chichester: Horwood Publishing.

Izard, J. (1997). Assessment of complex behaviour as expected in mathematical projects and investigations. In K. Houston, W. Blum, I. Huntley, & N. Neill (Eds.). *Teaching and Learning Mathematical Modelling-Innovation, Investigation, and Applications.* (pp.109-124). Chichester: Albion Publishing.

Izard, J., Haines, C., Crouch, R., Houston, K., & Neill, N. (2003). Assessing the impact of teaching mathematical modelling. In S Lamon, W Parker, & K Houston (Eds). *Mathematical Modelling: A Way of Life* (pp. 165-178). Chichester: Horwood Publishing.

Lingefjard, T., & Holmquist, M. (2001). Mathematical modelling and technology in teacher education – visions and reality. In J. Matos, S.K. Houston, W. Blum, & S. Carreira (Eds). *Modelling and Mathematics Education: Applications in Science and Technology* (pp. 205-215). Chichester: Horwood Publishing .

London McNab, S., Moss, J., Woodruff, E., & Nason, R. (2004). Investigating fairness in ranking Commonwealth games performance: collaborative mathematical modelling in a grade 5/6 classroom. In H-W. Henn, & W. Blum (Eds). ICMI Study 14: *Applications and Modelling in Mathematics Education (Pre-conference volume)* (pp.167-174). Dortmund: University of Dortmund.

O'Carrol, M., Hudson P., & Yeats, A. (1987). A comprehensive first course in modelling. In J.S. Berry, et al., (Eds.) *Mathematical Modelling Courses* pp 113-121. Chichester: Ellis Horwood.

Pollak, H. (1969). How can we teach applications of mathematics? *Educational Studies in Mathematics*, 2, 393-404.

Stillman, G. (2002). *Assessing Higher Order Mathematical Thinking through Applications* (PHD thesis). Brisbane: The University of Queensland.

Stillman, G. A., & Galbraith, P. L. (2003). Towards constructing a measure of complexity of application tasks. In S. Lamon, W. Parker, & K. Houston (Eds.*). Mathematical Modelling: A Way of Life* (pp. 179-188). Chichester: Horwood Publishing .

Turner, R. (2004). Modelling and Applications in PISA. In H-W. Henn, & W. Blum (Eds). ICMI Study 14: *Applications and Modelling in Mathematics Education (Pre-conference volume)* (pp.273-278). Dortmund: University of Dortmund.

---

[1] International Conference on the Teaching of Mathematical Modelling and Applications

Chapter 2.6

MODELLING FOR LIFE: MATHEMATICS AND CHILDREN'S EXPERIENCE

Brian Greer, Lieven Verschaffel and Swapna Mukhopadhyay
Graduate School of Education, Portland State University, Email: brian1060ne@yahoo.com / Center for Instructional Psychology and Technology, University of Leuven, Belgium, Email: lieven.verschaffel@ped.kuleuven.be / Department of Curriculum and Instruction, Portland State University, USA, Email: swapna@pdx.edu

Abstract: We make the case for introducing fundamental ideas about modelling early, in particular through reconceptualizing word problems that describe real-world situations as exercises in modelling. Further, we argue for modelling as a means of giving children a sense of agency through recognizing the potential of mathematics as a critical tool for analysis of issues important in their lives.

1. EARLY AND AUTHENTIC CONNECTIONS

In this contribution we will first make the case for the early introduction in schools of fundamental ideas about modelling and for laying the foundations of a mathematical disposition, in particular through reconceptualizing word problems that describe real-world situations as exercises in modelling (Verschaffel, 2002). Further, and more broadly, we will argue for modelling as a means of giving children a sense of agency through recognizing the potential of mathematics as a critical tool for analysis of issues important in their lives, their communities, or society in general (Mukhopadhyay & Greer, 2001).

Our plea to take mathematical modelling seriously already at the elementary school level is a reaction against the fact that investigations of, discussions about, and instructional efforts for, mathematical modelling take place, almost exclusively, at the higher secondary and tertiary level. However, as argued convincingly by Usiskin in his contribution to this book, the generally accepted definitions of *mathematical model* and the *modelling process* do not require mathematics at such a (high) level. Indeed, these two terms are often described in such a way that makes mathematical modelling synonymous with what might be termed real-world problem solving. This realization suggests that modelling might begin as early as the very first years of the primary school.

Before going further, we want to emphasize that, especially among researchers and designers working at the primary school level, the term mathematical modelling is not only used to refer to a process whereby a situation has to be problematized and understood, translated into mathematics, worked out mathematically, translated back into the original (real-world) situation, evaluated and communicated. Besides this type of modelling, which requires that the student has already at his disposal at least some mathematical models and tools to mathematize, there is another kind of modelling, wherein model-eliciting activities are used as a vehicle for *the development* (rather than the application) of mathematical concepts. This second type of modelling is called 'emergent modelling' (Gravemeijer, 2004). Although it is very difficult, if not impossible, to make a sharp distinction between the two aspects of mathematical modelling, it is clear that they are associated with different phases in the teaching/learning process and with different kinds of instructional activities. However, in this contribution the focus will be on the first aspect of modelling.

2. MODELLING WITH ELEMENTARY ARITHMETIC

2.1 Suspension of sense-making

There are many examples of responses by children to word problems that show an apparent willingness to ignore things that they know about the world, language, and logic (see the first chapter of Verschaffel, Greer, & De Corte (2000) for a survey). The most dramatic and well-known example is probably the French research (prompted by a satirical reflection from Gustave Flaubert, see Verschaffel et al., 2000) in which elementary school children were posed questions of the type:

- There are 26 sheep and 10 goats on a ship. How old is the captain?

A large majority gave a numerical answer, while only a small minority questioned whether an answer is possible.

Intrigued by this example and other manifestations of "suspension of sense-making" (Schoenfeld, 1991, p. 340), we carried out in parallel in Northern Ireland and in Flanders pencil-and-paper studies with upper elementary and lower secondary school students, using a set of somewhat different problems including those listed below (Greer 1993; Verschaffel, De Corte, & Lasure, 1994):

- A man wants to have a rope long enough to stretch between two poles 12 meters apart, but he only has pieces of rope 1.5 meters long. How many of these would he need to tie together to stretch between the poles?
- Carl has 5 friends and Georges has 6 friends. They decide to give a party together. They invite all their friends. All friends are present. How many friends are there at the party?
- Bruce and Alice go to the same school. Bruce lives at a distance of 17 km from the school and Alice at 8 km. How far do Bruce and Alice live from each other?
- John's best time to run 100 meters is 17 seconds. How long will it take him to run 1 kilometer?

Note that, while all problems mentioned above are about sense-making, the four items from the latter list differ from the former "captain's problem" in that they admit sensible answers. We termed these latter items "problematic" (P-items, for short), in the sense that a proper answer requires (at least from our point of view) the application of judgment based on real-world knowledge and assumptions, rather than the straightforward application of arithmetical operations on the given numbers (as in standard word problems).

Both studies showed that the vast majority of children responded by apparently assuming that the situations described in the problems could be neatly mapped on to simple arithmetic operations. For example, on the four sample problems above, 100, 89, 95 and 97 % of children tested gave the non-realistic answers '8 pieces', '11 friends', '9 km' or '25 km', and '170 seconds', without any further qualification (Verschaffel et al., 1994).

Our studies were replicated in many other countries, using the same P-items, the same testing conditions and the same scoring criteria. Findings were strikingly consistent with our results, sometimes to the great surprise and disappointment of these other researchers, who had anticipated that the 'disastrous' picture of the Northern Irish and Flemish pupils would not apply to their students (see Verschaffel et al., 2000).

Our first reaction to such findings was shock. How could it be that the results of some years of mathematics education could be the willingness of children to collude in negating their knowledge of reality? We came to realize that this apparent "suspension of sense-making" can be construed as sense-making of a different sort, namely a strategic decision to play the "word problem game". As expressed by Schoenfeld (1991, p. 340):

> ... such behavior is sense-making of the deepest kind. In the context of schooling, such behavior represents the construction of a set of behaviors that results in praise for good performance, minimal conflict, fitting in socially etc. What could be more sensible than that?

Students' strategies and beliefs develop from their perceptions and interpretations of the didactical contract (Brousseau, 1997) or socio-mathematical norms (Yackel & Cobb, 1996) that determine – mainly implicitly – how to behave in a mathematics class, how to think, how to communicate with the teacher, and so on. More specifically, this enculturation seems to be mainly caused by two aspects of current instructional practice, namely (1) the nature of the problems given and (2) the way in which these problems are conceived and treated by teachers. Support for the second factor comes from a study by Verschaffel, De Corte, and Borghart (1997), wherein pre-service elementary school teachers were asked, first, to solve a set of problems themselves, and, second, to evaluate alternative answers from (imaginary) pupils to the same set of problems. The results indicated that these future teachers shared, though in a less extreme form, students' tendency to suspend sense-making.

As a result, students learn to play the "Word Problem Game", the rules of which include:

- Any problem presented by the teacher or in a textbook is solvable and makes sense.
- There is a single, correct, and precise numerical answer which must be obtained by performing one or more arithmetical operations with numbers given in the text.
- Violations of your knowledge about the everyday world may be ignored.

2.2 Word problems as modelling exercises

A minimal and rather easily achievable goal is to improve the quality of word problems as applications in numerous ways that have been suggested over many years, such as:

- Break up the expectation that *any* word problem can be solved by adding, subtracting, multiplying, or dividing the numbers given in the problem, or a simple combination thereof, by varying the quantitative relationships, including irrelevant data, and so on.
- Weed out word problems that supposedly describe real-world situations but are not realistic (we do not argue for getting rid of puzzle-like or "whimsical" word problems).
- Valorize forms of answer other than single, exact numerical answers.

More radically, we recommend a modelling perspective whereby arithmetic operations should be mindfully evaluated as candidate models for a given situation presented verbally or otherwise, with many examples to help students discriminate between cases where application of the operations provides a model that is (a) precise, (b) approximate, (c) inappropriate. To put it another way, we are arguing that what we have termed "implicit modelling"

(Greer & Verschaffel, this volume), carried out through routine expertise, be replaced by "explicit modelling" necessitating adaptive expertise.

A number of design studies have been carried out by researchers to take the modelling approach seriously, in particular an intervention involving the creation and evaluation of a learning environment for mathematical modelling and problem solving in upper elementary school children. Without any attempt to be exhaustive, we list a few design studies that have been set up recently according to this modelling perspective:

- Several so-called 'developmental research' projects by the Freudenthal Institute in The Netherlands (Gravemeijer, 2004; Van den Heuvel-Panhuizen, 2004),
- The Jasper studies of the Cognition and Technology Group at Vanderbilt (1997),
- The numerous design experiments with model-eliciting activities summarized in the recent book by Lesh and Doerr (2003), and
- The learning environment for mathematical modelling and problem solving in upper elementary school children that we developed, implemented and tested a few years ago in Leuven (Verschaffel, De Corte, Lasure, Van Vaerenbergh, Bogaerts, & Ratinckx, 1999).

Characteristics common to these experimental programs include:

- The use of more realistic and challenging tasks than traditional textbook problems,
- A variety of teaching methods and learner activities, including expert modelling of the strategic aspects of the modelling process, small-group work, and whole-class discussions, and
- The creation of a classroom climate conducive to the development of the elaborated view of mathematical modelling and of the accompanying beliefs.

Generally, these studies have produced positive outcomes in terms of performance, underlying processes, and motivational and affective aspects of learning. After reviewing the available research evidence, Niss (2001, p. 8) concluded that "application and modelling capability can be learnt – and according to the above-mentioned findings has to be learnt – but at a cost, in terms of effort, complexity of task, time consumption, and reduction of syllabus in the traditional sense".

3. MODELLING SOCIAL ISSUES

3.1 The political nature of mathematics education

> This is a great discovery, education is politics! After that, when a teacher discovers that he or she is a politician, too, the teacher has to ask "What kind of politics am I doing in the classroom?" (Freire, 1987, p. 46)

Generally, it is implicit that mathematics education, like mathematics, is politically neutral and thus exempt from Freire's declaration. According to Apple (2000, p. 243): "It is unfortunate but true that there is not a long tradition within the mainstream of mathematics education of both critically and rigorously examining the connections between mathematics as an area of study and the larger relations of unequal economic, political, and cultural power". However, there are signs of change, building on a major shift within the discipline of mathematics education from a mainly psychological and pedagogical perspective towards recognition of the historical, cultural, and social contexts of both mathematics and mathematics education. This shift is encapsulated in the phrase "mathematics as a human activity" whence the acknowledgment of the political situatedness of mathematics education is a natural outgrowth (Mukhopadhyay & Greer, 2001; for an excellent review of the emergence of this perspective see Vithal, 2003, Chapter 1).

Particularly in the United States, there are powerful conservative counter-forces at work. In the Mathematics Framework for California Public Schools (California Department of Education, 2000, p. 157) there is a discussion of the following "extreme example":

> The 20 percent of California families with the lowest annual earnings pay an average of 14.1 percent in state and local taxes, and the middle 20 percent pay only 8.8 percent. What does that difference mean? Do you think it is fair? What additional questions do you have?

The following comments are made (p. 157):

> ...a proper understanding of the difference in the two figures of 14.1 percent and 8.8 percent would require a strong background in politics, economics, and sociology... Moreover, the idea of "fairness" is a difficult one even for professional political scientists and sociologists. Formulating a *mathematical* transcription of this elusive concept in this context is therefore beyond the grasp of the best professionals, much less that of school students. Since it is impossible to transcribe the problem into mathematics ... this is therefore not a mathematical problem.

In effect, the argument defines as non-mathematical any act of modelling that does not uncontroversially lead to a precise set of equations that can be solved to yield a single answer.

3.2 Modelling as a tool for critical analysis

Here we consider three examples of mathematical analysis applied to aspects of contemporary US culture.

The first example, from the work of Mukhopadhyay (1998), starts with the simple question: "What would Barbie look like if she was the height of an average woman?". Barbie dolls, perhaps the ultimate icon of US culture, have given rise to a very extensive literature analysing the remarkable associated cultural phenomena from sociological and other perspectives. The investigation begins as an exercise in proportional reasoning. In order to dramatize the contrast between the doll that is often idealised as having a "perfect" human body, and the individual chosen for comparison, the contour of that individual is sketched. The projected Barbie, using the computed measures of her relevant body-parts is then superposed on the full-size contour drawing. The obvious differences in body shape (for example, Barbie's waist is so narrow she could not bear children) lead into discussions of issues of body-image and eating disorders. The conditions under which the dolls are manufactured in, for example, China and Brazil lead to another area of discussion based on data about the economics of sweatshop labor.

The second example is from Eric Gutstein, a mathematics educator at the University of Illinois, Chicago. As part of his work, he teaches middle school mathematics in a public school situated in a low-income, Mexican immigrant community. In a recent project, he used an article from the *Chicago Tribune* as the basis for a three-week project on whether there is racial discrimination in the allocation of mortgage loans. The resulting discussion was intense and open. One student wrote as follows (Gutstein, 2003a, p. 36):

> [It would seem that racism] is a factor because white applicants no matter what their income was, they were always denied less times than African Americans and Latinos. And it is also a factor because the ratio of applicants denied between African Americans and whites is 5:1 and between Latinos and whites is 3:1. That data shows that racism is a factor.
>
> There are always two sides to a story. Racism is not a factor because we do not know whether or not those people had bad credit or were unemployed. It could be possible that a lot of those people could have been in debt. Even though the banks want to make loans they also want to make sure that they get paid.

> So with the data provided it is very hard to conclude whether or not racism is a factor when it comes to obtaining a mortgage loan in the Chicago area.

This is a sophisticated response that belies the condescending attitude of the statement cited above that children cannot be expected to understand such complex issues.

The third example is from a description by Tate (1995a) of the pedagogy of a teacher in a predominantly African American urban middle school. Students are asked to pose a problem negatively affecting their community, to research it and develop strategies to tackle it, and to resolve the problem by implementing these strategies. As a particularly striking example, the students identified the presence of 13 liquor stores within 1000 feet of their school as a problem, developed a plan to move them away, and carried out that plan by various direct actions, including lobbying the state senate. Mathematical modelling was one tool used in this real-problem-driven exercise. For example, the students analyzed the local tax and other codes that led to financial advantages for the liquor stores and reconstructed this incentive system to protect their school community. As Tate (1995a, p. 170) comments:

> This required the students to think about mathematics as a way to model their reality... Percentages, decimals, and fractions became more than isolated numbers as the students tried to mathematically manipulate these different, yet related, symbol systems and to link them to real problem solving and decision making.

3.3 Diverse realities

If a decision is made to mathematize situations and issues that connect with students' lived experience, then it brings a further commitment to respect the diversity of that experience across genders, classes, and ethnicity (Cooper & Dunne, 2000; Gutstein, 2003b). A very clear example is the following:

> It costs \$1.50 each way to ride the bus between home and work. A weekly pass is \$16.00. Which is the better deal, paying the daily fare or buying the weekly pass?

When African American students were asked about their responses, it was discovered that they "transformed the 'neutral' assumptions of the problem – all people work 5 days a week and have one job – into their own realities and perspectives" (Tate, 1995b, p. 440). In their experience, as opposed to white middle-class experience, a job (such as cleaning) might mean making several bus trips every day, not just two, and working more than 5 days a week. If items of this type are used for assessment, and assumptions are made about the "right" answers, the implications for inequity are clear, given that, as Tate (1995b, p. 440) puts it: "the underpinnings of school mathematics curriculum, assessment, and pedagogy are often more closely aligned with the idealized experience of the White middle class".

4. CONCLUDING COMMENTS

In typical elementary schools worldwide, the teaching of early arithmetic is predominantly focused on computational proficiency. Even word problems that putatively link mathematics and aspects of the real world are often no more than thinly disguised exercises in the four basic operations. Given that many adults claim inability to do, and fear of, mathematics, we may conjecture that such a regime establishes early a negative and narrow disposition towards mathematics in many children. An alternative vision, doubtless Utopian, sees early arithmetic as an opportunity to lay the foundations for a positive and productive mathematical disposition, including a grasp of the relationship between aspects of reality and mathematical structures as mediated by modelling acts, and a belief in the power of mathematics as a sense-making and critical tool.

REFERENCES

Apple, M. W. (2000). Mathematics reform through conservative modernization? Standards, markets, and inequality in education. In J. Boaler (Ed.), *Multiple perspectives on mathematics teaching and learning* (pp. 243-259). Westport, CT: Ablex.

Brousseau, G. (1997). *Theory of didactical situations in mathematics*. Dordrecht, The Netherlands: Kluwer.

California Department of Education (2000). *Mathematics framework for California public schools: Kindergarten through Grade 12*. Sacramento, CA: Author.

Cognition and Technology Group at Vanderbilt (1997). *The Jasper project: Lessons in curriculum, instruction, assessment and professional development*. Mahwah, NJ: Lawrence Erlbaum Associates.

Cooper, B., & Dunne, M. (2000). *Assessing children's mathematical knowledge: Social class, sex and problem-solving*. Buckingham, England: Open University Press.

Freire, P. (with Shor, I.). (1987). *A pedagogy for liberation*. Westport, CT: Bergin & Garvey.

Gravemeijer, K. (2004). Emergent modelling as a precursor to mathematical modelling. In H.-W. Henn, & W. Blum (Eds), *ICMI Study 14: Applications and modelling in mathematics education. Pre-conference volume* (pp. 97-102). Dortmund: Universität Dortmund.

Greer, B. (1993). The modelling perspective on wor(l)d problems. *Journal of Mathematical Behavior, 12*, 239-250.

Gutstein, E. (2003a). Home buying while brown or black. *Rethinking Schools*, Fall, 35-37.

Gutstein, E. (2003b). Teaching and learning mathematics for social justice in an urban, Latino school. *Jorunal for Research in Mathematics Education, 34*, 37-73.

Lesh, R., & Doerr, H.M. (2003). *Beyond constructivism. Models and modelling perspectives on mathematics problem solving, learning, and teaching*. Mahwah, NJ: Lawrence Erlbaum Associates.

Mukhopadhyay, S. (1998). When Barbie goes to classrooms: Mathematics in creating a social discourse. In C. Keitel (Ed.), *Social justice and mathematics education: Gender, class, ethnicity, and the politics of schooling* (pp. 150-161). Berlin: Freie Universität.

Mukhopadhyay, S., & Greer, B. (2001). Modelling with purpose: Mathematics as a critical tool. In B. Atweh, H. Forgasz, & B. Nebres (Eds.), *Socio-cultural aspects in mathematics education* (pp. Mahwah, NJ: Lawrence Erlbaum Associates.

Niss, M. (2001). Issues and problems of research on the teaching and learning of applications and modelling. In J.F. Matos, W. Blum, S.K. Houston, & S.P. Carreira (Eds.), *Modelling and mathematics education. ICTMA 9: Applications in science and technology* (pp. 72-89). Chichester, U.K.: Horwood.

Schoenfeld, A. H. (1991). On mathematics as sense-making: An informal attack on the unfortunate divorce of formal and informal mathematics. In J.F. Voss, D.N. Perkins, & J.W. Segal (Eds.), *Informal reasoning and education* (pp. 311-343). Hillsdale, NJ: Erlbaum.

Tate, W. F. (1995a). Returning to the root: A culturally relevant approach to mathematics pedagogy. *Theory into Practice, 34*, 166-173.

Tate, W. F. (1995b). School mathematics and African American students: Thinking seriously about opportunity-to-learn standards. *Educational Administration Quarterly, 31*, 424-448.

Van den Heuvel-Panhuizen, M. (2004). Schifts in understanding: The didactical use of models in mathematics education (pp. 285-290). In H.-W. Henn & W. Blum (Eds), *ICMI Study 14: Applications and modelling in mathematics education. Pre-conference volume* (pp. 97-102). Dortmund: Universität Dortmund.

Verschaffel, L. (2002). Taking the modelling perspective seriously at the elementary school level: Promises and pitfalls (Plenary Lecture). In A. D. Cockburn & E. Nardi (Eds.), *Proceedings of the 26th Annual Meeting of the International Group for the Psychology of Mathematics Education, Vol. 1* (pp. 64-80). Norwich, England: University of East Anglia.

Verschaffel, L., De Corte, E., & Borghart, I. (1997). Pre-service teachers' conceptions and beliefs about the role of real-world knowledge in mathematical modelling of school word problems. *Learning and Instruction, 4*, 339-359.

Verschaffel, L., De Corte, E., & Lasure, S. (1994). Realistic considerations in mathematical modelling of school arithmetic word problems. *Learning and Instruction, 4*, 273-294.

Verschaffel, L., De Corte, E., Lasure, S., Van Vaerenbergh, G., Bogaerts, H., & Ratinckx, E. (1999). Design and evaluation of a learning environment for mathematical modelling and problem solving in upper elementary school children. *Mathematical Thinking and Learning, 1*, 195-229.

Verschaffel, L., Greer, B., & De Corte, E. (2000). *Making sense of word problems*. Lisse, The Netherlands: Swets & Zeitlinger.

Vithal, R. (2003). *In search of a pedagogy of conflict and dialogue for mathematics education*. Dordrecht, The Netherlands: Kluwer.

Yackel, E., & Cobb, P. (1996). Sociomathematical norms, argumentation, and autonomy in mathematics. *Journal for Research in Mathematics Education, 27*, 458-477.

Chapter 2.7

MODELLING IN LOWER SECONDARY MATHEMATICS CLASSROOM – PROBLEMS AND OPPORTUNITIES

Gabriele Kaiser and Katja Maaß
Faculty of Education, University of Hamburg, Germany, Email: gabriele.kaiser@uni-hamburg.de / Mathematical Institute, University of Education, Freiburg, Germany, Email: Katja.Maass@ph-freiburg.de

Abstract: The paper deals with the gap between the relevance of applications and modelling in didactical discussions and its minor importance in everyday mathematics teaching. Results of our own empirical studies that describe mathematical beliefs of teachers and students as being central obstacles are presented. Further, the studies demonstrate the possibility to change these beliefs as well as ways to promote modelling competencies.

1. INTRODUCTION

Empirical studies have shown that applications and modelling only play a minor role in everyday mathematics teaching. This situation has not yet changed decisively, although many teaching materials have been developed over the last decades and are available. This is not exclusively a typical German situation. A number of empirical and international comparative studies indicate that applications and modelling are less significant in everyday school life in many countries, although there are country specific differences.

In the following, we present the results of our own studies which point out problems and opportunities of integrating modelling and application in lower secondary mathematics classroom. The results of the first study refer to students (Maaß, 2004), while the results of the second study focus on teachers (Kornella, 2003; Ross, 2002).

2. RESULTS OF AN EMPIRICAL STUDY FOCUSSING ON MATHEMATICS STUDENTS

This study deals with the following questions:

1. How far do students' mathematical beliefs change when modelling problems are included in ordinary mathematics lessons?
2. How far do such lessons enable students to carry out modelling processes on their own?
3. What kind of connections exist between the students' mathematical beliefs and their modelling competencies?

2.1 Theoretical frame

To answer these questions, the theoretical approach focuses on discussions about applications of mathematics and discussions about beliefs.

First to the discussion of modelling and modelling competencies: We describe a modelling process as a process in which a non-mathematical problem is solved through the application of mathematics (cf. Blum, 1996). Competencies for modelling include abilities to model problems as well as the willingness to implement them.

Secondly, we refer to the conception of mathematical beliefs which are described as an individual's stable knowledge of certain objects and affairs as well as of corresponding attitudes and emotions (Pehkonen & Törner, 1996). Considering possible connections between beliefs and learning processes, the search for methods to change beliefs is a central problem which has not yet been solved. An important contribution to characterize students' beliefs was given by Grigutsch (1996). He categorizes students' beliefs mainly by four aspects of mathematical belief systems which refer to mathematics as a field of science. Mathematics can be understood as a science which mainly consists of problem solving processes (aspect of process), a science which is relevant for society and life (aspect of application), an exact, formal and logical science (aspect of formalism) or a collection of rules and formulas (aspect of scheme). The first two aspects are called dynamic beliefs, the last two static beliefs.

Based on this theoretical background, this study aimed to show the effects of modelling lessons on students in a comprehensive manner.

2.2 Methodological approach

Classroom setting

During the data collection period of 15 month (April 2001 – June 2002), six modelling units were integrated into two parallel classes, age 13 – 14, in a Gymnasium (i.e. school type for higher achieving students). For example, in three of these units the students had to answer the following questions:

1. How large is the surface of a 'Porsche'?
2. How can different rates of various mobile phone contracts be compared depending on customers' habits?
3. Is it possible to heat the water required in Stuttgart-Waldhausen by solar collectors on the roof of houses?

Theoretical basis and methods in data collection

Aiming at an explanation of complex relations in the context of everyday life and at a contribution to an empirically founded theory, the study is a qualitative study which theoretically starts off mainly from the Grounded Theory (Strauss & Corbin, 1998). Furthermore, a long-lasting incorporation of modelling tasks into everyday mathematics teaching practice became possible because in this study the researcher and the teacher were represented by the same person according to Action Research (Altrichter & Posch, 1998).

In order to meet the complexity of the research's objectives, a variety of methods in data collection were used (questionnaires, interviews, learner's diaries, tests, concept maps). Based on computer-aided data evaluation, typologies were created to explain interrelations between phenomena. The main tool to elucidate the results was the construction of ideal types as described by Weber (see Kelle & Kluge, 1999).

2.3 Results of the study

The following results are based on the data-evaluation of 35 students and their development during the whole period of the study. First, the reconstructed types of reaction will be explained. Then we will refer to the problems and possibilities of integrating modelling into mathematics classes.

Types of reaction

Mathematical belief systems

The results of this study show, on the one hand, that the aspects reported by Grigutsch exist in the students' mind. On the other hand, it became clear, that those aspects do not sufficiently describe the students' mathematical belief systems. Many students seemed to have no idea how to characterize mathematics as a science. Their thinking primarily concentrated on the lessons taught and their own role in these lessons. These beliefs will be called 'non subject-based'.

Among others, the following beliefs were reconstructed:

Cognitive shaped non subject-based beliefs:

- Beliefs about the short duration of teaching units within the mathematics classes: *'Exercises in maths lessons should only last one hour.'*
- Beliefs about a minor importance of words in the exercises: *'You have to write in German, in mathematics you have to calculate'.*
- Beliefs about the necessity of learning: *'Either one is able to deal with mathematics or not, learning is useless.'*

Affective shaped non subject-based beliefs:

- Beliefs about teaching methods: *'I liked the lesson because we were allowed to work in groups.'*
- Beliefs about the atmosphere within mathematics classes: *'It is absolutely shit that some students cannot respect Mrs Maaß' orders. Then she gets angry and the atmosphere gets really bad.'*
- Beliefs about understanding: *'Today I understood everything. And when you understand something, you like it.'*

These non subject-based beliefs seem to be so important for some students that subject-based beliefs, as described by Grigutsch (1996), could not be reconstructed. Altogether, the reconstruction of students' individual mathematical belief systems shows a complex structure of different beliefs. However, frequently one subject-based or non subject-based aspect turned out to be the most important. Furthermore, the results indicate that almost all components of the belief system are responsible for the way students act in a typical manner. The interrelation between belief system and the students' actions can be described by six ideal types.

Ideal type A+B: Students with a process-oriented or an application-oriented mathematical belief system have a positive attitude towards modelling examples. The application-oriented beliefs increased during the study.

At the end of the study, one student with a process-oriented belief system who, at the beginning, had hardly any application-oriented beliefs answered to the question "Do you think that you can use the things you learned during the modelling units?" *'I think it can't do any harm, because knowledge is always regarded positively. Moreover, I have learned to react independently and to see whether anybody wants to cheat me.'*

Ideal type F: Students with an affective-shaped non subject-based mathematical belief system who have also the impression to understand the content quite well, regard the modelling examples as positive. They develop application-oriented beliefs.

One student answered the question 'What did you learn from modelling examples?' as follows: *'A lot! 1. Mathematics is everywhere. 2. Maths lessons can be fun. 3. Everybody needs mathematics. 4. Which mobile phone contract I have to choose... 5. Well, many important things.'*

Ideal type C+D+E: Students with a scheme-oriented, a formalism-oriented or a cognitive-shaped non subject-based belief system reject modelling examples in an emotional way. No or only very few application-oriented beliefs are developed until the end of the study.

Students answered the question about what they have learned, as follows: *'Modelling examples don't belong into maths lessons because you don't have to calculate... I have learnt nothing.' 'I have learnt that maths lessons can be horrifying.'*

On the question "When do you need mathematics in life?" one student answered: *'It depends on the profession. As an engineer you need geometry, as a shop assistant you need plus, minus, times and divided by.'*

Modelling competencies

Reaction patterns of the students can be reconstructed from mathematical competencies as well as mathematical beliefs which have great influence on the acquisition of modelling competencies. In an idealized way, four types of modellers can be distinguished:

Reality-distant modellers have a positive attitude towards context-free mathematics and reject modelling examples. As consequence an affective barrier is set up which mainly results in a lack of competency to solve problems closely connected to context-related mathematics which means that they have problems with the construction of real models, with their validation and partially also with the interpretation of the results.

Mathematics-distant modellers clearly give preference to the context of real-world problems and show only low performance in mathematics lessons. These students are very enthusiastic about modelling examples. They are able to construct real models and validate solutions quite well. Lack of ability is found in constructing mathematical models, in finding a mathematical solution and in interpreting complex solutions.

Reflected modellers have positive attitudes towards mathematics itself as well as towards modelling examples. They show an appropriate performance in mathematics. Deficits within the modelling process are hardly to be found.

Uninterested modellers are neither interested in the context of real-world problems nor in mathematics itself. They have deficits in mathematical competencies. While dealing with modelling problems, problems occur in every part of the modelling process.

Problems and opportunities

Which problems may occur?

The negative reactions of those students whose belief system can be characterized as scheme- or formalism-orientated or as cognitive shaped non subject-based might prevent many teachers from integrating modelling problems in their classes after a first effort. Thus, students' beliefs might even prevent a broad implementation of realistic tasks in everyday mathematics teaching.

What opportunities are offered by the integration of modelling examples in daily school routine?

The integration of modelling examples in mathematical lessons can lead to the development of students' application-oriented beliefs as we have seen above.

Students at lower secondary level are able to develop modelling competencies which include meta-knowledge of modelling processes. Therefore, students become qualified to model unknown real world problems by themselves and to question critically already accomplished modelling. At the end of the study almost every student was able to deal with simple modelling tasks even when the context of the task was unknown to him/her. Many of them were even able to deal with complex modelling problems.

Modelling problems provide an important educational contribution to mathematics lessons which meet the individual abilities of (many) more students (than in usual mathematics lessons). The open formulation of modelling problems and the necessity to simplify the complex reality enables students to develop solutions by themselves, according to their capabilities. The results of this study show that strong students choose more challenging models while weaker students prefer simpler ways to achieve their final solution.

The positive attitude towards modelling examples evoked by the connection to reality and the unusual success of weaker students allows an affective access to mathematics and, from a long-term perspective, may positively improve the acquirement of mathematical competencies.

3. RESULTS OF AN EMPIRICAL STUDY FOCUSSING ON MATHEMATICS TEACHERS

The second study gives insight to the teachers' perspectives. Many teachers think it desirable to discuss contextual and modelling problems in lessons, but a look at teaching practice makes it clear that contextual and modelling problems play only a rather minor role. For this reason, within the framework of our study we will examine the question 'what are the mathematical beliefs of teachers towards applications and modelling tasks?'

3.1 Frame and design of the study

The study was conducted within the evaluation of a pilot programme by the German government together with the federal states which was aimed at increasing the efficiency of mathematical and scientific teaching. This innovative programme, carried out during the period 1998 – 2003, aimed at fundamental changes in mathematics teaching: namely, a change of the tasks as practised in lessons and a change of the dominating learning and teaching structures focused on a stronger integration of applications and modelling examples. Over the whole time teachers were offered further education programmes, both by internal and external initiatives. Furthermore, the participating teachers were asked to try out already existing material and to develop new material through teamwork. Teachers were given access to a great amount of material – developed all over Germany within this innovation programme – through a special server: (http://blk.mat.uni-bayreuth.de).

The study, whose results will be described, is restricted to the evaluation of this programme at the six participating schools in Hamburg. Due to organisational constraints, the evaluation is limited to a period of only one year. This short time implied that really great effects of change could not be expected.

The evaluation study started when the students of the 6 participating schools attended year 7 and 8 and ended when they were in year 8 and 9. The study is divided up into different components: In the first the development of mathematical literacy as well as students' beliefs are examined. Due to lack of space, we do not refer to this part of the study (see Kaiser & Willander, 2005). In the second qualitatively oriented component the mathematical belief systems of the teachers involved were examined.

The theoretical approach of his study, like that of the first study, uses the approach of Grigutsch (1996) about the classification of mathematical belief systems, that is; process-oriented and application-oriented mathematical beliefs as dynamic belief systems; and formalism-oriented and scheme-oriented mathematical beliefs as static mathematical belief systems (for details concerning teachers see Grigutsch, Raatz, & Törner, 1998). Beliefs are characterised as stable patterns of conviction.

Results of empirical studies show how strongly mathematical beliefs about mathematics and mathematics teaching control the pedagogical behaviour of teachers. Mathematical innovations like the introduction of applications and modelling bring up the question how much there is a possibility for change.

There exists nearly no empirical study which investigates the difficulties in changing beliefs. However, well known studies that analyse the difficulties of changing attitudes – a psychological construct closely related to the concept of beliefs – show that beliefs are not easily modifiable. Ambrose

(2004) points out for example that the changes in the belief systems of teachers are more incremental than monumental. In this context, Pehkonen (1994) distinguishes between "surface beliefs" which are not deeply rooted within the belief system, and "deep beliefs" that are functioning as central anchor points. Pehkonen (1994) points out that these deep beliefs need to be changed because they are motivating teachers during their mathematics lessons.

Methodologically, the study is qualitatively oriented and applying methods from qualitative social science. Furthermore, the applied empirical methods concerning choice of sample, data analysis and data interpretation are based on the theoretical attempts of the Grounded Theory (Strauss & Corbin, 1998).

In this study, all teachers involved in mathematics teaching of year 7 and 8 students of the six participating schools have been asked about their mathematical beliefs at the beginning of the project and after one year. This has been done in written form via open and closed items. Altogether 41 teachers participated at the beginning and 29 at the second questioning. With 8 teachers, who were chosen for certain theoretical criteria, partly standardised interviews were done, 4 at the beginning and 4 at the end of the study (for details see Ross, 2002; Kornella, 2003).

3.2 Results of the study

The written questioning at the beginning of the study shows a clear dominance of static beliefs about the nature of mathematics, in other words for these teachers mathematics mean exact mathematical thinking and exact ways of working as it is described in the formalism-oriented approach.

In one of the in-depth interviews a teacher describes his view about mathematics as follows: *Mathematics is at first a 'formal language', in contrast to colloquial language 'not redundant', 'precise' and 'logical'.* According to this teacher's opinion there is only a weak relation between mathematics and everyday teaching: *'For me mathematics is ... not always, sometimes yes, ... has also a relation to life.'*

Besides the formalistic position there are also scheme oriented understandings. Within these understandings, mathematics is reduced to the accumulation of rules and formulae. Mathematics is – as expressed in an interview – *'the logical sequence of formulae'*. Non-mathematical applications do not form a constitutive part of mathematics. In mathematics lessons students learn *'the basic conditions of mathematics', 'and everything else comes from the other subjects, there one continues to calculate.'*

This goes along with the fact that there is only seen a weak relation between mathematical subject knowledge and the real world. In the interview, one teacher explained that mathematics might even be replaced by playing chess because mathematics is aimed at developing thinking competencies.

Mathematical beliefs for which the aspect of application plays a central role, could only seldom be reconstructed. In one of the in-depth interviews it becomes clear that for teachers with such an orientation the aspect of application has a fundamental meaning: *'What shall I do with mathematics, if I cannot apply it somehow for my life?'* However, not its usefulness, but the training of 'critical questioning' is as important as the training of thinking abilities.

Beliefs about the nature of mathematics teaching were also dominated by static understandings. Likewise, the beliefs about the goals of mathematics teaching are predominated by schematic aspects. Dynamic ideas only prevail with beliefs about the learning of mathematics.

Taken together, it becomes obvious that for the whole group of questioned teachers applications and modelling play only a minor role in their beliefs about mathematics and mathematics teaching.

In the follow-up study conducted one year later, only slight changes could be observed: Altogether, the beliefs about the nature of mathematics and the nature of mathematics teaching changed slightly towards a greater relevance of application and modelling examples. The results of the in-depth interviews are as follows: Teachers with mathematical beliefs, in which the aspect of application only plays a minor role, interpreted application oriented beliefs about the nature of mathematics or the nature of mathematics teaching in a way by which they became appropriate for their own mathematical beliefs. In detail: Teachers with a process oriented understanding of mathematics and mathematics teaching stress the many chances which exist for developing solutions and reduce applications and modelling to this aspect. In contrast to that, teachers with schematic mathematical beliefs restrict applications and modelling to examples that enable easy mathematisations or lead directly to a formula. For teachers with formalistic beliefs, the context nearly does not play any role.

4. PROSPECTS AND POSSIBLE CONSEQUENCES

On the one hand, the studies demonstrate that it is possible to integrate modelling examples into mathematics lessons, on the other hand, they make clear that it is extremely important to do so in a consequent way. The intense disapproval reaction of students with scheme and formalism oriented mathematical beliefs, as described in the first study, demonstrate how relevant it is to tackle applications and modelling examples as an integral part of mathematics teaching, starting at primary level. Furthermore, the studies show that the students' partly disapproval reaction must not lead to the neglect of applications and modelling. If one adapts oneself to these reactions, one can meet them in an adequate manner. In addition to a change in the mathematical beliefs of the students, a very positive development of the students' modelling competencies could be observed. Students developed especially a high level of meta-cognitive competencies which are generally regarded as an undeniable component of modelling competencies.

The results of both studies support the findings and assumptions from other studies. Specifically the result that teachers and their beliefs about mathematics must be regarded as essential reasons for the low realisation of applications and modelling in mathematics teaching. In order to promote real world and modelling examples within the mainstream mathematics education, it will be necessary to integrate real world examples and modelling courses in pre-service and in-service education for teachers.

REFERENCES

Altrichter, H., & Posch, P. (1998). *Lehrer erforschen ihren Unterricht.* Bad Heibrunn: Klinkhardt.

Ambrose, R.(2004). Initiating change in prospective elementary school teachers' orientations to mathematics teaching by building on beliefs. *Journal of Mathematics Teacher Education*, 7, 91-119.

Blum, W. (1996). Anwendungsbezüge im Mathematikunterricht – Trends und Perspektiven. In G. Kadunz et al. (Eds.), *Trends und Perspektiven.* (pp. 15-38) Wien: Hölder-Pichler-Tempsky.

Grigutsch, S. (1996). *Mathematische Weltbilder von Schülern, Struktur, Entwicklung, Einflussfaktoren.* Unpublished doctoral dissertation, Gerhard-Mercator-Universität, Gesamthochschule Duisburg, Duisburg, Germany.

Grigutsch, S., Raatz, U. , & Törner, G. (1998). Einstellungen gegenüber Mathematik bei Mathematiklehrern. *Journal für Mathematik-Didaktik, 19*, 3-45.

Kaiser, G., & Willander, T. (2005). Development of mathematical literacy: results of an empirical study. *Teaching Mathematics and its Applications, 24*(2-3), 48-60.

Kelle, U., & Kluge, S. (1999). *Vom Einzelfall zum Typus* Opladen: Leske und Budrich.

Kornella, M. (2003). *Beliefs als Barrieren von Innovationen im Bildungswesen.* Pehkonen, E. (1994). On teachers' beliefs and changing mathematics teaching. *Journal für Mathematik-Didaktik, 15*, 177-209.

Pehkonen, E., & Törner, G. (1996). Mathematical beliefs and different aspects of their meaning. *Zentralblatt für Didaktik der Mathematik, 28*, 101-108.

Ross, N. (2002). *Lehrerkonzeptionen von Aufgabenkultur im Mathematikunterricht Hamburg.* Unpublished master thesis, University of Hamburg, Hamburg, Germany.

Strauss, A., & Corbin, J. (1998), *Basics of Qualitative Research.* Newbury Park: Sage.

Chapter 2.8

MATHEMATICAL MODELLING – A CONVERSATION WITH HENRY POLLAK

Henry Pollak
Teachers College, Columbia University, New York, USA, Email:hpollak@adsight.com

Foreword to Henry Pollak's paper

Henry Pollak is, without any doubt, one of the pioneers in the field of applications and modelling in mathematics education. As early as in the sixties of the last century he pleaded for an integration of applications and modelling into mathematics teaching (for instance, Pollak, 1969). He was able to do that particularly competently and credibly since he was not part of the educational system itself but a leading member of Bell Laboratories. The educational scene of the sixties and the early seventies was shaped by the New Maths movement which had, in contrast to its own intentions, led in many countries to an emphasis on intra-mathematical aspects (see Pollak, 2003, for the situation in the USA). Internationally, Henry's engagement for applications and modelling became particularly visible at ICME-3, 1976, where he gave the survey lecture on "The Interaction between Mathematics and Other School Subjects" (Pollak, 1979). Also in the first ICTMAs, Henry was active on a prominent position, for instance as a plenary speaker at ICTMA-3, 1987 (Pollak, 1989).

How is the situation of applications and modelling in mathematics education today, thirty years after ICME-3 and twenty years after ICTMA-3? Especially due to Henry Pollak's strong voice and his enormous influence, applications and modelling now occupy in many countries far more important curricular positions than at those times. Mathematics education now deals much more intensively with questions of teaching and learning modelling and applications, as can be seen, among other things, by the series of ICME and ICTMA conferences and not least by the present ICMI Study. For personal reasons, Henry Pollak was not able to come to the Study Conference. However, he was present in a plenary session through a video that was produced for this special purpose, and which showed excerpts from interviews with him.[1] The following paper is a transcript of that video. By this, we hope to convey the authenticity and livelyness of Henry's contribution as far as possible.

Mathematics education owes a great deal to Henry Pollak, and this chapter of the Volume ought also to be seen as homage to this pioneer in our field.

1. INTRODUCTION

The important conferences on the teaching of applications and modelling which have been going on for twenty years now have also been a major force in the spreading of those ideas and I am delighted to have a chance to participate once again in such a conference – it's been a long time since I have been able to come to one.

I think the question you have to ask yourself is: What is a complete mathematics education, that is, what are all the things that should be going on when kids are learning mathematics?

2. MATHEMATICS EDUCATION

Mathematicians may often feel that it is just the importance and the beauty of the subject that obviously justifies all this time – not all the students feel that way. There are going to be students – I think many of them – who are more interested in the usefulness of mathematics and being able to apply it to something else that they are interested in – rather than becoming professional mathematicians.

You have to remember roughly what the data are: roughly speaking, if you start at 9th grade and go through graduate school, you lose about half the students each year from the year before and so you end up with millions of students in the 9th grade and in the United States we end up with about two per year per million population getting a PhD in mathematics.

We have lost all the others along the way some way or the other, and many of those are going to be interested in how the subject gets used.

And so, I myself feel it should be an essential part of mathematics education to learn about how one uses mathematics. Now you will say: 'Well, we do that. Because, after all, our textbooks are full of word problems and all those uses of maths' and the answer is quite frequently: they are not. They are simply using words from some other discipline in the hope that those will keep the students interested.

But basically in the use of mathematics you have to formulate problems from outside of maths as well as solving them after you succeeded in formulating them. And you go back and forth between these two things a good deal, and this whole process of beginning with the situation that you are trying to (or that you hope to) understand mathematically until you finally get to a picture of it, and a formulation of it that will allow you to get some answers – that whole process is what we call mathematical modelling.

First thing you do is to identify something in the real world that you want to know or to do or to understand so the result at the end of step one is a question in the real world. Then we select particular objects within that question that seem important to the real world question and we identify the relations among them. So at this stage we have identified key concepts in the real world situation. Three – we then decide what we will keep and what we'll ignore about the objects and their interrelations. You simply cannot take everything into account. And the result then is an idealized version of the original question. Then, once we have this idealized version, we translate it into mathematical terms and we obtain a mathematical formulation of this idealized question. This is what is called a mathematical model.

There are many disciplines in which modelling has been the game for centuries, physics and chemistry, there are branches of engineering, and more recently there are branches of the social sciences. It is their great challenge to take situations in their field and try to understand them in a quantitative theoretical kind of way, in order to explain what they see and to make suggestions and predictions and forecasts for the future.

The question I am asking is not when did model building ever appear, or when did models ever appear? The question I am asking is when did mathematics educators get interested enough and think it was important enough so that they started to pay some attention to it in the teaching of mathematics?

The feeling has been, well, what we need to do is to teach some pure mathematics and let the other people apply it. And the great difficulty with this is that you lose your students. You are asking them – and it is one of the many different ways in which we ask them to do something like this – we are asking them for delayed clarification. We do this over and over again.

'Why do I have to learn this?'

'Well, you will see, you are going to need it. That discipline needs it. Maybe this will become clear to you later.' 'Why do I have to learn this?'

'Because it is going to be on the test.'

'But, how does that tie in?'

'Well, ok, you need this for another subject, you need it for a job that you may have. You may need it for intelligent citizenship. You may need it to get ahead.

But right now, you need it because of the test. This test is going to allow you to get into a good college, and that college then will allow you to get the good job and become an intelligent citizen and get ahead.'

So the gratification is always delayed.

What do I think? I don't think you can motivate very many students by the beauty of mathematics alone without seeing the usefulness.

3. INTER-RELATIONSHIPS

It is very natural for some people to think that, look, the existing system created me and I am a success, so obviously that's the best possible system. I don't know the extent to which that is true. I suspect but can't prove that almost any way that you had taught me mathematics I would have become a mathematician, simply because I liked the subject so much and I learnt and investigated it on my own in various ways no matter what anybody said or what they did.

According to my father's diaries I used to be looking for patterns in things that I did not understand already when I was five years old or something like that. So there was an instinct there. But how did I make the transition into being also interested in the connection between mathematics and the rest of the world?

I can remember one particular incident which happened at the supper table when I was a senior in high school. My father asked me a question. He didn't usually worry much about what I was doing in school and didn't ask about it, but he said to me: 'What are you studying in physics right now?' I said: Well, currently what came up is the subject of adhesion and of cohesion – that is substances sticking together. And he said: "Well, what is that?" I said: "Well, one of them is where one substance sticks to itself and the other one is where the substance sticks to another one." And I said: "I can't remember which is which." He said: "Well, I know," and he proceeded to tell me. I said: "You never studied physics, how do you know that?" He said: "Well, I don't know any physics, but I know Latin." Now that was a hit below the belt, because I had also studied Latin all the way through high school and was pretty good at it. But it had never occurred to me that you could use subjects on each other. That was not part of my education. Nobody had ever said that you can figure out which is adhesion and which is cohesion because 'co' means together (it's 'cum') and 'ad' means one to the other, next to each other or towards each other, so adhesion is two different substances and cohesion is the same substance.

That was a real eye opener, and it was at that point that I started thinking hard about how in fact different subjects were related to each other, rather than just trying to be good at each one. You start thinking about what the relation between them is, and how they affect each other. And it is at that point that I certainly began wondering, why didn't we ever talk about how mathematics is related to everything else? This was relating physics to Latin, so how is everything else related to everything else? So that was a key incident to take me in that particular direction.

Throughout my career – certainly throughout much of it – I maintained some interest in education. I got my PhD in 1951, so more than a half a century ago – and went to work at Bell Labs, and eventually, of course, after 35 years of this, I retired. I had struggled for decades with the fact that I was interested both in mathematical research and in mathematics education. I wore those two hats in different official capacities and yet I had to wear those two hats on the same head. And I struggled with the consistency between these two. And the consistency between these two hinges tremendously on an understanding of what applications of mathematics are really all about. And that was our business in Bell Labs.

4. THE NEW MATHS

In the 1950s we all experienced the excitement engendered by Sputnik and by the competition for outer space that gripped our leaders, and we decided we needed more students getting good at mathematics and science. The New Maths was born. I had been a student as an undergraduate of Ed Begle who founded the School Mathematics Study Group in 1958, and Al Tucker of Princeton was one of its chief advisors.

And so in the winter of 57/58, when they formed the first team to start working on SMSG –the School Mathematics Study Group - they remembered me, and they said, well, let's give it a try, and see whether this "takes" . And so I was invited back to Yale where I had been an undergraduate, and where the first four-week summer session of the School Mathematics Study Group began.

A major aim of SMSG was to make sense of school mathematics - as we said before. To anchor an apparently endless series of apparently isolated tricks to a structure within which these tricks become realizations of a small number of important mathematical principles. That was what SMSG was all about.

People have criticized the new maths a lot and have said it was a failure. I don't think that it was a failure – at least not in my terms. Mathematics as a percentage of all college majors reached an all time historical high in 1970 which was the peak of people coming out of the wave of New Maths.

It was – as I recall the figure – close to 5% of all college majors in 1970. Now mathematics as a major is back down to less than 1%.

It was the quality of what they were learning in terms of the understanding, making sense, that I think attracted so many students into mathematics.

Now, the great trouble is, when it comes to applications and to the usefulness of mathematics, that when we taught it we taught it without making sense - only this time we taught it without making sense in two ways rather than one. That is, we didn't try to make sense out of the mathematics that turned out to be involved, and we also didn't make sense of the external situation to which we claimed we were applying the mathematics . So it was twice as bad as what we were doing before.

You know, the typical event in a calculus course is: 'Today we are going to study the centre of mass, or moment of inertia or something.' What are they? Well, here are the formulae. What do the formulas mean? We don't know, it doesn't matter, just do the calculus. Worse than that, go further along: 'Ok, students, today we are going to study the Coriolis effect. Consider the following partial differential equation.' And then you go ahead and work with that equation. So what does it mean? What do the variables mean? How do you know it's right? What did you keep and what did you throw away in writing down that equation?

So the reason that the teaching of applications at that time often was so bad, was that people didn't bother to understand them as well as not bothering to understand the mathematics. My feeling was that when we turned towards teaching the usefulness of mathematics, teaching the modelling was needed in order to make sense of the application, and that combined with making sense of the mathematics would then produce a first rate curriculum.

In teaching modelling, there are obviously two things that you have to do, very basic kinds of questions. You have to take some models that have been created and have been known to be successful and students have to study those models, and understand what makes them work, and think about what went into their creation and the way they were formulated and their success.

Also, students have to take situations for themselves and start creating models of those situations, make decisions of what you have to keep and what you can afford to ignore, and how you are going to test whether you really succeeded. Find something that needs to be done, bite off a piece that you think you can chew – and then chew it. And incidentally, find out if you have succeeded. A very basic question that you have to answer is: which of these two aspects of teaching modelling do you do first?

The same kind of question was asked years ago about learning computer programming. You have to read other people's programs and see why they worked, and you also have to create programs on your own and make mistakes and learn from them.

How do you balance those two? Research on just understanding the simplest aspects of this question has been begun, for example by Jerry Lege, where you begin to study the effects of each of these activities, of looking at other people's models and the effects of creating some of your own. So, we need a research base we don't have, but which is in the process of beginning to be created.

5. PROBLEM FORMULATION

When is problem formulation over? When have you finished in fact the key stage of problem solving? And the answer is: When you get to the point where the mathematics itself is familiar. So that's a moving target – as we hope – as our students get older and better, they know more and more mathematics, and there is less and less that you have to do in formulating before you get to the point at which it's routine. But the really exciting part of problem formulation is over at the moment that you have got to familiar mathematics. When we have done this we translate it back to the real world and what we now have is a theory of the idealized question. And then you carry out the mathematics that is indicated.

After that's done, comes one of the key points – now comes the reality check. Do we believe what's being said? Are the results practical, are the answers reasonable, are the consequences acceptable? If yes, the real world problem solving has been successful. Our next job, namely to communicate with potential users, is both difficult and extraordinarily important.

If I can go back to my earlier existence for just a minute, one of the hardest things that I had to do at Bell Labs was to work with everybody on exposition, as my bosses had worked with me when I first came. People have to learn to write things down in such a way that the person who needs to understand what's being said can do so. It's just amazing to me how many people will do an interesting piece of work and get it all down in their head and their notebook, and then be utterly unwilling to write it down for the public in such a way that somebody else can understand it. It's absolutely no good to do a piece of work, get it all neatly written down in your head or in your notebook, and be unable to communicate to anyone else as to what it is that you have done. So certainly people who are going to do quantitative, structural, systematic – i.e. mathematical work on their jobs have got to learn to explain what they have done.

6. PROBLEM SOLVING

Polya was a tremendous educator. I was very fortunate to know him and know him very fairly well. It's a very amusing sideline that our birthdays were exactly 40 years apart. The analysis and the ideas that he contributed towards good teaching were phenomenal. It is also true that he was very interested in applications and did a lot with them. For example, if you look at his little book on mathematical methods in science which is absolutely gorgeous, this is a book that he wrote specifically to use with teachers to give them a good feeling for how mathematics is used in science in their grade, but he is always very careful to stop while the subject is still beautiful and before it gets hard. He will do the particular pieces which are particularly nice. And then the next piece that one has to do in going further into the understanding of the subject gets harder, and he stops and goes on to something else. Well, of course, it becomes an exciting course for teachers and there are plenty of beautiful things that you can do without getting into a mess.

I think that is a little wrong. I have taught that little book a number of times, and I always insist on doing a few messy things that go beyond what's in the book because I want the teachers to understand that while there is a lot of beauty here, and a lot of lovely ways of getting understanding, sometimes you've got to work harder than that, and I want them to have seen it.

The role of technology is really very large. Technology has the possibility of making certain subjects possible. For example, there are many concepts related to data analysis and to understanding what a set of data is trying to tell you. Often, to get the understanding that you want, you need to process more numbers than you have time to do. The pedagogic problem is the following: Suppose, you have a thousand numbers and you need to know what the structure of that set looks like. By the time the class can agree on what this is, they have long since forgotten why they were interested in the problem. And so technology has made it possible to teach certain subjects and to get involved in some things which are very interesting and very important – you just couldn't do them without technology.

Technology has made certain things more necessary, that is, for example, in understanding what goes on in computers and connecting them with their applications, you need to think much harder about discrete mathematics, you need to think much harder about what an algorithm is, and what you expect from success than you had ever needed to look at before.

Technology has certainly also changed the relative importance of various topics. As we have said, it has made some topics more important – it has made other topics less important. I know that there is a lot of disagreement about this but it is just true that the technique of division isn't as important as it was before we had the way in which everybody now does it. And it simply isn't worth as much time as it used to have.

But probably the biggest thing that technology has done is to change the image and the practice of mathematics itself. Mathematicians have never really admitted how they practice their subject. We show students lovely results and interesting developments, and they don't necessarily find out how difficult it was to get all of these things, and the thing that the technology has done is to force us to come out in the open with unspeakable practices that we used to indulge in with our doors closed and our blinds drawn, and which nobody ever knew about and that is - we experiment. Our job is not just: 'here is a problem – solve it' or 'here is a theorem – prove it'. Our job also is: 'here is a situation, can you figure out what the hell is going on?'

This is experimental work. And what technology has done is to return, after an absence of maybe a hundred or two hundred years, return mathematics back into the full spectrum of science. It is also an experimental subject and we do experiments to find out what's going on, and then we try to see whether we are right or not. And we are now in a position where it is respectable to admit this.

7. THE OXFORD SEMINARS

I know mostly about what goes on in the United States and don't claim to try to do things thoroughly as far as the world is concerned. I couldn't possibly do that. The Oxford Seminars began in 1968. Now let me describe what the Oxford Seminars were. People from private industry or from government owned industry in the British Isles could bring difficult problems, in classical applied mathematics typically, that is mathematical physics problems - on which they needed help. And they and some professors and some graduate students at Oxford would spend a week together, no cell phones, no interruptions, no nothing. They lived together for a week and worked on those problems.

What I was told is, that about a third of those problems led to complete solutions to the enormous satisfaction of the chemical industry, or transmission, or various others. About a third of the problems had partial solutions and on about a third of them they got nowhere, and incidentally a number of graduate students got very good jobs -which is nothing to be sneezed at. The reports of the Oxford Seminars reflected back on how modeling is or is not done in education. Because this was a brand new direction. The method of presentation of applied mathematics in secondary and tertiary education has on the whole changed very little changed since 30 years ago. When the majority of mathematicians went on to teach in traditional ways at school or university, the training they received was not too bad. But more and more are seeking jobs outside teaching, and are looking for jobs in industry, commerce etc., where they will be expected to use their mathematical expertise to solve problems which are of interest to the engineers, scientists, and others around them.

The employers complained bitterly that we produce many graduates who are incapable of this. I would like to quote a pen portrait of the typical mathematics graduate: 'He is good at solving problems, but not so hot at formulating. The graduate is not particularly good at planning his work nor at making a critical evaluation of it when completed.'

The evidence from schools, universities and the employers leads one to the conclusion that we must look again at the teaching of applications of mathematics at all levels. This could well involve the consideration of new situations outside science to which mathematics can be applied - but this was England after all. But this should also include new stimulating treatments of scientific topics. At the moment in many cases too much emphasis is put on the elegance of the solution and too little on the significance in relation to the original problem. Too little time is spent on developing a model from a real situation.

The Oxford Seminars were very early, but the early efforts were very similar in the United States. Oxford had a recognized influence on what happened at the Claremont Colleges in California, which were one of the very early practitioners of modelling in the United States.

8. NEGLECTED TOPICS IN MATHEMATICS EDUCATION

People are always interested in how to predict, and prediction is one of the most nearly impossible as well as one of most interesting things that you can do if you think you may be good at it. I was in a position to try it.

I realized that topics in the mathematical sciences that we at Bell Labs found to be particularly important, but which our current and future employees were very unlikely to have learnt in school, were very fine candidates for future importance. What I found by the time I got to Bell Lab in the 50s and from there on is, that there were two things that were very much in the forefront of research which people typically did not learn - one was the subject of data analysis and the other was the subject of discrete mathematics. I felt that if those were so important to the work of what was then the huge unified Bell System, that those were things that really ought to be kept in mind for the future of education – because, by god, they were important. Certainly in the fifty years that I have been involved, those two subjects have grown tremendously. Several of the topics in data analysis that were invented at Bell Labs after I got there have now found their way into elementary school!

One of our problems is that we do teach some successful data analysis in elementary school, and you have to figure out how to keep it alive until you get to college because it is so difficult to get data analysis used in the high school curriculum. I didn't mean to go off onto that tangent – but as people will learn as they listen to me, my talk consists entirely of tangents.

9. TEACHING MODELLING

So, I came back in thinking about how do we get to the teaching of modelling. I think that is so tremendously important because more students will stick with us if we have this kind of mathematics. We have to teach our mathematics in such a way that it keeps people together as long as possible, so that no matter what background you come from you have the longest possible time in which the directions that you would like to go in, and the capabilities that you have, can show themselves.

So its very natural to ask yourself what aspects of mathematics do everybody have to learn, will be good for everyone to be good at. Everybody has to use mathematics in everyday life, and everybody has to understand how mathematics relates to the rest of the world that they live in. And so modelling can become a unifying force, it can be something that if you spend time on it early and often, it helps to keep the kids together. And the longer you can keep them together the better it is for our society. We can't afford to be a place where you decide at the end of 4th grade what somebody's future is going to be. You want to provide experiences that everybody should have, so that everybody can be kept together just as long as possible – and modelling is one of the subjects that allows you to do that. We want to get additional topics that are important for modelling, we want to get modelling itself, into the curriculum. We have been working on trying to show samples of curricula that take modelling much more seriously than we have in the past, and finding out how successful they are with how many students and how well this works in motivating kids and keeping them interested.

The major difficulty that is often cited is that if you are going to model phenomena from outside of mathematics, you of course, as we have said before, have to understand those phenomena. This requires people who teach mathematics to understand things that they normally are not required to understand. But if you say to yourself, 'look, my intention is that some of the time is to be spent on the kids learning how to model,' then you can find loads and loads of subject matter areas where they know all about the subject involvcd, and they can indeed practice modelling. Modelling as an activity doesn't have to be learned in areas which the teacher and the students don't understand!

Maybe we can think of the possibility that society will get sick and tired of the stone-walling against modelling at school level. Maybe society might decide that traditional mathematics is simply not worth being taught every year to everybody, and they might choose to use some of the time for real world problem formulating and problem solving. This might contain lots of different important and interesting mathematics.

Suppose this were to happen – that is one seriously tries to say 'Ok, you don't want to let us into traditional courses, very well, we will try to set up some new different courses in mathematical modelling.'

Now what would a curriculum look like? How would you educate teachers for this? How would you interact and connect with traditional mathematics? How would you connect with statistics and science and computer science? How would you convince a skeptical world that you had succeeded with something important, that is, how would you assess the students in this new subject?

If I were at your wonderful meeting in Dortmund, I would be listening for ideas, no, I would be pestering you for ideas. And so it is probably a very good thing that I am not there. Thank you very much!

REFERENCES

Pollak, H. (1969). How can we teach applications of mathematics? *Educational Studies in Mathematics 2*(2/3), 393-404.

Pollak, H. (1979). The interaction between mathematics and other school subjects. In UNESCO (Eds.), *New Trends in Mathematics Teaching IV*, (pp. 232-248). Paris.

Pollak, H. (1989). Recent applications of mathematics and their relevance to teaching. In W. Blum et al. (Eds), *Applications and Modelling in Learning and Teaching Mathematics,* (pp. 32-36. Chichester: Ellis Horwood.

Pollak, H. (2003). A history of the teaching of modelling. In G. Stanic & J. Kilpatrick (Eds.), *A History of School Mathematics,* (pp. 647-671). Reston: NCTM.

Polya, G. (1977). Mathematical Methods in Science. *New Mathematical library* (Vol. 26) Washington: Mathematical Associaion of America.

Raynor, M.E. (1973). Mathematical Applications in Science. *New Aspects of Mathematical Applications at School Level* (pp. 209-216). Eich sur Alcette, Luxembourg: Imprimerio Victor S.A.

[1] Our sincere thanks go to Brian Cohen who conducted the interviews and in particular to Jere Confrey who wrote the questions and edited the interviews into a coherent piece.

Part 3

Issues in Applications and Modelling

Section 3.1

EPISTEMOLOGY AND MODELLING

Edited by Jere Confrey

Chapter 3.1.0

EPISTEMOLOGY AND MODELLING – OVERVIEW

Jere Confrey[1]
Washington University in St. Louis, USA, Email: jconfrey@wustl.edu

Abstract: This section introduces the issues of epistemology, including how to relate modeling and mathematical growth, the implications of complexity, and the limitations of the approach.

1. INTRODUCTION

Modelling emerges as a major theme in mathematics education from several perspectives. It seems a natural historical and evolutionary descendant of constructivism, and includes central connections among disciplines, develops the importance of multiple representational systems, and incorporates elements of students' prior knowledge, strategies, representations, inscriptions, and reflections. Modelling can highlight fundamental issues from a socio-cultural point of view by emphasizing iterative and emergent development of complex ideas, subject to critical discussion and judgment, and honed through interactions. The strongest arguments for modelling are based on the view that it will be advantageous for the development of student thinking. Many papers in this section try to unpack this perspective, from an epistemological point of view.

Many mathematics curricular initiatives have valued the use of situations in mathematics education (realistic maths, didactical engineering, etc.). Supporting this orientation is the view that drawing on situations strengthens relationships among mathematics and (1) students' prior knowledge, (2) their knowledge outside of schooling, and (3) their subsequent applications of mathematics to other contexts. However, modelling's value to students goes beyond inviting them more effectively to enter mathematics or preparing them to draw connections beyond the mathematical arena. Modelling-based changes in epistemology can permeate instruction during the whole of the course of mathematics education. This section's authors illustrate a wide range of impacts on instructional programs and learning.

Two contrasting positions provide a useful heuristic for considering the role of modelling in mathematics education: (1) advocating *modelling for mathematics*--modelling as one of a number of means to demonstrate views and uses of mathematics, and (2) supporting *mathematics for modelling*--mathematics viewed as a tool critical to the modelling enterprise. The former, should be recognized and valued, especially as a transition strategy for reform and change. However, authors in this section examine the more encompassing perspective that mathematics acts as a tool for building models. Within this perspective it is still important to attend to the place for learning the skills and concepts of mathematics, developing structures, and developing multiple methods of proof. To be clear, the group did not call for a "watered down" view of mathematics, but rather for a more flexible and substantial foundation that can support stronger uses of mathematical prediction, explanation, and justification.

For some, modelling is breathing new life into the ideas of mathematical literacy, and should affect mathematics as taught to a broad cross-section of students. This perspective also argues that more people must understand how mathematics transforms the modelling process, shapes the claims that can and should be made, and influences societal resources and conceptions. Scholars increasingly speak of the need to unpack the black boxes of choices and decisions, based on mathematical models, which shape human enterprises but are otherwise inaccessible to citizens' critical understanding. Finally, in addition to considering the epistemology of mathematics as taught through modelling, it is important to explore how that vision might be realized in relation to particular strategies in instruction. Three key issues arise that merit thoughtful attention: (1) How can one relate student modelling activity and mathematical growth? (2) What epistemological ramifications result from intensified use of complexity? (3) What are the limitations of this point of view? Each contribution addresses these issues in different ways.

2. PAPER SUMMARIES

The plenary by Confrey and Maloney set the stage for addressing issues of student modelling and instructional growth. We presented modelling as a process of transforming an indeterminate situation to a determinate one, (the model), emphasizing a key role for inquiry and reasoning in the process of using technology. Inquiry stimulates the process of transformation; reasoning permits one to use the known structures and tools of mathematics to create and justify a set of key artifacts. Epistemologically we stressed that the artifacts (observations, measurements, interactions, indicators and descriptors) are the data by which the model is assessed, and that the student looks through the model to recognize salient, explanatory, or predictable patterns in his/her world. Student investigation of a spring with a motion detector illustrated the potential of a single modelling task to provide appropriate content at different levels of student mathematical sophistication. This suggests that modelling provides an effective means to differentiate instruction and to manage heterogeneity.

Arzarello, Pezzi, and Robutti, in discussing how real and fictive motions support students' interpretations of graphs, further illustrate how modelling activity may evolve from the personal to the imagined to the abstract. Their description of cognitive activity comprises a complex evolution, starting in bodily experience; continuing with the evocation of the just-lived experience through gestures and words; developing further by connecting it with the data representation, and culminating with the use of algebraic language to record the relationships among the quantities involved in the experiment. In this work, modelling permeates the development of mathematical thought.

Lehrer's invited paper articulates the teaching of modelling as promoting students' progressive mastery of expert and complex inscriptions. His theory of modelling successively incorporates physical and mechanical models, representations, syntactical systems, and, finally, hypothetical-deductive and emergent systems. His summary states key epistemological assumptions: "...challenges include recognizing that models edit, rather than copy, the world. Models amplify phenomena by specifying relationships that one might not have otherwise considered, [and] suggest qualities of the world to modelers. One must learn to look through the model into the world." Further he recognizes a key role of competition or comparison of models: judgments of the quality of a model "...rely on entertaining alternative models. ...[W]ithout opportunity to invent and revise models, this epistemic quality is largely hidden from view." (this volume, p. 160) The process, refinement, and critique of modelling will be needed to bring to the foreground the epistemological changes, implied by the approach.

Gravemeijer identifies a concept of emergent modelling wherein, instead of trying to concretize abstract mathematical knowledge, the objective is to try to help students model their own "informal mathematical activity" (this volume, Chapter 3.1.2). Following the work of Latour and Lehrer and Schauble, he discusses a "cascade of inscriptions of a chain of signification." Like Confrey and Maloney's transformational process, and Arzarello et al.'s sequence of graphing experiences, Gravemeijer emphasizes the role of progressively reorganizing situations. One can suggest that all three of these authors recognize that modelling permits complexity into the classroom, to be treated as leading to successive reorganization of content, to account for varied aspects of the problem. This can change the classroom dynamics, replacing the quest for "truth" by more nuanced views of shades of meaning, successive approximations, and continuous reassessment and review.

Lesh and Yoon articulate principles for designing *model-eliciting activities*, which include refinement and revision in the problems, the expression of current and changing beliefs, and developing proficiency with sharable and reusable tools. For them, models draw on conceptual systems, rely on a variety of representational systems, and involve intermediate design cycles. It is of particular importance that these aspects of modelling lead to methods of research and assessment that encourage greater participation among students and bring forth a diversity of approaches.

Sträßer contributes another perspective, key to how mathematics is experienced in this complex world. He recognizes the dilemma that with increasing automation, the role of mathematics seems to hide, yet people simultaneously argue for its importance in a technological society. He traces how technologies of selling leave sellers only keying in unit prices or identification codes, but that in non-routine or breakdown situations mathematics' essential character reemerges. His argues the importance of professional situations which entail a dialectic between concrete and abstract, and, in classrooms exploring these relationships in a demystifying manner.

Finally, Hanna and Jahnke demonstrate a role for modelling even with respect to proving, that most sacred of mathematical activities. They show that mathematical justifications can be used as a form of explanation, and can draw upon familiar ideas from, for example, physics, such as balance and centers of gravity. Then, these "tools" can be a means to generate proofs for geometric theorems. This leads to the use of a physical system to represent and manipulate a non-physical, mental system such as geometry.

These papers provide state-of-art analyses of epistemological issues underlying modelling, and argue that modelling benefits the student and more accurately portrays the role of mathematics in technology and society. They do not simply assert the global value of the approach, but rather provide examples and analyses of how it played out in everyday instructional settings, and will require careful and explicit attention to describing changes in student thought over time.

---

[1] Valued contributions to a discussion that formed a basis for the introduction to the chapter were provided by: Michèle Artaud, Michèle Artigue, Ferdinando Arzarello, Stephen (Sen) Campbell, Jere Confrey, Koeno Gravemeijer, Gila Hanna, Eva Jablonka, Niels Jahnke, Gabriele Kaiser, Susan Lander McNab, Katie Makar, Alan Maloney, Geoffrey Roulet, Ralph Schwarzkopf, Heinz Steinbring, Rudolf Sträßer, Marji van den Heuvel-Panhuizen, Wim Van Dooren, Igor Verner.

Chapter 3.1.1

MODELLING BODY MOTION: AN APPROACH TO FUNCTIONS USING MEASURING INSTRUMENTS

Ferdinando Arzarello, Giovanni Pezzi and Ornella Robutti
Departement of Mathematics, University of Torino, Torino, Italy,
Email: ferdinando.arzarello@unito.it / Liceo "Torricelli", Faenza, Italy,
Email: gpezzi@racine.ra.it / Departement of Mathematics, University of Torino, Torino, Italy,
Email: ornella.robutti@unito.it

Abstract: The paper presents an approach to modelling in secondary schools where technological instruments are used for measuring and modelling motion experiences. In all cases one or more sensors measure various quantities and are connected to a calculator. In some examples we study pupils (9-th grade) who run in the class and see the Cartesian representation of their movement produced by a sensor in real time. In others, pupils (11-13-th grade) go on switchbacks or other similar merry-go-rounds and use instruments to measure some quantities (speed, acceleration, pressure), which are recorded on graphs and tables. In both cases, pupils discuss what has happened and interpret the collected data. Within a general Vygotskian frame, the authors use different complementary tools to analyse the situations: the embodied cognition by Lakoff and Núñez, the instrumental approach by Rabardel, the definition of concept by Vergnaud. In particular the role of the perceptual-motor activity in the conceptualisation of mathematics through modelling is stressed.

1. THE THEORETICAL FRAMEWORK

It is well known that pupils have difficulties in conceptualising the function concept. According to the current research, their difficulties concentrate in interpreting graphs, particularly those in which a variable is time-dependent, as for example space-time or velocity-time graphs. In fact, two main misinterpretations have been pointed out in the literature. One is the graph-as-picture interpretation, in which students expect the graph to be a picture of the phenomenon described. In kinematics, this can result in the students interpreting a graph of space versus time as if it were a road map, with the horizontal axis representing one direction of the motion rather than representing the passage of time (Clement, 1989). Another common misinterpretation is the slope/height confusion, in which students use the height of the graph at one point, when they should use the slope of the line tangent to the graph at a point, and vice-versa.

To overcome such difficulties, we have designed a teaching project where the function concept can be approached within suitable experience fields (Boero et al., 1995b) so that its meaning can be built up by students in a proper way. To this end, we use a motion sensor and a symbolic-graphic calculator, with which students create graphs and number tables to model different kinds of motion (either of their body or of other objects). The didactical aim of the teaching experiment is the construction of the concept of function as a tool for modelling motion. Our particular goal with these activities is that the students can reach competencies in describing mathematically a function, both from a global and a local point of view, starting from their perceptions and experiences with the sensor. At a more advanced level, they can use such competencies to interpret more complex situations, e.g. the motion on a switchback.

The research aim is the analysis of students' cognitive processes involved in the construction of meanings for the mathematical objects, through modelling representations. Specifically, our investigations focus on their mental dynamics while they interpret the different representations of data (tables, graphs) in order to grasp their meaning with respect to the concrete experiment of motion. This analysis is made by the observation of all the students' activities, including their gestures, language, and interactions with the instruments.

Hence our research can be framed within the challenge of Issue 1 of the Discussion Document (see Blum et al. on p. 515). Specifically, it makes some contribution to the following questions:

- What are the processes of modelling? What is meant by or involved in each?
- What is the meaning and role of abstraction, formalisation and generalisation in applications and modelling?
- How much extra-mathematical context must be familiar and understood to undertake applications and modelling?

The general framework of our research is Vygotskian: the emphasis is on the social construction of knowledge and on the semiotic mediation given by cultural artefacts (Bartolini Bussi et al., 1999). The social dimension is given by the recourse to the 'mathematical discussion', orchestrated by the teacher (Bartolini Bussi, 1996); the artefacts are represented by the symbolic-graphic calculators, by the sensors and by the switchbacks.

To describe the crucial cognitive aspects of pupils' learning processes in interaction with technological instruments, we use three analysis tools:

- The embodied cognition approach by Lakoff & Núñez (2000) (see also: Arzarello, 2000a; Arzarello et al. 2003);
- The instrumental analysis by Rabardel (1995) and others (Artigue, 2001, Vérillon et al., 1995);
- The definition of concept given by G.Vergnaud (1990)[1], in particular the notion of operating invariant.

We think it is possible to integrate the instrumental approach with new results from cognitive science, in particular embodied cognition. These two approaches help us to analyse the students' activities from a new point of view. In fact, if the instrumental approach can give us a framework to analyse the use of technologies by students, in terms of schemes of use, it is not sufficient for interpreting their mental activities, especially during the conceptualisation processes. On the other hand, cognitive science is perfectly aimed to study pupils' mental activities; however, its approach to conceptualisation processes in mathematics focuses on some fundamental aspects but does not explain all of the theoretical and symbolic features of the mathematical thinking. Hence we find it useful to embed our analysis within the framework of Vergnaud's definition of concept.

2. THE TEACHING EXPERIMENTS

A main problem for students who are requested to interpret graphs or numerical tables (which model situations) regards their static features (see Kieran, 1994; Boero et al., 1995a), which risk blocking their mental dynamics, hence inhibiting a fruitful exploration (Boero et al., 1995b). In fact, to cognitively grasp the meaning of a function one needs complex dynamic activities; for example so called fictive motion (Talmy, 1996), produced when the subject interprets a graph in a dynamic and oriented way, as if it were produced by a moving trajector. Such an activity can be observed through the words and gestures of subjects (see Lakoff & Núñez, 2000, pp. 31 and 37). From this point of view it is interesting to observe how a graph is generated on the screen of a graphic calculator, which represents data on-line measured by a sensor (CBR[2]). The observer looks at a genuinely oriented generation of the points in time, which is a sensibly different experience from perceiving a graph given in a holistic way. Such a dynamic graph is easier to interpret by subjects, when compared with a static one. This is the starting point for our first working hypothesis: suitable fields of experience (see Boero et al., 1995a) where students experience real and fictive motions, can support pupils while interpreting graphs. Such a field is our "Real data in real time", where pupils live some concrete experience (e.g., running); in the meanwhile some data are relived by an on-line measurement tool and represented in real time on the screen of a graphic calculator. Successively, pupils are asked to interpret the graphs and tables on the screen, exploiting what these mean with respect to their lived experience. In the end they are asked to analyse some of their specific features and to represent them using suitable algebraic language. Our second working hypothesis is that body, language, and instruments mediate and support the transition of students from the perceptual facts to the symbolic representation, e.g. the algebraic one: in fact they can stimulate the production of an intense cognitive activity, which is marked by rich language and gesturing activity, for example with production of grounding metaphors. The purpose of our proposal is to describe the development of students' cognitive activities from bodily (e.g. perceptual, kinetic, ..) to theoretical features. In such a development a crucial point is the genesis of the meaning for mathematical objects through modelling activities exploiting temporal explorations towards their just past experience and anticipating hypothesis and conjectures. Words and gestures reveal crucial insights within this activity; in particular language provides students with a fruitful cognitive activity based on their just lived kinetic and visual experiences. This genetic process allows students: (i) to produce a mathematical sense for the graphs they see on the screen and (ii) to start and support their transition to the algebraic register. For a wider discussion see the Research Forum at PME 27 (Nemirowski et al., 2003).

The teaching experiment is organised as a long-term intervention of activities during the year, each activity lasting for two-three one hour class sessions, and possibly including some open air activity, e.g. going on switchbacks in a funfair. During the sessions the students work in groups of three-four pupils and they use the tools of the activity (e.g. a measure instrument or a graphic calculator or a sheet of paper). In each activity they have to answer some questions on a working proposal form, related to the construction of the meaning of a mathematical object. The researcher, who is present during the activity has the role of observer (she records everything with a video-camera) and guides the final discussion.

3. SOME EXAMPLES OF MODELLING ACTIVITIES

3.1 Example 1

The experiment, organised by O. Robutti, consists in a sequence of activities scheduled as follows:

1. Analysing a graph and answering some questions about the points and

their co-ordinates;
2. measuring the length of objects with different tools (ruler, meter, ...) and finding regularities;
3. representing data in tables or graphs using a graphic calculator (a TI92, by Texas Instruments);
4. collecting time and distance data by using a sensor of position and analysing them on the graph and in the table of the calculator screen (Fig. 3.1.1-1);
5. constructing models of a phenomenon, knowing the rate of change of a quantity vs. time;
6. measuring data of a variable quantity vs. time and modelling the phenomenon.

Each activity is divided into three parts: in the first, the students (in small groups) explore a situation (using a proper tool or by paper and pencil); in the second the groups answer some written questions which ask them to use/build suitable data representations (tables, graphs) to interpret the situation in a mathematical way (within a pencil and paper or calculator environment); in the third and final part, the students participate in a class discussion, guided by a researcher.

3.2 Example 2

In our students' schools mathematics and physics are both taught by the same teacher. The idea here is to design activities within the pupils' field of experience "Real data in real time" and to use sensors to collect data on some physical quantities (speed, acceleration, pressure) while riding on a switchback or some similar machine, and then to use graphical and numerical representations to discuss the model so obtained. The goal is for pupils to enter more and more deeply into the physical concepts experienced while going on the machines, using the mediation of the mathematical model represented on the screen of the computer. The experiment is conceived with the same philosophy as that above, but requires more mathematical knowledge: in fact pupils are 2 – 3 years older than in the previous case. This part of the experiment has been designed by G. Pezzi and his equipe in Faenza. Fig. 3.1.1-1 (next page) shows the sensor-kit organised to measure the physical quantities (courtesy of Texas Instruments): the kit is assembled in a bag, which can be fastened to the experimenter's body or directly to the machine. Fig. 3.1.1-3 illustrates one of the machines (the Thunder Sierra): it is a switchback with a height difference of 32.5 m, whose structure and interesting aspects are sketched in Fig. 3.1.1-2.

Using pressure measures, a profile of the road has been drawn. Moreover an accelerometer has been used to record data concerning the acceleration of the coach: the diagrams (Fig. 3.1.1-4) have been obtained using the program Graphical Analysis 3.0, using the smoothing function in order to eliminate the noise from the acceleration graphics.

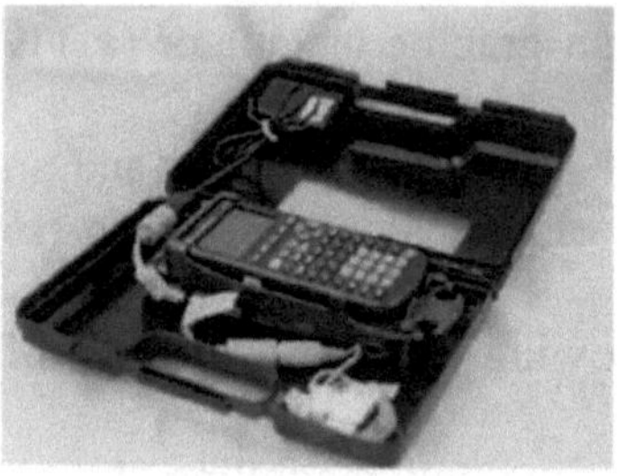

Figure 3.1.1-1. Sensor kit

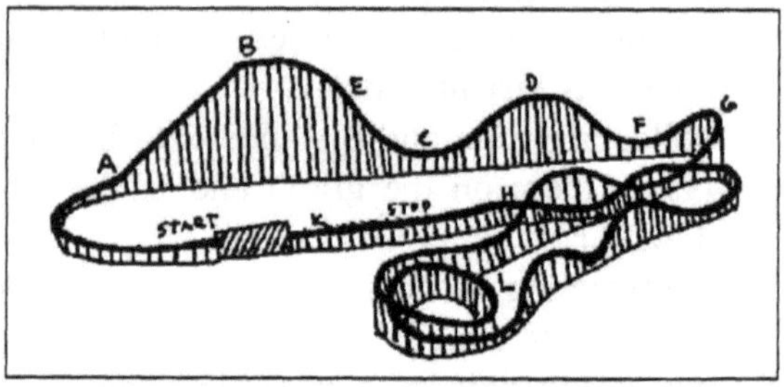

Figure 3.1.1-2. Structure of Thunder Sierra

Figure 3.1.1-3. Thunder Sierra

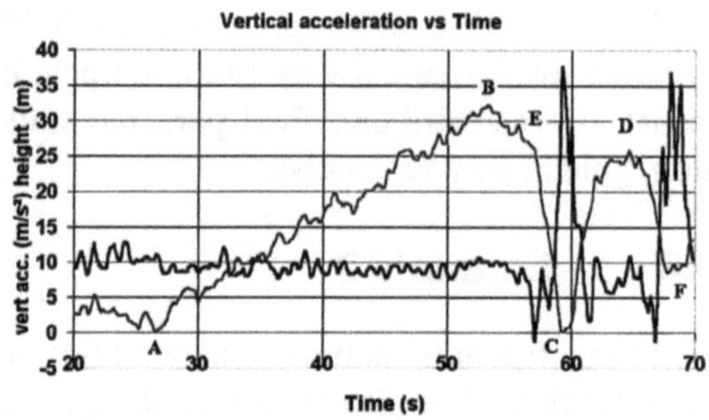

Figure 3.1.1-4. Measurements on board Sierra

4. SOME PARTIAL CONCLUSIONS

The written protocols of all the students show that most of them have good linguistic production and a flexible co-ordination among different registers: verbal, graphical, algebraic. Moreover there is an interesting genesis of the mathematical concepts through metaphors, fictive motion and managing of the inner times (Varela, 1999; Arzarello et al., 2001). We can observe this intense cognitive activity through their gestures and linguistic productions.

It is interesting to observe that the students' cognitive activity passes through a complex evolution, which starts with their bodily experience; goes on with the evocation of the just lived experience through gestures and words; continues by connecting it with the data representation; and culminates with the use of algebraic language to write down the relationships between the quantities involved in the experiment. The recalling process has a double nature: from the one side words and gestures start the generative action function towards a suitable representation of what they have done (i.e. with tables, graphs, functions); from the other side, it allows a meaningful interiorisation of their experience. In fact, there is dialectic between mathematical concepts (for example, a function), and their representations (for example, its graph), which develops through the generative action function supported by language and gestures.

Some didactical conclusions can be drawn from our experience and may possibly be confirmed by the research, which is going on in the meanwhile. a) The approach to functions in the school often inhibits or curtails experiences that encourage the productions of fictive motions schema. For example, the graphs in books and exercises generally have a static and holistic aspect. But new technology allows teachers to design experiences where graphs can be presented in a dynamical and genetic way. b) Using grounding metaphors seems to facilitate such functions as the generative and generalising ones, which can support students in the transition to a meaningful managing of algebraic language. In fact metaphors are based on common cognitive activities that all people can do. However, grounding metaphors may be not always appreciated in the class of mathematics, since they have not a rigorous flavour. On the contrary, encouraging their production by students can facilitate the understanding of formal aspects of mathematics. As a by-product, our findings suggest that a genetic structure appears in the way metaphors are produced, which intertwines deeply with inner times of pupils. Their cognitive activity shows a continuous dynamic movement from the present to the past (their lived experience) and to the future (the hypothesis or the de-timed sentences). The analysis of connections between inner times, rhythms and metaphors reveals investigations in Mathematics Education as a promising field from the point of view of research (genesis of mathematical objects), as well as practice (which cognitive activities can the teacher encourage to facilitate pupils' understanding of mathematics?).

REFERENCES

Artigue, M. (2001). *Learning mathematics in a CAS environment: the genesis of a reflection about instrumentation and the dialectics between technical and conceptual work.* Paper presented at the 2° CAME Symposium. Utrecht, The Netherlands.

Arzarello, F., & Robutti, O. (2003). Approaching Functions through Motion Experiments. In Nemirowsky et al. (Eds.), *Perceptuo-Motor Activity and Imagination in Mathematical Learning*. Hawai: Research Forum, Proceedings of PME 27.

Arzarello, F., & Paola, D. (2003). *Mathematical objects and proofs within technological environments: an embodied analysis.* WG9, CERME, Bellaria.

Arzarello, F., Bartolini Bussi, M.G., & Robutti, O. (2001). Time(s) in Didactics of Mathematics. A Methodological Challenge. To appear in: L. English, M. Bartolini Bussi, G. Jones, R. Lesh, & D. Tirosh (Eds.), *Handbook of International Research in Mathematics Education.* USA: LEA.

Arzarello, F. (2000a). *Inside And Outside: Spaces, Times And Language in Proof Production.* Proc. of PME 24, 1, 23-38.

Bartolini Bussi M.G., Boni M., Ferri F., & Garuti R. (1999). Early Approach to Theoretical Thinking: Gears in Primary School. *Educational Studies in Mathematics, 39*, 67-87.

Bartolini Bussi, M. (1996). Mathematical discussion and perspective drawing in primary school. *Educational Studies in Mathematics, 31*, 11-41.

Boero, P., Garuti, R., Lemut, E., Gazzolo, T., & Lladò, C. (1995b). *Some Aspects of the Construction of the Geometrical Conception of the Phenomenon of the Sun's Shadow.* Proc. PME 19, 3,3-10.

Boero, P., Dapueto, C., Ferrari, P., Ferrero, E., Garuti, R., Lemut, E., Parenti, L., & Scali, E. (1995a). *Aspects of the Mathematics-Culture Relationship in Mathematics Teaching-Learning in Compulsory School.* Proc. PME 19, 1, 151-166.

Clement, J. (1989). The concept of variation and misconceptions in Cartesian graphing. *Focus on Learning Problems in Mathematics, 11*(1-2), 77-87.

Kieran, C. (1994). *A functional approach to the introduction of algebra: some Pros and Cons.* Proc. of PME 18, 1, 157-175.

Lakoff, G., & Núñez, R. (2000). *Where Mathematics comes from.* New York: Basic Books.

Nemirowsky, R. (2003). Three Conjectures concerning the Relationship between Body Activity and Understanding Mathematics. In Nemirowsky et al. (Eds.), *Perceptuo-Motor Activity and Imagination in Mathematical Learning.* Hawai: Research Forum, Proceedings of PME 27.

Núñez, R. (2000). *Mathematical idea analysis: What embodied cognitive science can say about the human nature of mathematics.* Proc. of PME 24, 1, 3-22.

Rabardel, P. (1995). *Les hommes et les technologies.* Paris: Colin.

Talmy, L. (1996). Fictive motion in language and perception. In P. Bloom, M. Peterson, L. Nadel, & M. Garrett (Eds.), *Language and Space.* Cambridge: MIT Press.

Varela, F.J. (1999). A Dimly Perceived Horizon: The Complex Meeting Ground between Physical and Inner Time. *Annals of the New York Academy of Sciences, 879*, 143-153.

Vergnaud, G. (1990). La théorie des champs conceptuels. *Recherches en Didactique des Mathématiques, 10,* 133-170.

Vérillon, P., & Rabardel, P. (1995). Artefact and cognition: a contribution to the study of thought in relation to instrumented activity. *European Journal of Psychology in Education*, vol. IX, n°3.

---

[1] According to Vergnaud, a concept consists of: (i) *reference* ("l'ensemble des situations qui donnent du sens au concept"); (ii) *operating invariants* ("invariants opératoires": they allow the subject to rule the relationship between the reality and the practical and theoretical knowledge about that); (iii) external representations (language, gestures, symbols…).

[2] Calculator Based Ranger, Texas Instruments. For a technological description, see http://education.ti.com/educationportal/sites/US/productDetail/us_cbr_2.html or http://www.vernier.com/calc/cbr.html

Chapter 3.1.2

EMERGENT MODELLING AS A PRECURSOR TO MATHEMATICAL MODELLING

Koeno Gravemeijer
Freudenthal Institute, & Langeveld Institute, Utrecht University, the Netherlands,
Email: koeno@fi.uu.nl

Abstract: This chapter discusses the relation between 'emergent modelling' and 'mathematical modelling'. The former that has its roots in RME theory constitutes the main theme of this chapter. It is argued that mathematical modelling requires a preceding learning process, since it requires abstract mathematical knowledge to construe a mathematical model. The emergent-modelling design heuristic offers a means for shaping a series of modelling tasks that may foster the development of that abstract mathematical knowledge. The emergent-modelling heuristic is illustrated with an instructional sequence on data analysis.

1. INTRODUCTION

Students often seem to have difficulties with applying the mathematics they have learned. This problem may be described in various ways. One may describe it, for instance, in terms of mathematical modelling: The problem solver has to translate the given contextual problem into a mathematical problem to make it assessable for mathematical tools and procedures. In doing so, he or she construes a 'mathematical model' of the situation. In primary-school mathematics, solving word problems, offers a typical example of this type of modelling (Verschaffel, Greer, & De Corte, 2002). This modelling process can also be described as 'abstraction'. It may be useful to note, however, that abstraction, or abstracting, may refer to two very different situations, (a) situations that concern the activity of solving a given problem, and (b) situations that concern the long-term process of developing more abstract mathematical knowledge. In the former case students have to put more formal, abstract knowledge to use by making connections between the problem situation and that abstract knowledge. Here one often speaks of *'reduction'*, or, 'cutting bonds with everyday-life reality'. In the latter case, that of the long-term process, however, the central activity is that of *'construction'*. We may link the latter to the notion of 'emergent modelling' – which will be the topic of this contribution.

2. EMERGENT MODELLING

In contrast with the observed problems of students with mathematical modelling, there are also many reports that students are very inventive and successful when asked to solve novel, engaging, contextual problems. We may mention in this respect, the work of Lesh (Lesh & Harel, 2003) on model-eliciting activities, where the activity of the students is not so much that of applying mathematical ideas but of developing new mathematical ideas. The emergent modelling approach taps into the same potential, but with a focus on long-term learning processes, in which a model develops from an informal, situated model into a more sophisticated model. These emergent models are seen as originating from activity in, and reasoning about situations. From this perspective, the process of constructing models is one of progressively reorganizing situations. The model and the situation being modeled co-evolve and are mutually constituted in the course of modelling activity.

Although emergent modelling is an activity of the students, the term emergent modelling has it roots in the description of an instructional design heuristic within the domain-specific instruction theory for realistic mathematics education (RME). The 'emergent-modelling' design heuristic (Gravemeijer, 1999) was initially developed as an alternative for the common use of what we may call 'didactical models', manipulative materials and visual models that are meant to make abstract mathematics more accessible for the students. Especially at the primary and lower secondary level, manipulative materials and visual models are typically used as embodiments of mathematical concepts and objects in mathematics education. The problem with this kind of models, however, is that external representations do not come with intrinsic meaning. From a constructivist perspective, it may be argued that the meaning of external representations is dependent on the knowledge and understanding of the interpreter. This implies that in order to interpret these models correctly, students should already have at their disposal, the knowledge and understanding that is to be conveyed by the concrete models (Cobb, Yackel, & Wood, 1992).

The emergent-modelling design heuristic tries to circumvent this dilemma, by aiming at a dynamic process of symbolizing and modelling, within which the process of symbolizing and the development of meaning are reflexively related. The idea is that students start with modelling their own informal mathematical activity. Then, in the process that follows, the character of the model should change for the students. The model of their informal mathematical activity is expected to gradually develop into a model for more formal mathematical reasoning. In its latter form, the model may function in a manner as was intended for the didactical models, but now as a model that is rooted in the experiential knowledge of the students.

Mark that the model we are referring to is more an overarching concept than one specific model. In practice, 'the model' in the emergent-modelling heuristic is actually shaped as a series of consecutive *sub-models* that can be described as a cascade of inscriptions or a chain of signification. From a more global perspective, these sub-models can be seen as various manifestations of the same model. So when we speak of a shift in the role of the model in the following, we are talking about *'the model'* on a more general level. On a more detailed level, this transition may encompass various sub-models that gradually take on different roles.

The label 'emergent' refers both to the character of the process by which models emerge within RME, and to the process by which these models support the emergence of formal mathematical ways of knowing. According to the emergent-modelling design heuristic, the model first comes to the fore as a *model of* the students' situated informal strategies. Then, over time the model gradually takes on a life of its own. The model becomes an entity in its own right and starts to serve as a *model for* more formal, yet personally meaningful, mathematical reasoning.

In relation to this, we can discern four different types or levels of activity (Gravemeijer, 1999):

1. *activity in the task setting*, in which interpretations and solutions depend on understanding of how to act in the setting
2. *referential activity*, in which models-of refer to activity in the setting described in instructional activities
3. *general activity*, in which models-for derive their meaning from a framework of mathematical relations
4. *formal mathematical reasoning*, which is no longer dependent on the support of models-for mathematical activity.

These four levels of activity illustrate that models are initially tied to activity in specific settings and involve situation-specific imagery; at the referential level, models are grounded in students' understandings of paradigmatic, experientially real settings. General activity begins to emerge as the students start to reason about the mathematical relations that are involved. As a consequence, the model looses its dependency on situation-specific imagery, and gradually develops into a model that derives its meaning from the framework of mathematical relations that the students construe in the process. The transition from model-of to model-for coincides with a progression from informal to more formal mathematical reasoning that is interwoven with the creation of some new mathematical reality – consisting of mathematical objects (Sfard, 1991) within a framework of mathematical relations. Thus, the model-of/model-for transition is not tied to specific manifestations of the model, instead, it relates to the student's thinking, within which 'model-of' refers to an activity in a specific setting or context, and 'model for' to a framework of mathematical relations.[1]

3. DATA ANALYSIS AS AN EXAMPLE

The emergent-modelling heuristic is elaborated in various research projects on a variety of topics. We will take one of those research projects to illustrate the emergent modelling with a concrete example. This example concerns a teaching experiment on data analysis, carried out by Cobb, Gravemeijer, McClain and Konold in a 7th-grade classroom in Nashville (USA) (see Cobb, 2002). Our point of departure was, that although user-friendly data analysis software packages may seem to be the self-evident accessories for exploratory data analysis, this is only true for experienced data analysts, and not for students who still have to learn about data analysis. In order to be able to use such software in a proficient manner, one has to be able to anticipate what kind information one might be able to deduct from a certain way of representing the data. Working with such data analysis software packages therefore rather signifies an end point of the intended learning process, than a means of supporting it. We therefore turned to designing software tools that can be used for exploratory data analysis on an elementary level. In fact, these so-called 'minitools' are so designed, that they can support a process of progressive mathematization by which conventional statistical concepts and representations are reinvented. What is especially aimed for, is that the activity of structuring data sets with the minitools will foster a process by which the students come to view data sets as entities that are distributed within a space of possible values.

The visualizations offered by the minitools can be seen as manifestations of the same overarching model, which we may be describe as *a graphical representation of the distribution of the data values*.

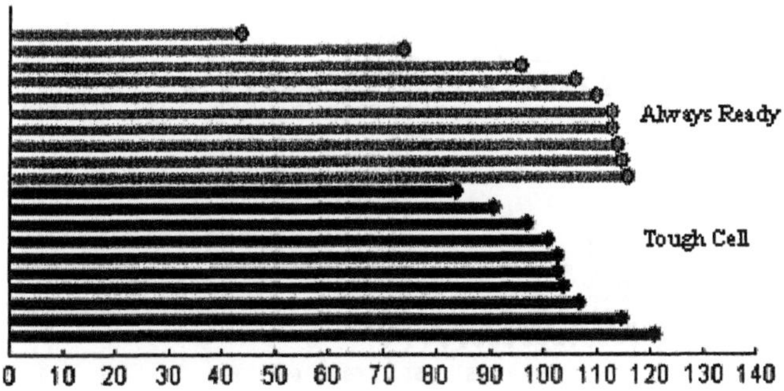

Figure 3.1.2-1. The life span of two brands of batteries.

The starting point is in visualizing the *measures*, or magnitudes, that constitute the data set. With minitool 1, *magnitude-value bars* (Fig. III.1.2-1) are introduced, where each value bar signifies a single measure. This tool has various tool options that can be used when analyzing data sets, such as a vertical value bar to mark certain values, or to split the data set, and various options for sorting the data.

One of the first tasks concerns the comparison to the life spans of two brands of batteries, Though Cell and Always Ready. The lif-span measures of ten batteries of each brand are presented as value bars in the minitool (Fig. 3.1.2-1). When confronted with this problem, the 7th-grade students introduced the term 'consistency' to argue that they 'would rather have a consistent battery (...) than one that you just have to try to guess'.We may interpret this argument as referring to the shape of the distribution, which is visible in the way the endpoints of the value bars are distributed in regard to the axis. In relation to this, we may speak of a graphical representation of the distribution as a *model of* a set of measures.

In the discussions on distributions represented by value bars, the students started to focus on the end points of value bars. As a consequence, these end points came to signify the lengths of the corresponding value bars for them. This allowed for the introduction of a line plot as a more condense (local) model, that leaves out the value bars, and only keeps the end points (Fig. 3.1.2-2 next page).

In Minitool 2 various tool options are made available to help the student structure the distribution of data points on a line plot. One of the tool options partitions a set of data points into four quartiles. The corresponding inscription is in principle similar to the conventional box plot (see Fig. 3.1.2-3 next page).

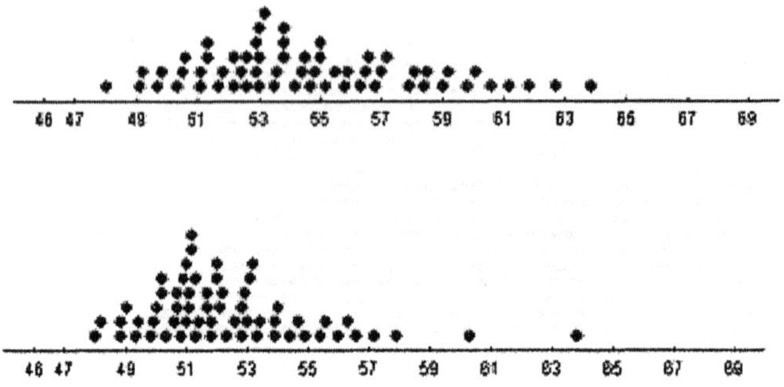

Figure 3.1.2-2. Data on the speeds of cars before and after a speed trap

While working with the second minitool, the students started to use the term ‘hill’ to denote the shape of the distribution. They did so for the first time when they discussed the effect of a speed trap on the basis of data on the speeds of cars before and after the speed trap (see Fig. 3.1.2-2). One of the students used the following argumentation: ‘If you look at the graphs and look at them like hills, then for the before group the speeds are spread out and more than 55, and if you look at the after graph, then more people are bunched up close to the speed limit which means that the majority of the people slowed down close to the speed limit.’

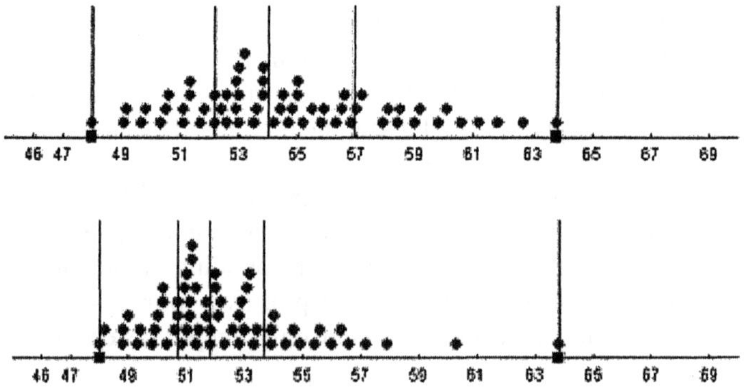

Figure 3.1.2-3. Four equal groups as precursor for the box plot.

Eventually the students started to use the four-equal-groups display of the second mintool to reason about shape and density. The distance between two vertical bars that mark a quartile were interpreted as indicating how much the data are ‘bunched up’. Moreover, the median started to function as an indicator of 'where the hill is', for unimodal distributions. Finally, the students started to treat distributions as entities with certain characteristics. In this regard, we may describe the four-equal groups display as a graphical representation of the distribution that started to function as a *model for* reasoning about distributions.

In the sequence, the model initially comes to the fore as a *model of* a set of measures. At first, the density-function aspect is rather implicit, although the shape of a sorted magnitude-value-bar graph of minitool 1 can be interpreted as signifying variation in density. Gradually, however, density comes more to the foreground, and in this manner, the model can become a *model for* reasoning about various types of distributions. Not only does the distribution become an entity with certain characteristics, but the students also begin to see relations between these characteristics. The normal distribution can be taken as a typical example; the students may learn eventually that a normal distribution is symmetrical, and that as a consequence, mean, median, mode, and midrange coincide.

4. CONCLUSION

We started this chapter with the observation that students experience difficulties when they are expected to apply the mathematics they know, but are good at tackling applied problems, if they feel challenged to invent novel solutions. We believe that we can resolve this paradox by using emergent modelling to shape mathematics education that prepares students for mathematical modelling. The emergent-modelling instructional design heuristic is based on the idea of sequencing modelling tasks in order to support a long term process of 'abstraction-as-construction', within which students construe mathematical knowledge that is grounded in their earlier informal experience, and which is meaningful, and applicable. In addition, the implied modelling activity familiarizes them with a mathematical approach to every-day-life situations. In this sense, modelling serves both as an instructional goal and as a means of helping students reinvent mathematics, and preparing them for 'applications' and 'modelling'.

REFERENCES

Cobb, P. (2002). Modelling, symbolizing and tool use in statistical data analysis. In: Gravemeijer, K., Lehrer, R., van Oers, B., & Verschaffel, L. (Eds.), *Symbolizing, modelling and tool use in mathematics education,* (pp. 171-195). Dordrecht, the Netherlands: Kluwer Academic Publishers.

Cobb, P., Yackel E., & Wood T. (1992). A constructivist alternative to the representational view of mind in mathematics education. *Journal for Research in Mathematics Education, 23*(1), 2-33.

Gravemeijer, K. (1999). How emergent models may foster the constitution of formal mathematics. *Mathematical Thinking and Learning.1*(2), 155-177.

Lesh, R., & Harel, G. (2003). Problem solving, modelling, and local conceptual development. *Mathematical Thinking and Learning, 5*, 157-190.

Sfard, A. (1991). On the dual nature of mathematical conceptions: Reflections on processes and objects as different sides of the same coin. *Educational Studies in Mathematics, 22*, 1-36.

Verschaffel, L., Greer; B., & De Corte, E. (2002). Everyday knowledge and mathematical modelling of school word problems. In Gravemeijer, K., Lehrer, R., van Oers, B., & Verschaffel, L. (Eds.), *Symbolizing, modelling and tool use in mathematics education*, (pp. 257-276). Dordrecht, the Netherlands: Kluwer Academic Publishers.

---

[1] Within the context of the emergent modelling heuristic, the model-of/model-for terminology is only used when this transition is linked to the constitution of a framework of mathematical relations.

Chapter 3.1.3

PROVING AND MODELLING

Gila Hanna and Hans Niels Jahnke
Ontario Institute for Studies in Education, University of Toronto, Canada, Email: ghanna@oise.utoronto.ca / Fachbereich Mathematik, Universität Duisburg-Essen, Germany, Email: njahnke@uni-essen.de

Abstract: This paper discusses the complementary roles of modelling and proof. The two are inseparably linked, and the authors argue that this should be reflected in teaching. Two examples are discussed. The first describes a teaching unit using arguments from statics to prove geometrical theorems, the second discusses the role of thought experiments in general and a specific thought experiment for deriving Pick's formula.

1. MODELLING AS A VALUABLE TOOL IN TEACHING PROOF

With today's stress on teaching mathematics as a coherent, reasoned activity and on communicating its elegance and power, teachers are increasingly being encouraged to focus on the explanation of mathematical concepts, and students are being asked to justify the steps in their problem-solving. This would seem to be precisely the right classroom climate in which to make use of proof, not only as the ultimate form of mathematical justification, but also as an explanatory tool.

It is well known, however, that not every valid proof of a conjecture provides satisfying reasons why the conjecture is true. Indeed, mathematicians, mathematics teachers, philosophers of mathematics, and mathematics education researchers have all come to distinguish between proofs that explain and proofs that do not (Blum, & Kirsch, 1991; Hanna, & Jahnke, 1993; Hersh, 1993; Mancosu, 2001; Rav, 1999; Rota, 1997; Thurston, 1994). For this reason, there is a need to identify the types of proof that can best make clear why a conjecture is true.

Unfortunately there is no single account of what makes a proof explanatory, or even of what mathematical explanation is. The great variety of criteria has been the subject of recent explorations in the philosophy of mathematics (Mancosu et al., 2005; Manin, 1998; Rav, 1999; Rota, 1997). In any case, what might suffice as an explanation among mathematicians is quite different from what would be desired as an explanation in school mathematics.

Clearly, an explanatory proof in school mathematics, as in any other context, must be one that not only demonstrates the truth of its assertions, but also helps one understand why the assertions are true. The aim of such a proof is always to bring to light underlying relationships that place its assertions in a broader mathematical context. In the classroom, however, an explanatory proof must rely upon the more limited mathematical knowledge of students and make use of the properties of objects best known to them.

A number of different methods have been employed to this end, such as the judicial use of visualisations (Hanna, 1990), explorations with dynamic software (de Villiers, 2002), pencil-and-paper proofs appropriate to the cognitive development of the student (Tall, 1998), or the use of arguments from physics (Hanna, & Jahnke, 2002). The present paper will explore further the use of arguments from physics, looking in particular at the light it sheds on the relationship between proving and modelling.

The following example will help explain how the use of arguments from physics can provide both a proper deductive proof of a theorem and greater insight into why the theorem is true (Hanna, & Jahnke, 2002; Polya, 1981; Uspinskii, 1961). Ideas from physics that are already familiar to students, such as the concepts of balancing objects and of the centre of gravity, are presented as tools. The students are then prompted to use these tools to prove a geometrical theorem. That is, students are encouraged to build a mathematical proof by taking as postulates one or more principles of physics.

Let us take as postulates the following three principles of statics:

P1: The uniqueness of the centre of gravity (each system of masses has one and only one centre of gravity).

P2: The lever principle (the centre of gravity of any two masses lies on the straight line joining the masses, and its distances from the masses are inversely proportional to them).

P3: The principle of substitution (if any two individual masses are replaced by a single mass equal to the sum of the two masses and positioned at the centre of gravity of the two masses, then the location of the centre of gravity of the total system of masses remains unchanged).

These three principles can then be used in proving the following geometrical theorem:

The medians of a triangle intersect at a single point and on each median this point is located two-thirds of the way from the vertex. The point of intersection is the centre of gravity of the triangle, or the *centroid.*

This rather simple proof proceeds as follows. Consider the vertices of the triangle as loaded with equal masses of unit weight and connected by rigid weightless rods. The centre of gravity of any side is the midpoint of that side (by P1 and P2). Therefore the unit masses at its ends can be replaced by a mass of weight 2 at its midpoint (by P3) without altering the centre of gravity of the entire triangle. If we then connect this midpoint with the third vertex to form a median, we can conclude that the centre of gravity of the whole triangle must lie on this median, and by P2 the median must be divided in the ratio 2:1 (Fig. 3.1.3-1 and 3.1.3-2). Since this construction can be repeated with the other two sides, the three medians must meet in one and the same point, the centre of gravity. Of course this proof can be extended to the more general Ceva theorem, of which this *centroid* theorem is a corollary.

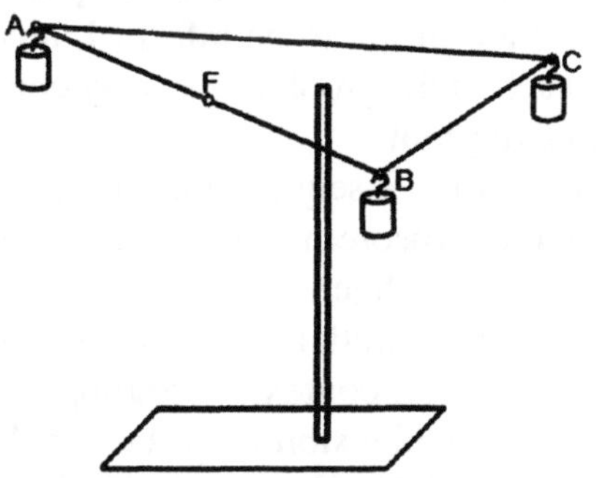

Figure 3.1.3-1. Triangle with equal masses at the vertices, A, B and C

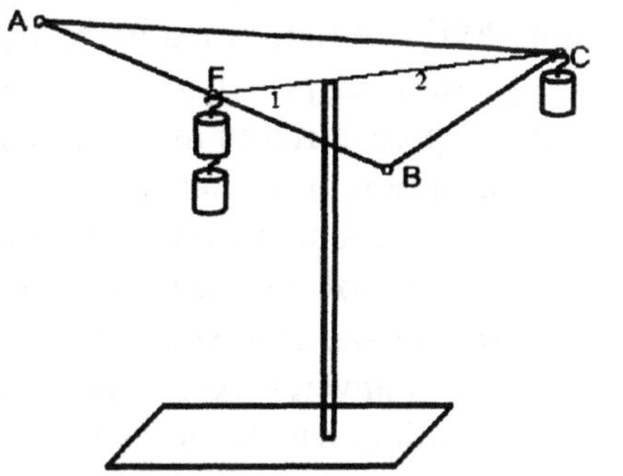

Figure 3.1.3-2. Masses at A and B are moved to the midpoint F of AB.

The fact that this proof deals with abstractions from physical objects (unit masses) does not make it any less of a deductive proof. It is not an induction based upon an experiment or series of experiments, but rather draws logical implications from the three stated postulates. It is more compelling to students than a purely geometric proof, however, because the initial postulates, coming as they do from statics, appeal to physical intuition. This has the advantage of increasing the plausibility of the conclusion in their eyes.

One can regard this proof as a very useful application of modelling, in which the system of masses serves as a model of the triangle. It is true that modelling often has to do with creating a non-physical representation of a physical system.

These non-physical representations are of interest primarily because, once calibrated, they can be manipulated to represent states of the physical system that are impossible or too expensive to study directly. So it is perhaps more common for a physical system to be modelled and a non-physical one to be the model.

But this assignment of roles is arbitrary, or rather it is function of which system, the physical or the non-physical, is more easily manipulated and which is the subject of study. When a physical system is used in the classroom to help convey an understanding of a geometric relationship, as in the proof above, the usual situation is reversed: the physical system is the model, and it is the geometric relationship that is being modelled. The physical system is the one that is easiest for students to grasp and to manipulate in their minds, while it is the non-physical system, in this case the geometric relationship, that is the real subject of study and the one the teacher is eager to convey.

Thus teaching mathematical proof using physical principles is a successful use of modelling, and in fact can be complemented in the classroom, where appropriate, by the actual manipulation of physical objects. This way of teaching proof has some similarity to the use of "reality-related" proofs (Blum, 2003). Whereas reality-related proofs are meant to be informal, however, a proof using principles from physics, such as the proof above, may enjoy the same degree of rigour as any other deductive proof.

By definition, a deductive proof of a theorem is a sequence of steps that shows, using accepted rules of inference, that the theorem is a logical consequence of a set of premises (postulates or axioms). Such an argument is no less legitimate and compelling mathematically if the initial postulates happen to come from physics. At the same time, the physical context, appealing as it does to intuition, has the advantages of making the plausibility of the conclusion more readily apparent (Hanna, & Jahnke, 2002; Uspenskii, 1961), of offering more appeal to the students, and of fostering better mathematical understanding.

The median theorem mentioned above has been actually the subject of two teaching experiments carried out in an Ontario school with Grade 12 students. The research sought to determine the extent to which arguments from physics might make the proof explanatory. Indeed, the vast majority of students did find that the physical arguments helped them understand *why* the theorems were true. Typical comments were: "the physical arguments helped because they provide the visual aspect" and "Help understanding of why they are true… mainly because we know that the whole centre of gravity thing is true from real life experience" (Hanna, DeBruyn, Sidoli, & Lomas, 2002). These experiments as well as experiences with Grade 9 students in Germany showed that the necessary concepts from physics are known and well understood by a majority of students.

2. THOUGHT EXPERIMENTS: THE COMPLEMENTARY ROLES OF MODELLING AND PROOF

The use of concepts and arguments from physics in mathematical proofs is valuable to students not only because of a possible gain in intuitiveness. It also provides them with the important new insight that one can look at one and the same statement from different points of view. In mathematics and science, as in everyday life, there may be different explanations for the same situation, every explanation starting from different premises and exhibiting new aspects of the statement being considered.

In the teaching of proof it is well known that it is often just as demanding to identify the premises from which one can derive a certain statement as it is to find the chain of logical steps – the proof itself – leading from the premises to the statement. If students are working within a domain already familiar to them, one in which they have gained a certain degree of experience, setting up premises is simply an act of *selection*. The students might say to themselves "This theorem is about angles in a circle, so what do we know about this topic?", and try to remember what theorems they have already proved concerning angles in a circle. Selecting known truths from a limited domain is the classical paradigm of proving in the classroom, in geometry and in other fields of mathematics.

In general, however, identifying adequate premises for a proof of a statement is most often a more open-ended and *creative act*, and the entire process, encompassing this creative act and the subsequent rigorous proof, is in fact the same as that involved in *modelling*. Examining the situation in more detail, as we do below, shows that an argument can be made that proving itself is an example of modelling.

When proving a statement we have to do two things: (1) find the "right premises" and (2) devise the chain of deductive steps leading from the premises to the statement. Looking only at the completed proof, however, the first activity is presumed to have been done already, or to lie outside of mathematics, and it is the second activity, the chain of deductive steps, that is seen as the interesting one, the one that matters. Thus the second of these two activities has become most closely associated with the notion of proof, while the first activity has been downplayed.

In a real-life situation, on the other hand, the relationship is exactly the other way around. What then matters most is setting up the premises, which is exactly what in the applied sciences is called modelling. Only after we have built a model can we go on to perform inferences within it, that is, to move on to the second activity of constructing chains of deductive steps. Necessary as it is, the applied scientist takes this manipulation of the model for granted, and thus in this context it is the second activity that is downplayed.

Of course it is stressed in the theory of modelling that modelling is a circular or spiral process of setting up a model, drawing conclusions, modifying the model, drawing conclusions, and so on. However, the same is true for the creation of a proof by practicing mathematicians. Proving, far from being limited to the chain of deductions one sees in the finished product, is a spiral process of selecting a set of premises, drawing conclusions, modifying the set of premises, drawing conclusions, and so on.

It is only under an artificial division of labour, therefore, that modelling and proving appear to be separate activities. They are inseparably linked, in fact, and this should be reflected in teaching. Unfortunately, though, teachers often have the idea that there is a deep gap between modelling and proof. One day they work on applied problems, the next they do rigorous proof. Because of the complementary relationship between proving and modelling, however, the approach of using arguments from physics in mathematical proof deserves a position of special importance in the teaching of proof.

An especially concise form and a paradigm for the complementary relation of modelling and proving is what physicists have called a "thought experiment" ("Gedankenexperiment") since Galileo's times. It is a concept with which every student should be confronted, because thought experiments embody the most elegant use of arguments from physics in mathematical proofs.

As an example, we present in concluding our paper a thought experiment recently devised by Christian Blatter (1997) from the Federal Technical University of Zurich (ETHZ) to prove a famous theorem of Georg Pick (1859 – 1942). The theorem is suitable for the early secondary level. Pick's theorem is as follows:

> Let P be a simple lattice polygon, i.e. a polygonal Jordan domain in the plane whose vertices have integer coordinates. Then, its area $m(P)$ is given by
>
> $$m(\mathrm{P}) = i + {}^{b}\!/_{2} - 1$$
>
> where i denotes the number of points in the interior of P and b the number of points on its boundary.

The formula gives the area as multiples of the unit square. Thus, the area can be determined by simply counting the numbers of points in the interior and on the boundary of P. In Figure 3 we have $i = 16$ and $b = 11$, therefore, $m(P) = 16 + {}^{11}\!/_{2} - 1 = 20 + {}^{1}\!/_{2}$.

There are numerous proofs of Pick's area theorem. In one of them the area of P is dissected into minimal triangles, and then Euler's polyhedron formula is applied (see Aigner, & Ziegler, 2002). Blatter's thought experiment using an argument from physics runs as follows (Aigner, & Ziegler, 2003). We assume that at time $t = 0$ a unit of heat is concentrated at each lattice point of the plane. This heat will be distributed over the whole plane by heat conduction so that at time $t = \infty$ the heat is equally distributed on the plane with density 1. Therefore, the area $m(P)$ is equal to the amount of heat in the interior of P.

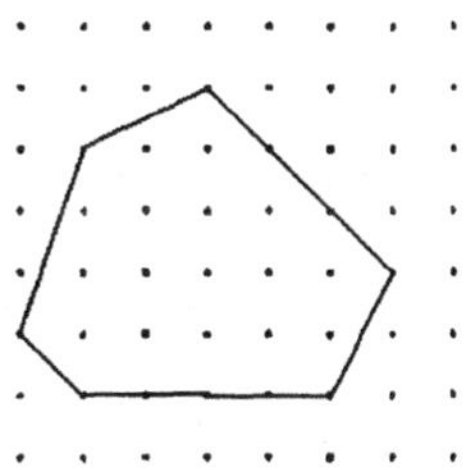

Figure 3.1.3-3. Pick's area theorem

In the following we do not need to make any assumption about how heat conduction works, and we do not use differential equations. The only assumption is that an equal amount of heat will flow from every point in every direction, i.e. that the lattice point situation is rotationally symmetric.

Now, where does the heat in the interior of P come from? First of all, we consider an edge of the polygon and observe that the lattice points lie symmetrical to its midpoint M.

Therefore, to every point Q on one side of the edge belongs a point Q' on the other side lying symmetrical to M. Thus Q and Q' will send an equal amount of heat across the edge of the polygon from different sides. Thus the total heat flux across the edge is 0. This implies that the heat in the interior of P comes only from the lattice points inside of P and from the points on its boundary. Each inner point contributes heat of amount 1, whereas each point on the boundary sends heat into the polygon proportionally to the interior angle of which the point is the vertex. To calculate the sum of the interior angles we walk along the boundary in, say, the counter-clockwise direction. A lattice point on the boundary which is not a vertex sends half its heat into the interior. The heat coming from a vertex is half minus the turning angle of the boundary at that vertex, measured in units of 2π. The sum of all turning angles for a simple polygon is one full turn, and, from this follows Pick's formula.

The specific merit of this thought experiment is that it gives every term in Pick's formula a clear and intuitively plausible interpretation. This is especially true for the $b/2$ and the mysterious -1. Thus one could think about presenting this thought experiment as an enrichment activity after one has treated Pick's theorem in a purely geometrical fashion. As we said, no special knowledge about heat conduction is required.

As suggested in this paper, the careful identification of premises, including the search for alternative premises, as well as the use of principles from physics, are examples of potentially successful uses of modelling and should play a greater part in the teaching of proof than has been the case.

REFERENCES

Aigner, M., & Ziegler, G. (2002). *Das BUCH der Beweise*. Heidelberg/Berlin: Springer Verlag.

Aigner, M., & Ziegler, G. (2003). *Brillanten für das „BUCH der Beweise"*. ftp://ftp.math.tu-berlin.de/pub/combi/ziegler/WWW/lmu.pdf

Blatter, C. (1997). Another Proof of Pick's Area Theorem. *Mathematics Magazine, 70*(3), 200.

Blum, W., & Kirsch, A. (1991). Preformal proving: Examples and reflexions. *Educational Studies in Mathematics, 22*(2), 183-203.

Blum, W. (2003). *On the Role of "Grundvorstellungen" for Reality-Related Proofs Examples and Reflections.* http://wwwdidactique.imag.fr/preuve/CERME%203% 20papers/TG4-Blum.pdf

de Villiers, M. (2002). *Developing Understanding for Different Roles of Proof in Dynamic Geometry.* http://mzone.mweb.co.za/residents/profmd/homepage.html

Hanna, G. (1990). Some Pedagogical Aspects of Proof. *Interchange 21*(1), 6-13.

Hanna, G., & Jahnke, H. N. (1996). Proof and Proving. In Bishop, A. J. et al. (Eds.), *International Handbook of Mathematics Education* (pp. 877-908). Dordrecht: Kluwer.

Hanna, G., & Jahnke, H. N. (1993). Proof and application. *Educational Studies in Mathematics, 24*(4), 421-439.

Hanna, G., & Jahnke, H. N. (2002). Arguments from physics in mathematical proofs: An educational perspective. *For the learning of mathematics, 22*(3), 38-45.

Hanna, G., DeBruyn, Y., Sidoli, N., & Lomas, D. (2004). Teaching proof in the context of physics. *Zentralblatt für Didaktik der Mathematik, International reviews on mathematical education, 36*(3), 82-90.

Hersh, R. (1993). Proving is convincing and explaining. *Educational Studies in Mathematics, 24*(4), 389-399.

Mancosu, P. (2001). Mathematical explanation: Problems and prospects. *Topoi, 20*, 97-117.

Mancosu, P., Jørgensen, K. F., & Pedersen, S. A. (Eds.). (2005). *Visualization, Explanation and Reasoning Styles in Mathematics.* Dordrecht: Springer.

Manin, Yu. (1998). Truth, rigour, and common sense. In H. G. Dales, & G. Oliveri (Eds.), *Truth in mathematics* (pp. 147-159). Oxford: Oxford University Press.

Polya, G. (1981). *Mathematical discovery: On understanding, learning and teaching problem solving* (2 vols.; combined ed.). New York: John Wiley & Sons.

Rav, Y. (1999). Why do we prove theorems? *Philosophia Mathematica, 7*(3), 5-41.

Rota, G-C. (1997). The phenomenology of mathematical proof. *Synthese, 3*(2), 183-197.

Tall, D. (1998). *The cognitive development of proof: Is mathematical proof for all or for some.* Paper presented at the UCSMP conference. Chicago, USA.

Thurston, W. P. (1994). On proof and progress in mathematics. *Bulletin of the American Mathematical Society, 30*(2), 161-177.

Uspenskii. V. A. (1961). *Some Applications of Mechanics to Mathematics.* New York: Balisdell Publishing Company.

Chapter 3.1.4

A DEVELOPMENTAL APPROACH FOR SUPPORTING THE EPISTEMOLOGY OF MODELING

Richard Lehrer and Leona Schauble
Vanderbilt University, Nashville, Tennessee, USA, Email: rich.lehrer@vanderbilt.edu / leona.schauble@vanderbilt.edu

Abstract: Across a wide spectrum of disciplines and forms of investigation, scientists invent and revise models. Although central to scientific practice, models cannot simply be imported into classrooms. Instead, pedagogy must be designed so that students can come to understand natural systems by inventing and revising models of these systems. Considering theories of analogical development, we suggest rooting first experiences of modeling in resemblance – physical microcosms – and in inscription – children's drawings and related writings. From these starting points, we seek to stretch inscription into mathematization, so that children describe natural systems by recourse to mathematical systems and structures. Engagement in these practices has epistemic consequences, fundamentally altering how children view the natural world.

1. INTRODUCTION

Across a wide spectrum of disciplines and forms of investigation, scientists invent and revise models. Scientific ideas derive their power from the models that instantiate them, and theories change as a result of efforts to invent, revise, and debate the qualities of competing models. Models are not only assemblies of ideas, but also mobilize a wider network of representational forms, institutions, material means, and specialized ways of talking (e.g., Bazerman, 1988; Latour, 1999; Pickering, 1995). Hence, we consider model-based reasoning and associated forms of practice as constituting the epistemic core of coming to understand natural systems. Our approach to understanding the growth and development of this form of reasoning is rooted in pedagogical design; we work with teachers to craft classroom ecologies that cultivate model-based reasoning across grades and years of schooling, and then study the practices and outcomes that result (Lehrer & Schauble, 2005). Among the desirable outcomes is an appreciation of the epistemic grounds of modeling. In this brief paper, we first describe an approach to pedagogy rooted in developmental theory and then outline a typology of models that we find useful to support growth and development of children's reasoning (see also, Lehrer & Schauble, 2006).

2. DEVELOPMENTAL ORIGINS OF MODELLING IN RESEMBLANCE

When thinking about the developmental roots for modeling, we find it useful to recall that at its most basic level, a model is an analogy. A familiar set of objects and relations stands in for a less familiar set of objects and relations. For example, the Bohr model metaphorically renders the atom's nucleus as the sun and its electrons as planets, with attendant implications of force and orbital paths. Familiarity also exacts a price: Analogies are not mere copies, so it is a matter of test to determine which aspects and relations of the more familiar system are pertinent for understanding the new system. Hence, one informational resource for pedagogical design is research that considers how analogical reasoning develops.

Gentner and Toupin (1986) suggest that there is a continuum in correspondence between familiar and less familiar systems. These range from literal similarity (copy) to pure relational structure. The most accessible correspondences for children are literally similar: A stands in for B because it resembles B in some way. For example, when we asked first graders to use springs, dowels, Styrofoam™, and other materials to build a model that "works like your elbow," the children's initial constructions copied perceptually salient features. The first-graders used round foam balls to simulate the "bumps" in their elbow joints and Popsicle sticks to simulate fingers, even though neither of these features captured anything about the way the elbow functions. However, resemblance served an important developmental role: It appeared to bridge the gap between form and function. Resemblance supported the very idea of representation – that is, that hardware originally designed for other purposes could be re-assembled and re-purposed to represent an elbow. It helped persuade classmates that there was something elbow-like about their assemblies, yet as students encountered range of motion and related constraints, these resemblances were eventually eliminated as unnecessary for persuading classmates about "good" models. Over the course of a few lessons, the ground of legitimate correspondence shifted away from resemblance toward satisfying constraints. This shift from literal similarity to mapping relations is a hallmark of analogical reasoning, according to Gentner's theory. Children's appreciation of the epistemic grounds of modeling appeared to be affected significantly by these experiences. During follow-up interviews, most suggested that models were not copies of reality. Instead, first-graders talked about how models edit reality to focus on the way things work. One even consoled a researcher who asked about the apparent lack of complete fidelity of the child's hardware model to an elbow by assuring him that "it's only a model."

The shift from literal similarity to analogical mapping of systems of relation sets the stage for mathematical expression. Mathematical models encapsulate the structure of natural systems, although the correspondence between mathematical systems and natural systems is always subject to some uncertainty. Kline (1980) referred to this process as the mathematization of nature, and it forms the cornerstone of our efforts to introduce modeling to children. For example, in our studies, third-grade children who first built physical models of elbows went on to explore relationships between the position of a load and the point of attachment of the tendon in a more complex elbow model. The model treated the arm as a third-class lever, with the elbow acting as the fulcrum. Students expressed the torque of the system mathematically as product of effort (supplied by the biceps) and distance (from elbow to point of attachment of biceps on the forearm). This mathematical description served not only as a summarization but also as a point of departure for further investigation. If distance and effort were related, how might different points of attachment trade off effort and range of motion? In our view, looking at nature through the lens of a model is an important hallmark of model-based reasoning. Models do not simply describe. They also suggest new avenues of vision.

3. A TYPOLOGY OF MODELS BASED ON ANALOGICAL MAPPING

Considering models as forms of analogy suggests prospective developmental pathways based on types of mappings between models and natural system. In our view, each kind of mapping represents an interpretative stance governed by intentions and purposes of modelers, not solely by the qualities of the models. Models, like all symbols and representations, have external features and qualities, but their status as a model relies on interpretation, rendering any typology of models approximate. For example, one might suggest that a pendulum is a model system for periodic motion. Yet, for most, the pendulum simply swings back and forth and does not stand in for anything other than itself.

3.1 Physical Microcosms

Mechanical models of the solar system, terraria, and model airplanes intentionally resemble the systems they are intended to represent, albeit at different scale. They are microcosms of natural systems, so physical models often prove apt entry points to practices of modeling. For example, one class of first-graders observed that in one corner of the schoolyard, a pile of tomatoes was changing from week to week. The onset of winter slowed change to a standstill, so students sought a way to continue their study. Compost columns were the means that they pursued to hold nature within their grasp. Children's choices of materials to include in the columns included moldy fruit, dirt and leaves, but also gum wrappers and stray pieces of Styrofoam packing material that they observed in the vicinity. Over time children's questions and interpretations were guided less by resemblance and more by systems of relation (e.g., How did moisture affect decomposition? Was mold alive?). Nonetheless, resemblance was critical for surmounting the initial hurdle of ensuring that the compost columns were accepted by all as legitimate models of the process occurring outside the walls of the school. Hence, physical microcosms afford early entrée to modeling via literal similarity, but they also typically provoke questions about relational structure as students consider what to include in a model system and why, and how to modify the model to account for new observations and data.

Although we have emphasized their role as starting points for modelling, microcosms are not merely "school stuff." In social studies of science, "irruptions" of objects into researchers' investigations have been essential to the pursuit of "modern" science (Latour, 1993). Partnership between artifacts and person has pedagogical implications. Consider, for example, sixth graders who conducted field studies of aquatic systems and then attempted to design a sustainable system in a one-gallon jar. Students initially regarded this task as unproblematic, to be solved by merely copying the elements (substrate, plants, animals) found in a nearby pond into their jar. To their surprise, sustainable systems proved elusive, and unexpected events, such as algal blooms and rapid increase in bacterial populations, transformed students' understandings of the functioning of aquatic systems and spurred new lines of inquiry. Resistances like these from the material world contribute to what Pickering (1995) describes as a "mangle of practice." No model specifies instrumentation and measurement in sufficient detail to prescribe practice, and this was a late-dawning but important insight for the sixth-graders attempting to design sustainable ecologies.

3.2 Representational Description

Models are typically expressed as systems of representation. Latour

(1999) suggests that systems of scientific representation (Latour calls them inscriptions) share properties that make them especially well suited for mobilizing cognitive and social resources in the service of establishing the plausibility of claims. The pedagogical challenge, then, is to support children's use and understanding of representation. Long before they arrive at school children have some appreciation of the representational qualities of pictures, scale models, and video representations. Emerging symbolic capacities are the foundation for engaging children in the invention and revision of systems of inscription – ways of representing – the natural world. This form of activity parallels that of the scientific community, notwithstanding significant differences in the worlds of scientists and children. In instruction, we attempt to place students in a position to invent inscriptions – to visually denote their commitments and conjectures about how a system functions – and to compare and contrast the affordances and constraints of different systems of inscription. Our rationale is that systems of representation do not simply communicate thought; they also shape it (Olson, 1994). For example, when third-grade children invented maps of their school's playground, initial maps were drawings of favorite pieces of playground equipment (Lehrer & Pritchard, 2002). As children shared their maps with classmates, they saw the need to take a less local view of the space. They then invented competing ways of representing distance and direction, and origin and scale. Contrasting their tentative maps provided children with a venue to re-consider the meanings of place and space and to re-create the playground symbolically, as a mathematical (polar coordinate) system.

It is important that children have repeated and extended opportunities to model. These same third-graders later attempted to describe the growth of plants. Each child grew his or her own plant. Children initially represented change by drawings and by pressed plant silhouettes. These relied on perceptual resemblance and mediated the later development of inscriptions that were further removed from phenomena. For example, plant silhouettes were supplemented by measurements and displays to create other descriptions of change, such as graphs depicting changing ratios (e.g., change in height to change in time). Yet measures and graphs did not simply overwrite or somehow incorporate silhouettes. Instead, silhouettes participated in a system of what Latour (1999) calls "circulating reference" (p. 72), in which they performed an important indexical function of relating measures and graphs to events in the world (the actual growth of the plants). This cascade of inscriptions transformed the conceptual terrain, so that students began to pose new questions about the plants. For example, while comparing plant root growth to shoot growth, students invoked their new technologies of display to argue both for and against the claim that roots grow like shoots. Mathematical description provoked new ways of seeing nature.

3.3 Syntactical Models

We just noted that representational systems often have their start in fundamental symbolic capacities of pretense or imitation and in basic inscriptional capacities, such as drawing. As representational systems stretch from resemblance into structural relations, they seem to begin to alter their character. In particular, they often summarize the essential functioning of a system but bear little resemblance to the system being modeled. For example, animal behaviors, such as food preference, can be investigated via a repeated process of flipping a coin to answer questions such as, are choices intentional, or simply a matter of chance? Note that one must first establish the correspondence between a model of chance, such as a coin flip, and events in the world. For example, a group of fifth-grade students investigated changes in the distribution over time of a collection of plants. During the course of their investigations, they sought to explore what might occur if they grew the plants again. What might be the shape of the data on the same day of growth if conditions were the same? Because they could not actually grow the plants again, students explored the behavior of repeated sampling of the collection as a stand-in for repeated growth. Yet the relation of the repeated process of sampling to the repetition of growth under the same conditions was purely a matter of structural isomorphy.

3.4 Emergent Models

Emergent models impose a further restriction on mapping between model and world: Relations between objects produce emergent behaviors that are not apparent in the description of either the object or the relation. For example, the value of a stock on the market emerges from the independent decisions of thousands or millions of investors. Resnick (1994) suggest that people often find emergent models implausible or even contradictory because they believe that complex systems must be orchestrated centrally. A common example (in the United States) contrasts evolutionary theory to intelligent design. It is difficult for people to believe that the watchmaker is blind. Penner (2000), who investigated these issues with middle school students, concluded that there are three challenges for developing appreciation of this form of modeling: (a) recognizing that there may not be a single cause (e.g., a central authority), (b) distinguishing between aggregate and individual levels of analysis, and (c) tracing the consequences of perturbations at the micro level to behavior at the macro level. Students often treat emergent levels as if they were hierarchies of inheritance, attributing properties of the individuals to the aggregate (Wilensky & Resnick, 1999). Chi (2005) further suggests that students may make ontological errors, confusing emergent systems with others that allow for more direct correspondences between model behaviors and world behaviors. Our explorations of emergence have been limited to introducing children to the mathematics of distribution. We rely on contexts of repeated measurement, because these contexts support viewing distribution as emerging from the collective activity of individual agent-measurers. Working in these contexts, children trace agent-level actions into the level of distribution (e.g., outliers might result from a measurer using a nonstandard method of measure). Moreover, descriptions of collective activity (the distribution) are readily distinguished from those of individual agent-measurers (the measures). Hence, some of the epistemic challenges noted by Penner (2000) and by Chi (2005) may be surmounted in some circumstances.

New modeling tools have been developed that may well afford more access to emergent modeling. What was once the province of differential equations can now be described by programming languages that allow models to be constructed as ensembles of independent agents acting in parallel, according to a comparatively simple set of relations Moreover, these forms of activity can be distributed across different types of computational media to model processes such as the spread of disease (Wilensky & Stroup, 1999).

4. SUMMARY

Modeling presents a series of epistemic challenges that we have sought to make visible and approachable by designing pedagogy informed by theories of the development of analogical reasoning. These challenges include recognizing that models edit, rather than copy, the world. Models amplify phenomena by specifying relationships that one might not have otherwise considered, so models suggest qualities of the world to modelers. One must learn to look through the model into the world. Our students have pointed out another challenge: How does knowing in mathematics relate to knowing in science? For example, when third-grade students represented constant ratios of the lengths of the sides of a rectangle (e.g., the long side is twice the short side), lines on the Cartesian plane constituted a closed system. That is, a line represented all possible rectangles with sides in a particular ratio, even those that students had not constructed. But when they later considered ratios of mass to volume (i.e., density), there was great consternation. Some students noticed that objects composed of what appeared to be the same material were not all represented on the same line. Eventually, they concluded that the most likely explanation was measurement error. Yet as one third-grade student put it, it was also possible that the objects were really made of different materials, and that would require a separate line for each. She concluded that they could not know for sure – a recognition of the fundamental uncertainty underlying scientific conclusions.

This recognition is an example of perhaps the greatest epistemic challenge, model competition. How does one judge the quality of a model? Such judgments rely on entertaining alternative models, so that without opportunity to invent and revise models, this epistemic quality is largely hidden from view. Unfortunately, it is our impression that school students are generally taught the virtues of particular models of nature. Competitors are either edited out of view in the interests of instructional efficiency, or perhaps even more provocatively, are recast as quaint relics of bygone eras, with their vital intellectual ferment evacuated. Students are taught models, or at least a model, but modeling remains hidden. Without opportunities to play modeling games from the beginning of education in mathematics and science, we fear that most students will reduce models to recipes—without ever understanding why and how they are generated, how they support claims and argument, and how they constitute explanations in reply to genuine questions.

REFERENCES

Bazerman, C. (1988). *Shaping written knowledge: The genre and activity of the experimental article in science.* Madison: University of Wisconsin Press.

Chi, M. T. H. (2005). Commonsense conceptions of emergent processes: Why some misconceptions are robust. *The Journal of the Learning Sciences, 14*, 161-199.

Genter, D., & Toupin, C. (1986). Systematicity and surface similarity in the development of analogy. *Cognitive Science, 10*, 277-300.

Kline, M. (1980). *Mathematics: The loss of certainty.* Oxford: Oxford University Press.

Latour, B. (1993). *We have never been modern.* Cambridge, MA: Harvard University Press.

Latour, B. (1999). *Pandora's hope: Essays on the reality of science studies.* London: Cambridge University Press.

Lehrer, R., & Pritchard, C. (2002). Symbolizing space into being. In K. Gravemeijer, R. Lehrer, B. van Oers, & L. Verschaffel (Eds.), *Symbolization, modelling and tool use in mathematics education.* (pp. 59-86). Dordrecht, Netherlands: Kluwer Academic Press.

Lehrer, R., & Schauble, L. (2005). Developing modeling and argument in the elementary grades. In T. Romberg, & T. P. Carpenter (Eds.), *Understanding mathematics and science matter*, (pp. 29-53). Mahwah, NJ: Lawrence Erlbaum Associates.

Lehrer, R., & Schauble, L. (2006). Cultivating model-based reasoning in science education. In R. Keith Sawyer (Ed.), *Cambridge Handbook of the Learning Sciences* (pp. 371-387). Cambridge: Cambridge University Press.

Olson, D. R. (1994). *The world on paper: The conceptual and cognitive implications of writing and reading.* New York: Cambridge University Press.

Penner, D. E. (2000). Explaining systems: Investigating middle school students' understanding of emergent phenomena. *Journal of Research in Science Teaching, 37*, 784-806.

Pickering, A. (1995). *The mangle of practice: Time, agency, and science.* Chicago: University of Chicago Press.

Resnick, M. (1994). *Turtles, termites, and traffic jams: Explorations in massively parallel microworlds.* Cambridge, MA: MIT Press.

Wilensky, U., & Stroup, W. (1999). *Learning through participatory simulations: Network-based design for systems learning in classrooms.* Computer Supported Collaborative Learning Conference, Stanford University, California.

Chapter 3.1.5

WHAT IS DISTINCTIVE IN (OUR VIEWS ABOUT) MODELS & MODELLING PERSPECTIVES ON MATHEMATICS PROBLEM SOLVING, LEARNING, AND TEACHING?

Richard Lesh and Caroline Yoon
Indiana University, Bloomington, Indiana, USA,
Email: ralesh@indiana.edu / csyoon@indiana.edu

Abstract: What is the nature of typical problem-solving situations where elementary-but-powerful mathematical constructs and conceptual systems are needed for success beyond school in a technology-based age of information? What kind of "mathematical thinking" is emphasized in these situations? What does it mean to "understand" the most important ideas and abilities that are needed in the preceding situations? How do these competencies develop? What can be done to facilitate development? How can we document and assess the most important achievements that are needed: (i) for informed citizenship, or (ii) for successful participation in the increasingly wide range of professions that are becoming heavy users of mathematics, science, and technology?

1. INTRODUCTION

The preceding questions are among the most significant that have shaped the models & modelling perspectives that are described in this paper. But, our views also were influenced strongly by the following observations which involve issues of equity as much as content quality.

Why do students who do not have histories of getting A's on tests and coursework often do exceptionally well beyond school? How can people be identified whose exceptional abilities do not fit the narrow and shallow band of abilities emphasized on existing textbooks and standardized tests? For example, when Purdue University's Gender Equity in Engineering Project assessed students using tasks that were designed to be simulations of "real life" problem solving situations, and when care has been taken to assess a broader range of understandings and abilities than those emphasized in traditional textbooks and tests, a broader range of students naturally emerged as having extraordinary potential. Furthermore, because such abilities were previously unrecognized, many of these students come from populations that are highly under represented in fields that emphasize mathematics, science, and technology.

Such observations generally are considered to be "common knowledge" among leaders in future-oriented fields ranging from aeronautical engineering to business management - where new levels and types of "mathematical thinking" are used in problem solving and decision making. But, these observations seldom make it clear what kind of instructional goals and experiences should be emphasized to help decrease mismatches between: (i) the narrow band of mathematical understandings and abilities that are emphasized in mathematics classrooms and tests, and (ii) those that are needed for success beyond school in the 21st century.

Many people assume that students simply need *more* practice with ideas and abilities that have been considered to be "basics" in the past. Others assume that *different* levels and types of understandings need to be developed – such as understandings that emphasize graphics-based or computation-based representational media. Still others assume that completely new topics and ideas (such as those associated with complexity theory, discrete mathematics, systems theory, or computational modelling) need to replace old conceptions of "basics." In this paper, our goal is not to resolve such issues. Our tentative answers have been given in a number of recent publications including Beyond Constructivism: Models & Modelling Perspectives on Mathematics Problem Solving, Learning & Teaching (Lesh & Doerr, 2003) and a special issue on models & modelling in the International Journal for Research on Mathematical Thinking & Learning (Lesh, 2003). The goal of this paper is simply to clarify some of the most important characteristics of what we refer to as models & modelling perspectives on mathematics problem solving, learning, and teaching.

2. POSSIBLE MISCONCEPTIONS TO DISMISS

Before describing foundations of our *models & modelling perspectives*, it is useful to dismiss several possible misconceptions. First, we are not advocating an "applied" mathematics courses – where traditional methods of instruction are used to teach about mathematical topics that someone considers to be "applied" (as opposed to "pure"). Instead, our perspectives are similar to those emphasized by John Dewey when he argued that the educational goal of *making science practical* is significantly different to *making practice scientific*. For example, in other publications (Lesh & Yoon, 2003a), we presented evidence showing why significantly different kinds of knowledge and processes tend to be emphasized depending on whether instructional activities are designed to: (i) guide students along (necessarily narrow) conceptual paths toward a textbook's (or teacher's) idealized portrayal of the meaning of a given idea, or (ii) put students in situations where they repeatedly express, test, and revise their own ways of thinking about the relevant concepts and conceptual systems. That is, *mathematizing reality* is significantly different than *realizing mathematics* – by pointing out applications of ideas and skills that are introduced.

Of course, such distinctions are not likely to seem significant to anybody who naively accepts the cliché: *It took brilliant mathematicians hundreds of years to develop most of the most powerful concepts in the elementary mathematics curriculum. It's not realistic to expect average ability children or adolescents to come up with such concepts in a few months or weeks – or during single problem solving episodes.* ... Models & modelling perspectives reject the notion that only a few exceptionally brilliant students are capable of developing significant mathematical concepts unless step-by-step guidance is provided by a teacher. Our research is filled with transcripts of model-eliciting activities in which the models (and conceptual tools) that students develop for making sense of the situations also result in significant developments of underlying constructs or conceptual systems (Lesh & Doerr, 2003). In fact, if the goal of instruction is to make significant changes in a student's underlying ways of thinking about important mathematical systems, then virtually the only way to induce significant conceptual change is to engage students in situations where they express > test > revise (or reject) their current ways of thinking.

Model-eliciting activities, as their name implies, are problem solving activities that elicit a model. That is, their solutions require students to express their current ways of thinking (i.e., their relevant models) in forms that are tested and refined multiple times. So, final solutions involve not only model development but also the development of constructs and conceptual systems that the models embody.

Principles for designing *model-eliciting activities* include the following. (i) Students must be engaged in problem solving activities in which they clearly recognize the need to revise or refine their current ways of thinking about the situation. (ii) Students must be challenged to express their current understandings in forms that they themselves can test and revise multiple times. (iii) The conceptual tools that students develop should be expected to be sharable (with others) and re-useable (beyond the immediate situation) beyond the specific situations in which they were developed (Lesh, et. al.,

2001). When these conditions are satisfied, our claim is that: *If the models involve mathematically significant concepts, then model development tends to involve significant forms of concept development; and, development often is achievable by students who have been labeled average or below-average in ability – as measured on traditional school tests and tasks.*

3. WHAT KIND OF MATHEMATICAL ABILITIES ARE EMPHASIZED IN MODEL-ELICITING ACTIVITIES?

According to *models & modelling* perspectives on mathematics, problem solving, learning, and teaching, "thinking mathematically" is about expression – interpretation, description, explanation, communication, argumentation, and construction – at least as much as it is about computation or deduction. And, mathematical interpretation is about quantification, dimensionalization, coordinatization, and systematization – or, in general, imposing structure on experience – as least as much as it is about deriving or extracting meaning from information that is presumed to be given. Therefore, when we ask what mathematics student have mastered, it is important to ask what kinds of situations they can describe at least as much as we ask what kind of data processing they can do. That is, mathematics learners and problem solvers are model developers at least as much as they are information processors; and, because models are the tools that mathematicians use to interpret experience, powerful, sharable, and reuseable models are among the most important cognitive objectives of mathematics instruction.

What is the nature of these models? First, models usually draw on concepts and conceptual systems from a variety of disciplines or textbook topic areas. Second, models usually are expressed using a variety of interacting representational media – each of which emphasize and de-emphasize somewhat different meanings of the underlying concepts and conceptual systems. Third, for models that are relevant to a given problem solving situation, most relevant constructs and conceptual systems are at intermediate stages of development. Fourth, model development usually involves a series of design cycles which involve different ways of filtering, organizing, and interpreting "givens" and "goals" in learning or problem solving situations. Primitive versions of a given model (or conceptualization) tend to be based on less refined and less complex relational/organizational systems; and unstable systems tend to: (i) notice only the most salient relationships and information in the problem situation, filtering out other important but less striking characteristics; and/or (ii) neglect to notice model-reality mismatches, thus imposing subjective and unwarranted relationships or interpretations based on "a priori" assumptions.

In general however, research on models & modelling has shown that the understandings and abilities that are critical for success in *model-eliciting activities* are similar to those that are emphasized in fields such as engineering or business management when expert job interviewers describe characteristics of students who are most sought-after in job interviews following the completion of their academic degree programs. That is, successful students are those who: (i) have histories of being able to make sense of complex systems, (ii) work well and communicate meaningfully within diverse teams of specialists, (iii) are skillful at planning, monitoring, and assessing progress within complex multi-stage projects; and, (iv) adapt rapidly to continually evolving conceptual technologies. ... Consequently, when we investigate the nature of mathematical problem solving situations that occur in future-oriented fields that are heavy users of mathematics, science, and technology, capabilities that emerge as being especially important are associated with the following problem characteristics.

- The mathematical products students need to produce generally involve more than easy-to-score answers to pre-mathematized questions.
- The products that are needed also usually involve sharable and re-useable conceptual tools (and underlying constructs and conceptual systems).
- The "problem solver" often is not simply an isolated individual. Instead, the "problem solver" often consists of a team of diverse specialists - who use a variety of rapidly evolving technical tools, and who represent a variety of different practical and theoretical perspectives.
- Because solutions involve the development of complex artifacts (or conceptual tools), design processes usually involve a series of iterative development > testing > revising cycles in which a variety of different ways of thinking about givens, goals, and possible solution steps are iteratively expressed, tested, and revised – or rejected.
- The context often involves too little time, too few resources, and conflicting goals (such as goals that are related to costs and benefits, completeness and simplicity, or quality and timeliness). Therefore, relevant ways of thinking usually draw on constructs and conceptual systems that come from a variety of disciplines and topic areas.

For people who are unfamiliar with the kind of mathematical thinking that is emphasized when groups of specialists use a variety of powerful technical tools to work on complex multi-stage projects, it often appears that the need for mathematical thinking surely must decrease (or become easier). *Don't tools and colleagues do some of the work for you? Don't complex or multi-stage projects enable you to focus on only a single component or stage of work that is needed? ...* In reality, tools and colleagues tend to create as many conceptual challenges as they eliminate; and, complex multi-stage projects tend to emphasize the need for higher-order understandings and abilities (Lesh, 2001).

4. MODELS & MODELLING VIEWS OF MATHEMATICS, PROBLEM SOLVING, LEARNING & TEACHING

In mathematics education research, it is common to characterize problem solving as a process of getting from givens to goals when the path is not obvious.

- The starting point is well defined. The relevant data are expressed in mathematical form; and, there seldom exist several plausible mathematical descriptions of relevant relations, actions, patterns or regularities (so that strengths and weaknesses of alternatives need to be considered). In other words, the mathematical description of the situation is not problematic.
- The desired end point is to produce some clearly specified type of mathematical "answer" – even though the purpose for which this answer is needed usually is not known. That is, it is not part of a tool whose usefulness needs to be verified outside the world of mathematics.
- The "problem" is simply to find a set of legal moves to get from "givens" to "goals" by moving along a path never needs to leave the world of mathematics.

Rather than being interested in "problem solving" for its own sake, *models & modelling perspectives* are interested in the development of meaning and usefulness for powerful mathematical concepts or conceptual systems. So, we focus on problems with the following characteristics.

- The situation to be understood involves some type of mathematically interesting system – which often (but not always) exists outside the world of mathematics. So, the most problematic aspects of tasks often involve developing useful ways to *think about* (describe, explain, interpret) relevant relationships, patterns, and regularities - or givens, goals, and possible solution paths – so that relevant mathematical tools can be used.
- The product that needs to be produced is not like a point (e.g., an answer such as "12 feet"). Instead, it is a complex artifact or a conceptual tool

(such as a spreadsheet with graphs) that needs to be developed for use in a variety of structurally similar situations.

- Development processes generally involve a series of express>text>revise "modelling cycles" (see Fig. 3.1.5-1) in which alternative ways of thinking are gradually sorted out, integrated, refined, or elaborated – or rejected. Consequently, solution processes tend to resemble genetic inheritance trees that describe the evolution of a community of conceptual systems – rather than progress along a conceptual path.

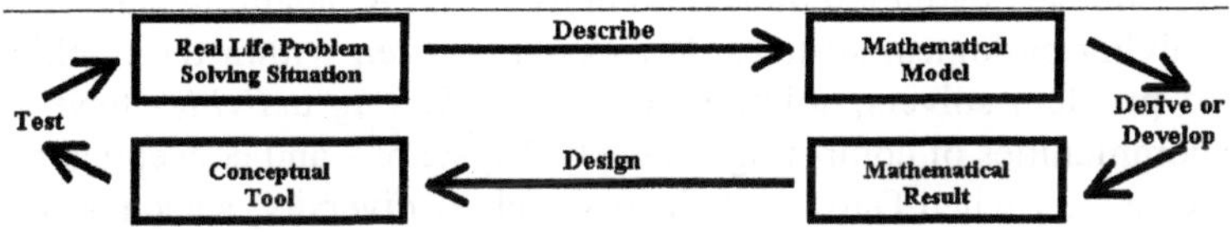

Figure 3.1.5-1. A Single Model Development Cycle

When traditional problem solving is characterized as *getting from (pre-mathematized) givens to (mathematical) goals*, then solution paths generally do not require students to leave the world of mathematics. So, solutions involve only moving from the upper right hand box to the lower right hand box in Fig. 3.1.5-1. Such problems can be thought of quarter-cycle activities.

Because of the nature of *model-eliciting activities*, it has been necessary to reconceptualize or redefine many popular heuristics and strategies to make them suitable for: (a) non-answer-giving stages of problem solving, (b) processes that are content-dependent, (c) solution paths involving multiple modelling cycles, and (d) conceptual models based on unstable organizational/relational (as well as procedural) systems.

(b) *The Nature of Learners and Problem Solvers:* In mathematics education research, it is common to characterize "problem solvers" as information processors – where "processing" generally emphasizes computation, and where "information" consists of pre-quantified data. But, in *model-eliciting activities*, data processing usually accounts for only a small portion of most problem solving episodes. Instead, the modelling cycles that problem solvers go through generally involve systematically rethinking the nature of givens, goals, and relevant solutions steps – or patterns & relationships that are attributed to surface-level data. Therefore, the most significant things that are being analyzed and transformed (or processed) are students' own ways of thinking about givens and goals – and patterns and regularities that are attributed to (rather than being deduced from) the information that is available.

Just as in more mature modern sciences, learning sciences are finding it necessary to recognize that most of the "subjects" (e.g., developing knowledge of students, or teachers, or learning communities) involve complex systems. Therefore, it has become necessary for theory development to move beyond machine-based metaphors and factory-based models to account for pattern and regularities in behavior. To develop models of students' behaviors in model-eliciting activities, *models & modelling perspectives* have moved away from industrial age machine metaphors (hardware), and beyond computer-age metaphors (software) toward metaphors grounded in an age of biotechnologies (wetware). In particular, the development of knowledge is viewed to be less like the construction of a machine or a computer program, and more like the evolution of a community of living, adapting, and continually evolving biological systems. For example, even within the thinking of isolated problem solvers, solutions to *model-eliciting activities* tend to involve communities of competing conceptual systems - and conceptual evolutions occur best when Darwinian factors such as diversity, selection, propagation, and conservation come into play. This is one reason why we often investigate problem solving where students work in teams. We use group-individual comparisons in much the same way that other researchers use expert-novice comparisons – or comparisons of gifted students to average ability students. In doing this, we go beyond emphasizing *the mind in society* (and *the mind of society*) to also emphasize *societies of mind* (Lesh & Yoon, 2003).

(c) *The Nature of Useful Mathematics:* In mathematics education research, it is common to assume that learning to solve "real life" problems involves three steps:

1. First, students should learn prerequisite ideas and skills (in decontextualized situations).
2. Next, students should learn certain problem-solving processes & heuristics in order to be able to use their ideas and skills effectively, as well as certain metacognitive processes, habits of mind, values, attitudes and beliefs (about the nature of mathematics, mathematical abilities, and problem solving) that facilitate decisions about when, where, and how to use these processes and heuristics.
3. Finally (if time permits), students should learn to use the preceding ideas, skills, processes and heuristics in messy "real life" situations.

Such views treat ideas, skills, heuristics, metacognitive processes, values, attitudes, and beliefs, as separate entities. In contrast, *models & modelling perspectives* considers models as including heuristics, metacognitive processes, values, attitudes, and beliefs which are inseparable from the constructs and conceptual systems they embody, all of which develop in parallel and interactively. They are not learned in isolation, they are learned as part of larger conceptual systems; and they are not learned in the abstract, they are learned in context – for specific purposes. It is only later that underlying abstractions are sorted out, isolated, analyzed, expressed using a single representational media, and integrated into elegant theories. In fact, even when students achieve sophisticated understandings of mathematical concepts, their thinking generally continues to organize knowledge around experience at least as much as around abstractions. That is, even ideas that are not logically connected often are psychologically connected because they have been used together in some familiar situation. Furthermore, just like the conceptual systems they embody, these metacognitive functions, values, attitudes, beliefs, and heuristics also vary in productivity from one situation to the next. A metacognitive function that is productive at one moment may be counter-productive at another moment. Therefore, no fixed, final, or inflexible profile of characteristics is likely to be productive across all circumstances. On the contrary, the best problem solving personality that a student can develop is one that can be adapted to suit changing circumstances.

(d) *The Nature of Useful Problem Solving Strategies & Heuristics*: When traditional mathematics education research treats problem solving as being about getting from givens to goals when the path is not immediately obvious (or it is blocked), heuristics usually have been thought of as being answers to the question: *What can you do when you are stuck?* But in *model-eliciting activities*, students seldom are stuck in the sense of having no ideas that are relevant to the situation. In fact, during early stages of students' work, several half-formulated (often logically incompatible) conceptualizations often operate simultaneously, each suggesting half-formulated solution procedures and/or alternative ways to select, filter, interpret, relate, organize, or synthesize information. Our instructional objective is to help average ability students use ideas that they do have, not to function better in situations in which they have none. In fact, many of the strategies and techniques that "good problem solvers" use to attack problems when they are "stuck" often are counter-productive when students work on *model-eliciting activities.* To identify what kinds of strategies *might* work, we treat substantive problem solving processes (or heuristics) developmentally, just as we treat the underlying constructs and conceptual systems they are associated with. For example, we ask: *What are primitive understandings of heuristics like "look for a similar problem", "draw a picture", or "clearly identify the givens and goals".*

During model-eliciting activities, the most beneficial decisions are seldom procedural; *What shall I do next?* In fact, students and groups who are preoccupied with "doing" (especially during early stages of solution attempts) typically do not do well compared with their peers (Lesh & Zawojewski, 2003). This is because solutions are constructed by gradually organizing, integrating, and differentiating unstable *conceptual* systems - in contrast to "arriving at answers" by linking together stable *procedural* systems (such as the condition-action rules associated with most information processing models of cognition). Therefore, finding a useful way to think about the situation (i.e., relationships among "givens," or interpretations of "givens" or "goals.") tends to be more important than rushing ahead to find a way to do it – or get from (prematurely conceived) "givens" to "goals."

In model-eliciting activities, many of the most effective activities facilitating solution attempt function not so much to help the problem solver amplify his/her problem solving abilities as they do to help the problem solver minimize cognitive characteristics associated with the use of unstable conceptual models. Consequently, the most important heuristics and strategies tend to be those dealing with: (i) how deficiencies in various models are detected, (ii) how to minimize the debilitating influences associated with the use of unstable conceptual models., (iii) how successively more complex and refined models are gradually constructed, and (iv) how competing interpretations are differentiated, reconciled, and/or integrated.

REFERENCES

English, L. (2002) (Ed.). *Handbook of international research in mathematics education.* Mahwah, NJ: Lawrence Erlbaum Associates.

Kelly, A. (2003). *Design research in education.* http://gse.gmu.edu/research/de/, Fairfax, VA: George Mason University. National Science Foundation funded project

Kelly, E., & Lesh, R. (Eds.) (2000). *Handbook of research design in mathematics and science education.* Mahwah, NJ: Lawrence Erlbaum Associates.

Lesh, R. (2001). What mathematical abilities are needed for success beyond school in a technology-based age of information? In M. Thomas (Ed.), *Technology in mathematics education.* Auckland, New Zealand: University of Auckland Press.

Lesh, R. (2001). Beyond constructivism: A new paradigm for identifying mathematical abilities that are most needed for success beyond school in a technology Based age of information. In M. Mitchelmore (Ed.), *Technology in mathematics learning and teaching: Cognitive considerations:* A special issue of the mathematics education research journal. Australia Mathematics Education Research Group. Melbourne Australia.

Lesh. R. (2003). Models and modelling in mathematics education. *Monograph for international journal for mathematical thinking and learning.* Hillsdale, NJ: Lawrence Erlbaum Associates

Lesh, R., & Doerr, H. M. (Eds.) (2003). *Beyond constructivism: Models and modelling perspectives on mathematics problem solving, learning, and teaching.* Mahwah, NJ: Lawrence Erlbaum Associates.

Lesh, R., Hoover, M., Hole, B., Kelly, A., & Post, T. (2000). Principles for Developing Thought-Revealing Activities. In A. Kelly & R. Lesh (Eds.), *Handbook of research design in mathematics and science education* (pp. 591-646). Mahwah, NJ: Lawrence Erlbaum Associates.

Chapter 3.1.6

EVERYDAY INSTRUMENTS: ON THE USE OF MATHEMATICS

Rudolf Strässer
Institutionen för matematik, Luleå tekniska universitet, Sweden, Email: rudolf@sm.luth.se

Abstract: Starting from a historical case study on weighing and pricing, the paper shows the growing integration of mathematics in instruments used in the workplace and on an everyday basis. Consequences for the teaching and learning of applications and modelling in vocational and general education are drawn – with special respect to the use of technology in mathematics.

1. POSITION IN THE ICMI STUDY ON APPLICATIONS AND MODELLING

This paper points to some aspects of applications and modelling of mathematics, which are closely linked to the first issue of the study entitled 'epistemology'. The case study in section 2 looks into an everyday situation (and the way it was / is normally lived during the last century and today). In doing so, a perspective on how mathematics is involved, sometimes also hidden in everyday situations is described – not only from the point of view of the 'normal' citizen, but also from the perspective of the professional user of the tools deeply involved in the situation. As a consequence of the case study, the conclusive remarks in section 3 do not only comment on epistemology, but elaborate some positions on the use of technology linked to applying / modelling (with the help of) mathematics and on issues related to professional use of mathematics and the teaching / learning of mathematics for professional use ('technical / vocational' education and training). The conclusions can also be read as ideas on the issue of "technological impacts" on applications and modelling in mathematics education.

2. CASE STUDY: A SHORT HISTORY OF HOW TO WEIGH AND PRICE

To make my position understandable, I present a historical case on the way mathematics was and is used (see also Strässer, 2002). I analyse the standard procedures of weighing and pricing in small and medium businesses and show the growing implementation of mathematics into various workplace tools (either material or organisational). The case study looks into a 'standard' everyday situation: weighing some three kilos of potatoes and telling the price of this merchandise.

2.1 Weighing 1: traditionally

In the past and even nowadays in marketplaces and old-fashioned shops, weighing was done with a pair of beam scales and normed weights (see Fig. 3.1.6-1; all illustrations courtesy of BIZERBA, one of the biggest producers of balances in Germany).

Figure 3.1.6-1. Traditional beam

The goods were placed in one scale and normed weights into the other (in complicated cases also into the one containing goods) to have the beam completely balanced. The weight of the goods can be read off the balance, adding and/or subtracting the weight in the respective scales. The price was then calculated separately (either mentally or in writing on a sheet of paper) by multiplying the unit price of the merchandise with the weight read from the beam scales. As for the mathematics involved, we see the partitioning of the weight according to the weights available, addition (and subtraction) to calculate the overall weight and a proportional model for the pricing.

Figure 3.1.6-2. Analogous weighing

2.2 Weighing 2: analogously

In Germany, since 1924, this way of weighing & pricing was slowly replaced by the introduction of a different analogous type of balance, in Germany called "Neigungsschaltwaage" (also: "Fächerkopf-Waage", a name related to the form of the balance; see Fig. 3.1.6-2 above). This type of balance still uses a simple proportional model. The partitioning and adding of weights was done "automatically" by the balance because the hand of the balance would move to the right proportionally to the weight of the merchandise – indicating the total weight on top of the scale. In addition, the pricing by multiplication was taken over by this new artefact: the price could be read off the scale at the correct place of the hand indicated by the unit price. The correct reading of prices and eventually adding the price of whole units of weight were therefore the only essential competences the seller should have. Mathematical "interpolation" was necessary in case of very large, very small or odd prices not on the hand and scale of the balance. Following information from BIZERBA, around 70% of the prices read off these balances were incorrect. Nevertheless, this type of artefact was widely used in Germany, until the 1980s at least.

2.3 Weighing 3: digitally

Nowadays, especially in larger shops and supermarkets, you would find digital balances (see Fig. 3.1.6-3 below) which directly offer prices for the merchandise put onto them with printouts of prices to be pasted to the goods – if the shop ever sells goods which are not pre-packed. The proportional model is still in use but hidden from the perception of the buyer. Reading weights and prices has become easy, while interpolation of large and small prices is unnecessary. "Odd" prices (like the "famous" 3.99 $ or DM) have only come into use with these balances or pre-packed merchandise.

Figure 3.1.6-3. Weighing & pricing digitally

What is left to be done by the seller or the buyer is keying in either the unit price or an identification number or symbol for the goods to be purchased. Normally, addition of several goods and identification of the individual seller is done automatically. Mathematics travels up the professional hierarchy to managers who decide on (quantity) discounts and special offers while (programmers of) computerised systems for checking the flow of goods in a company are responsible for a constant and realistic flow of information on the cash balance and economic success of the company.

2.4 Using Instruments and Mathematics

Looking back to the case described above, the most evident fact is the disappearance of mathematics from everyday perception – at least from the perception of the actual buyers and sellers. In fact, the mathematical ingredients of the situation of weighing and pricing (addition/subtraction of weights, calculation and addition of prices) are progressively turned into algorithms, automated and integrated into machines or "artefacts" (see below). Mathematics is hidden from the notice of those involved in the activity of weighing and pricing. If the situation (for the seller in most cases: the job) runs smoothly and routinely without unfamiliar and unforeseen events (today, the worst case would be the breakdown of the electricity supply), practitioners tend to rely on well-known routines for repetitive problems.

More generally, the routines are implemented in tools (like machines for calculating, scales to read, charts to fill etc., i.e. "primary" artefacts according to Wartofsky, 1979, p. 201ff). Difficulties when using mathematics tend to be simplified, if not totally avoided, by algorithms and routine activity flows. Book-keeping with its longstanding formalised set of concepts and practices (like discount and increase, recording of transactions by means of accounts, double entry book-keeping etc.) can serve as an additional illustration of how complicated workplace practices are made routine by "simple" algorithms which do not call for mathematical competences. As long as the workplace does not present unexpected situations, these instruments (for the concept of "instruments" as an entity made up of artefacts and 'utilisation schemes', in the French original: "schèmes d'utilisation" see Rabardel, 1995, more recently: Rabardel & Samurçay, 2001) go unrecognised and hide the mathematics they incorporate. Nevertheless it would be wrong to state that mathematics disappears altogether or becomes less important socially. On the contrary: the third phase of the weighing clearly shows the growing social importance of mathematics.

Is there a chance of "rediscovering" mathematics in the situations dominated by instruments hiding mathematics? Recent research on mathematics in vocational contexts offers a somewhat deceiving answer to this question. It is only in non-routine and non-standard situations, when usual practices fail or do not cover the situation to be faced at the workplace (the "breakdowns" or unfamiliar situations), that (even qualified) practitioners go back to unfamiliar, maybe innovative procedures. "They apply a fragment of professional knowledge, a half-remembered rule from school mathematics or a novel, though generally unsuccessful, use of a familiar tool" (Noss et al., 1998, p. 14; also Magajna, 1998; for the non-understanding of workplace mathematics cf. Hogan, 1996, p. 288 or Hogan & Morony, 2000). Here again the artefacts show up as one way to somehow manage non-routine problems. On the other hand, the "banking mathematics study" (cf. Noss & Hoyles, 1996) shows that computers and finely tuned software can even be used to offer a micro-world for exploring non-routine, unusual situations – if vocational training and education is explicitly targeted in this way.

To my knowledge, the problems of the disappearance of mathematics from societal perception for the 'ordinary citizen' up to now have only been studied in a very general way. Consequences of the implementation and integration of mathematics into ever more sophisticated instruments have not been analysed.

3. CONSEQUENCES FOR THE ICMI STUDY

At a general level, the above case study could caution, if not immunise against a too optimistic view on the societal appreciation of mathematics because mathematical knowledge is so widely used socially. Application of and modelling with the help of mathematics gradually disappears from societal perception by being hidden into ever more sophisticated instruments (artefacts together with standard utilisation schemes). Even if the range of applications of mathematics is growing at a breath-taking pace, this very development does not guarantee a growing social / political support for the teaching and learning of mathematics.

From the case study above, it is obvious that the last issue of the ICMI-study (entitled 'technological impacts') has to be considered not only as a means to enhance the "students' modelling abilities and to enrich the students' experience of … applications and modelling". Technology in itself is a major issue (a potential and a problem) for teaching and learning of mathematics because its use as everyday and professional instruments deeply changes the scope and way mathematics is used in society. Taking this into account, the ICMI-study could also look into issues like

- the role of mathematics in 'black boxes' created by technology,
- ideas on how / to what extent black boxes should be opened / 'de-greyed',
- how advanced technology can help to better understand the socially hidden mathematics.

For the teaching and learning of mathematics in institutionalised contexts, '*general*' education can only offer global mathematical concepts (like proportionality) to understand the vast diversity of instruments used in society. 'General' education will neither be able to study all the mathematical concepts used in the instruments available nor can it offer an analysis aiming at a deep and complete understanding of all black boxes using mathematics. As a consequence, the choice of mathematical concepts to be studied by every citizen is of utmost importance and has to be researched by Didactics of Mathematics.

In terms of application of and modelling with the help of mathematics, '*vocational / technical*' education (i.e.: education aiming at a defined area of work-related professional practice) faces an alternative which can be clearly identified: If vocational education only aims at the 'production' of workers who know which button to press in a given, pre-defined situation, ***no*** (!!) mathematics must be taught to these workers. The present, and even more so the future instruments will embody all the mathematics needed to operate the machines (and the situation). It could be considered a waste of time and money, it would be over-skilling the workers if they would be trained to understand the mechanisms built into the machines they handle (not control!).

There is a different picture of the worker handling the sophisticated instruments of today's (and future) workplaces: If the employer is / may be forced into looking for a qualified workforce which is able to control the highly integrated technology used in the workplace, it may be a good idea to have a workforce which knows about the concepts and algorithms built into the tools used in the workplace. Some of these concepts and algorithms are definitely of mathematical nature and can only be understood and competently controlled using mathematical knowledge.

Especially 'modern' information and communication technology (ICT) plays an ambivalent role in this complicated game: On the one hand, it can be used to hide the built-in mechanisms, to speed up the disappearance of mathematics from societal perception. On the other hand, when used in simulation modes, information and communication technology can be the first and best choice to better understand the role of mathematics when it is applied to model a situation.

REFERENCES

Hogan, J. (1996). A methodology for using classroom teachers to gather information about using mathematics from the work place. In P. Clarkson (Ed.), *Technology in Education: Proceedings of the 19th annual conference of the Mathematics Education Research Group of Australasia (MERGA)* (pp. 283-289). Melbourne: MERGA.

Hogan, J. & W. Morony (2000). Classroom Teachers Doing Research in the Workplace. In A. Bessot & J. Ridgway (Eds.), *Education for Mathematics in the Workplace* (pp. 101-113). Dordrecht: Kluwer.

Magajna, Z. (1998). Formal and informal mathematical methods in work settings. In A. Watson (Ed.), *Situated Cognition and the Learning of Mathematics* (pp. 59-70). Oxford: Departement of Educational Studies, University of Oxford.

Magajna, Z. & Monaghan, J. (1998). Non-elementary Mathematics in a Work Setting. *Psychology of Mathematics Education 22*, vol. 3, (pp. 231-238). South Africa: Stellenbosch.

Noss, R. & Hoyles, C. (1996). The visibility of meanings: Modelling the mathematics of banking. *International Journal of Computers for Mathematical Learning, 1*(1), 3-31.

Noss, R., Hoyles, C. & Pozzi, S. (1998). *Towards a Mathematical Orientation through Computational Modelling Project.* ESRC End of Award Report, Mathematical Sciences Group, Institute of Education, University of London.

Nunes, T., Schliemann, A. D., & Carraher, D. W. (1993). *Street mathematics and school mathematic.* Cambridge – New York – Oakleigh: Cambridge University Press.

Rabardel, P. (1995). *Les hommes et la technologie. Approche cognitive des instruments contemporains.* Paris: Armand Colin.

Rabardel, P., & R. Samurçay (2001). *From Artifact to Instrument-Mediated Learning.* New challenges to research on learning. International symposium organized by the Center for Activity Theory and Developmental Work Research, University of Helsinki.

Strässer, R. (no date.) Mathematics for Work - a Didactical Perspective. In C. Alsina, J. M. Alvarez, B. Hodgson, C. Laborde, & A. Pérez (Eds.). *8th International Congress on Mathematical Education - Selected Lectures* (pp. 427-441). Sevilla: S.A.E.M. 'Thales'. Also available at http://www.uni-bielefeld.de/idm/personen /rstraess/icme8.doc.

Strässer, R., & Bromme, R. (1992). The Description of Solids in Technical Drawing – Results from interviews of experienced draughtsmen. In W. Geeslin, & K. Graham (Eds.), *Proceedings of the Sixteenth PME* Conference, vol. 3, (pp. 43-50). Durham, New Hampshire, USA.

Strässer, R. & Zevenbergen, R. (1996). Further Mathematics Education. In A. Bishop, K. Clements, C. Keitel, J. Kilpatrick, & C. Laborde (Eds.), *International Handbook on Mathematics Education*, (pp. 647-667). Dordrecht: Kluwer.

Strässer, R. (2002). On the disappearance of Mathematics from society's perception. In H.-G. Weigand, N. Neill, A. Peter-Koop et al. (Eds.), *Developments in Mathematics Education in German-speaking Countries. Selected Papers from the Annual Conference on Didactics of Mathematics, Bern, 1999*, (pp. 124-133). Hildesheim: Franzbecker.

Wartofsky, M. W. (1979). *Models. Representation and the Scientific Understanding.* Dordrecht-Boston-London: Reidel.

Section 3.2

AUTHENTICITY AND GOALS

Edited by Peter Galbraith

Chapter 3.2.0

AUTHENTICITY AND GOALS – OVERVIEW

Peter Galbraith[1]
University of Queensland, Australia, Email: p.galbraith@mailbox.uq.edu.au

Abstract: This chapter samples a spectrum of views, informed by selected papers, and by input from a discussion involving participants from nine countries. It hence offers the reader an opportunity to engage with the diversity of thoughts expressed on these topics, and does not purport to attempt a definitive synthesis.

1. INTRODUCTION

Goals and authenticity are in practice inseparable, as the degree to which a task or problem meets the purposes for which it is designed is a measure of its validity from that perspective. Two broad theoretical conceptualisations can be identified in curricular implementation: referred to by Julie (this volume) as *modelling as vehicle* and *modelling as content*. In the former, real world problems are used to motivate and provide a basis for the development of particular mathematical content, and the needs of this curricular mathematics dictate the selection of problems to address. In the latter, developing the capacity of students to address problems located in the external world, and to evaluate the quality of their solutions are pre-eminent goals. Of course it is possible for a modelling program to embrace both versions to various degrees. However it appears that individuals, by and large, appear to identify themselves with one or other priority, and it is useful to bear this in mind when reading the contents of this section.

Papers included present alternative emphases, and issues raised are complemented by contributions from a discussion group featuring participants from nine countries. Three major themes emerged within the discussion – modelling at different levels of education; difficulties in the modelling process; and task selection. The following is representative of the range of viewpoints expressed, rather than attempting an impossible consensus.

Modelling at different levels of education: At school level mathematical modelling is a way of bridging the gap from reality to mathematics. On the one hand it is possible for children to use daily life experiences to understand mathematics, and on the other developing modelling competencies becomes a way to understand the world of reality, and to place mathematics in culture. At university level development of modelling competencies is more associated with learning to use mathematics as a tool.

Bridging the gap from mathematics to reality is very important, and we need to ask how this is affected by the age range – are younger children involved in mathematical modelling doing the same generic things as older students? Since mathematical modelling is important from the beginning there are implications for teacher education at all levels – asking what mathematics is, developing critical thinking as a basis for citizens to take a stand on issues, and so on.

Difficulties in the modelling process: Challenges to authenticity occur when portraying modelling as representative of the real world: for example, what is transferable; what can be discovered through manipulating models; how to avoid stereotypical models. Developing an ability to choose relevant information, and to look for missing data are important, as is a model viewed as an accountable personal construct of a problem situation.

A workable teaching and learning scaffolding process is needed: e.g. how to start the modelling process validly and proceed, and how to develop and maintain a proper balance with the use of technology. Mathematical modelling implies we learn something new in or about mathematics, and to model a new situation can complicate the view of the mathematics behind it. This raises learning and teaching issues concerning approach and purpose, and of course for the setting of valid goals that are also feasible for a given context.

Task selection: Here the question of authenticity becomes very significant. There is a need to ask of a potential task: Is it worth it? Does it really help us reach our goals? It is important to introduce *real* world modelling tasks, and in general realistic problems involve at least two models developed incrementally. Two aspects raised here are: the importance of using models based on experience (influenced by student background); and motivation e.g. looking to the world and to other disciplines for knowledge and problems.

It is central that choice of task should be consistent with avowed purpose. For example, if applications and modelling is included in mathematics education to attain goals such as 'students will experience school mathematics as useful for solving problems in real life outside the classroom' then students, to some extent, need to encounter tasks that are close parallels to comparable problem situations encountered outside the mathematics classroom. This issue is taken up in several of the papers that comprise this section.

2. PAPER SUMMARIES

Palm identifies goals that place value on learning and experiences located outside the classroom; including the development of application skills, a broadened conception of mathematics, and motivational benefits. He follows implications of these choices for the selection of teaching problems, inferring the need to include close simulations of real life situations. An authentic task is then one in which the situation described in the task (an event from real life that has occurred or may well occur) is truthfully described, and the conditions under which the task solving takes place in the real situation, are simulated with some reasonable fidelity in the school setting. For this purpose he proposes a modelling framework as a kind of operational blueprint: event – question – purpose – information/data – presentation – solution strategies, circumstances, solution requirements. This is essentially consistent with structures found in the various modelling diagrams that over time have served as guides to the solution process. Palm goes on to address the important subject of effectiveness, stressing that it is not enough to show that current practices have failed to achieve certain goals – there is a need for empirical evidence of the positive impact of *authentic* applications.

Jablonka draws attention to covert assumptions and values that permeate practice. She notes that different purposes result in different models of the 'same' reality, and taking note of inequalities that exist within and between countries, it is easy to imagine that problem contexts may reflect or advocate a lifestyle, which does not connect with many students' realities. In education mathematical practices are re-contextualised for purposes of enculturation, so that the selection of examples for applications and modelling has political content, and the resulting curriculum has social implications.

This means that identifying a core curriculum (and by implication an agreed set of core goals) for applications and modelling within general education is problematic. For if defined in terms of contexts, then given differences in cultural settings and between groups of students, it cannot be assumed that the same problems would be relevant to all. So the issue of goal setting reverts to time and place, and authentic (actual, not imitated, not false or adulterated) mathematical modelling takes place, when students and teachers are *bona fide* engaging in a modelling or application activity about an issue relevant to them or to their community.

In contrast to the *modelling as content* approaches of Palm and Jablonka Bonotto adopts an approach more akin to a *modelling as vehicle* perspective, in looking for modelling to motivate students with everyday life contexts but also "*to look for contexts that are experientially real for the students and can be used as starting points for progressive mathematization*" (Gravemeijer, 1999, p.158). In this approach everyday-life experience and formal mathematics, are not seen as two disjunctive and independent entities, but rather a process of gradual growth is aimed for, in which formal mathematics comes to the fore as a natural extension of the student's experiential reality. For Bonotto, goal realisation is the extent to which this process is deemed successful, authentic tasks are those that provide vehicles of safe conduct to this end, and she introduces an active duality into mathematics – real world linkages. Besides *"mathematizing everyday experience"* it is necessary to be *"everydaying mathematics"* In the classroom this can be implemented by encouraging students to analyze *'mathematical facts'* embedded in appropriate *'cultural artefacts'*; e.g. supermarket bills, bottle and can labels, a weekly TV guide. So the emphasis is on uncovering mathematics rather than on uncovering solutions to problem situations. Nevertheless this approach and those driven by the *modelling as content* perspective share overlapping sympathies. In this approach students are still expected to approach a problem as a situation to be mathematized, not primarily to apply ready-made solution procedures – the primary objective being to make sense of the problem. And in practice, moving back and forth between interpreting the problem and reviewing procedures or results, are processes consistent with similar iterative procedures celebrated within the *modelling as content* philosophy.

Schwarzkopf also examines word problem solving, noting that this involves interplay between two very different framings, namely "everyday-understandings" of a problems real-world context and "mathematical" framings. This interplay is complex in nature, because within different framings, participants are acting in different ways concerning the relevance of facts, the meaning of assertions, the acceptance of statements, rules for correct reasoning, and many other aspects. From this viewpoint the complexity of solving word problems originates from the *classroom interaction* and *not* from the *"internal" structure* of the task. The author reported the reconstruction of a variety of framings, drawing on very different and contradictory understandings of the task, and goes on to argue that the different framings raise problematic issues for interpretation, solution, and classroom interaction. By implication authenticity is strongly influenced by how well these respective components are orchestrated.

Together the papers and workshop comment cover substantial territory, while reflecting the interests, perspectives, and priorities of participants and authors. This section hence offers the reader an opportunity to engage with the diversity of thoughts and issues expressed by fellow scholars and practitioners, with the recognition that not all of the important matters associated with authenticity and goals have been (or can be) explored in this selection.

---

[1] Valued contributions to the discussion that formed a basis for this introduction to the chapter were provided by: Salett Biembengut, Morten Blomhoj, George Ekol, Djordje Kadijevik, Akio Matsuzaki, Susan McNab, Jarmila Novotna, Torulf Palm, Jacques Treiner.

Chapter 3.2.1

HOW TO REPLACE WORD PROBLEMS WITH ACTIVITIES OF REALISTIC MATHEMATICAL MODELLING

Cinzia Bonotto
Dipartimento di Matematica Pura ed Applicata, Università di Padova, Italy,
Email: bonotto@math.unipd.it

Abstract: In this contribution we want to discuss some classroom activities, whose overall aim is to change pupils' conceptions and beliefs of the role of real-world knowledge in mathematical classrooms activities, and to develop in them a positive disposition toward more realistic mathematical modelling. These activities make extensive use of cultural artefacts that could prove to be useful instruments in creating a new tension between school mathematics and the real world with its incorporated mathematics.

1. INTRODUCTION

In common teaching practice the habit of connecting mathematics classroom activities with everyday-life experience is still substantially delegated to word problems. But besides representing the interplay between formal mathematics and reality, word problems are often the only means of providing students with a basic sense experience in mathematisation and mathematical modelling. Recent researches have documented that the practice of word problem solving in school mathematics promotes in students the exclusion of realistic considerations and a "suspension" of sense-making, and rarely reaches the idea of mathematical modelling and mathematisation (see Verschaffel et al., 2000, for a review of these studies). Several studies point to two reasons for this lack of use of everyday-life knowledge: textual factors relating to the stereotypical nature of the most frequently used textbook problems, and presentational or contextual factors associated with practices, environments and expectations related to the classroom culture of mathematical problem solving. Furthermore the use of stereotyped problems and the accompanying classroom climate relate to teachers' beliefs about the goals of mathematics education.

This indicates a difference in views on the function of word problems in mathematics education. Researchers relate word problems to problem solving and applications, while student-teachers (and probably teachers in general) see word problems as nothing more, and nothing less, than exercises in the four basic operations which also have a justification and suitable place within the teaching of mathematics, though certainly not that of favouring mathematical modelling (Blum & Niss, 1991).

If we wish to establish situations of realistic mathematical modelling, in the sense of *"both real-world based and quantitatively constrained sense-making"* (Reusser & Stebler, 1997), in problem-solving activities, we have to: i) change the type of activity aimed at creating interplay between the real world and mathematics towards more realistic and less stereotyped problem situations; ii) change students' conceptions of beliefs about and attitudes towards mathematics (this means changing teachers' conceptions, beliefs and attitudes as well); and iii) change classroom culture by establishing new classroom socio-mathematical norms.

In this contribution we want to discuss how these changes can be realised at primary school level through classroom activities related more easily to the experiential world of the student, and consistent with a sense-making disposition. They make extensive use of cultural artefacts that could prove to be useful instruments in creating a new tension between school mathematics and the real world with its incorporated mathematics. We will show how suitable cultural artefacts and interactive teaching methods can play a fundamental role in this process.

2. CONNECTIONS BETWEEN CLASSROOM ACTIVITIES AND THE REAL WORLD

The connection between real world and classroom mathematics is not easy because the two contexts differ significantly. Just as mathematics practice in and out of school differs so does mathematics learning. Masingila et al. (1996) outlined three key differences between in- and out-of-school practices (goals of the activity, conceptual understanding, and flexibility in dealing with constraints). In out-of-school mathematics practice in particular, people may generalise procedures within one context but may not be able to generalise to another since problems tend to be context specific. Generalization, which is an important goal in school mathematics and an important aspect of the mathematisation process, is not usually a goal in out-of-school mathematics.

Although the specificity of both contexts is recognised, we think that the conditions that often make out of school learning more effective can and must be re-created, at least partially, in classroom activities. Indeed, while there may be some inherent differences between the two contexts, these can be reduced by creating classroom situations that promote learning processes closer to those arising from out-of-school mathematics practices. On the other hand the relationship between mathematics and the real world has always been both intricate and intriguing, as much complicated as interesting to deal with, and maybe we will never be able to analyse it completely and thoroughly. As a joke, we might say that it is a relationship of 'hate and love' since mathematics, although receiving nourishment from the real world, detaches from it as soon as possible, due to its special nature, only to come back to real experience in due time to pick up new problems and examples or to find new applications. As to didactics, the fact that this relationship is sometimes denied and at other times stressed, without any explanation of the reasons for these choices, makes it difficult for students to know whether or not it is permissible for them to exploit their everyday knowledge in approaching mathematical problems (Bonotto, 2001).

According to the Realistic Mathematics Education perspective we think that progressive mathematisation should lead to algorithms, concepts and notations that are rooted in a learning history which starts with students' informal experientially real knowledge. In our approach everyday-life experience and formal mathematics, despite their specific differences, are not seen as two disjunctive and independent entities. Instead, a process of gradual growth is aimed for, in which formal mathematics comes to the fore as a natural extension of the student's experiential reality. The idea is not only to motivate students with everyday-life contexts but also *"to look for contexts that are experientially real for the students and can be used as starting points for progressive mathematisation"* (Gravemeijer, 1999, p.158).

We stress that the process of bringing *"the real world into mathematics"* by starting from a student's everyday-life experience, is fundamental in school practice for the development of new mathematical knowledge. However it turns out to be necessary, but not sufficient, to foster for example a positive attitude towards mathematics, intended both as an effective device to know and critically interpret reality, and as a fascinating thinking activity. We contend that these educational objectives can only be completely fulfilled if students and teachers can bring mathematics into reality. In other words, besides *"mathematising everyday experience"* it is necessary to be *"everydaying mathematics"* (Bonotto, 2001). This can be implemented in a classroom by encouraging students to analyse *'mathematical facts'* embedded in appropriate '*cultural artefacts*'; there is indeed a great deal of mathematics embedded in everyday life.

The cultural artefacts we have introduced into classroom activities, e.g. supermarket bills, bottle and can labels, railway schedules, a cover of a ring binder, a weekly TV guide (see Bonotto, 2001; Bonotto & Basso, 2001; Bonotto, 2003), are concrete materials which children typically meet in real-life situations. We have therefore offered the opportunity of making connections between the mathematics incorporated in real-life situations and school mathematics, which although closely related, are governed by different laws and principles. These artefacts are relevant to children; they are meaningful because they are part of their real life experience, offering significant references to concrete situations. This enables children to keep their reasoning processes meaningful and to monitor their inferences.

We believe that immersing students in situations which can be related to their own direct experience and are more consistent with a sense-making disposition, allows them to deepen and broaden their understanding of the scope and usefulness of mathematics as well as learning ways of thinking mathematically that are supported by mathematising situations. This allows students to become involved in mathematics and to break down their conceptions of a remote body of knowledge. Obviously, usefulness and its pervasive character are just two of the many facets of mathematics that do not entirely capture its special character, relevance and cultural value; nonetheless these two elements can be usefully exploited from the teaching point of view.

3. ON THE USE OF CULTURAL ARTEFACTS IN CLASSROOM ACTIVITIES

The use of cultural artefacts in our classroom activities has been articulated in various stages, with different educational and content objectives.

First, the dual nature of the artefacts, that is belonging to the world of everyday life and to the world of symbols, to use Freudenthal's expression, allows movement from situations of normal use to the underlying mathematical structure and vice versa, in agreement with '*horizontal mathematisation*' (Treffers, 1987). For example using a receipt, which is poor in words but rich in implicit meanings, overturns the usual buying and selling problem situation, which is often rich in words but poor in meaningful references (Bonotto, 2001).

But these artefacts may also become real *"mathematising tools"* with some modification, e.g. removing some data that are present in the artefacts (Bonotto, 2001); in this way we can create new mathematical goals and provide students with a basic experience in mathematical modelling. In this new role, the cultural artefacts can be used as motivating stepping-stones to launch new mathematical knowledge, through the particular learning processes that Freudenthal (1991) defines as '*prospective learning*' or '*anticipatory learning*'. We think that this type of learning is better enhanced by a 'rich context' as outlined by Freudenthal, that is a context which is not only the application area but also a source for learning mathematics. The cultural artefacts and classroom activities we introduce are part of this type of context. These experiences have also favoured the type of learning *"retrospective"* that occurs when old notions are recalled in order to be considered at a higher level and within a broader context, a process typical of adult mathematicians. This different use of the artefacts also makes it possible to carry out '*vertical mathematisation*', from concept to concept.

The use of suitable artefacts allows the teacher to propose many questions, remarks, and culturally and scientifically interesting inquiries. The activities and connections that can be made depend, of course, on the students' scholastic level. These artefacts may contain different codes, percentages, numerical expressions, and different quantities with their related units of measure, and hence are connected with other mathematical concepts and also other disciplines (chemistry, biology, geography, astronomy, etc.). It could be said that the artefacts are related to mathematics (and other disciplines) as far as one is able to make these relationships. Furthermore we ask children to select other cultural artefacts from their everyday life, to identify the embedded mathematical facts, to look for analogies and differences (e.g. different number representations), to generate problems (e.g. discover relationships between quantities). In other words children should be encouraged to recognise a great variety of situations as mathematical situations, or more precisely *"mathematisable"* situations. In this way children are offered numerous opportunities to become acquainted with mathematics and to change their attitude towards mathematics, in contrast with the traditional classroom curriculum.

Besides the use of suitable cultural artefacts discussed above, the teaching/learning environment designed and implemented in our classroom activities is characterised by: i) the application of a variety of complementary, integrated and interactive instructional techniques (involving childrens' own written descriptions of the methods they use, individual and class discussions, and the drafting of a text by the whole class); and ii) an attempt to establish a new classroom culture through new socio-mathematical norms.

Regarding the first point, most of the lessons follow an instructional model consisting in the following sequence of classroom activities: a) a short introduction to the class as a whole; b) an individual written assignment where students explain the reasoning followed and strategy applied; c) a final whole-class discussion. We consider that the interactivity of these instructional techniques is essential because of the opportunities to induce reflection as well as cognitive and meta-cognitive changes in students. This process may be very important for teachers also, since it enables them to recognise and analyse individual reasoning processes that are not always explicit (corresponding to the individual written report). In the collective discussion, comparing different answers and strategies, noting children's first attempts at generalizing, and taking account of further remarks made during discussion, leads to collectively drawing up a text aimed at socialization of the knowledge acquired, which completes the activity.

As far as the second point is concerned, we expect students to approach a problem as a situation to be mathematised, not primarily to apply ready-made solution procedures. This does not mean that knowledge of solution procedures plays no part, but the primary objective is to make sense of the problem. In practice, it is often a matter of shuttling back and forth between interpreting the problem and reviewing possible procedures or results. At the same time, the teacher is expected to encourage students to use their own methods, exploring their usefulness and soundness with regard to the problem. The teacher should stimulate students to articulate and reflect on their personal beliefs, misconceptions and problem-solving strategies. Other possible strategies for solving the same problem when it appears next are emphasised, and students are encouraged to make comparisons between strategies.

4. DISCUSSIONS AND OPEN PROBLEMS

In this contribution we discuss some teaching experiences based on the use of suitable cultural artefacts, interactive teaching methods, and the introduction of new socio-mathematical norms. According to the results we can say that, contrary to the practice of word-problem solving in school mathematics, children did not ignore the relevant, plausible, and familiar aspects of reality, nor did they exclude real-world knowledge from their observation and reasoning. As found in previous studies, children exhibit flexibility in their reasoning processes by exploring different strategies, often sensitive to the context and quantities involved. Children were therefore able to resort to realistic considerations that are both real world based and quantitatively constrained sense making, in the sense of realistic mathematical modelling.

We do not suggest that the classroom activities described here are a prototype for all classroom activities related to mathematics, although in agreement with Verschaffel et al. (1999, p. 226), we think that *"the development of mathematical problem-solving, skills, beliefs, and attitudes should not emanate from a specific part of the curriculum but should permeate the entire curriculum"*, for example by following both a *"mixing approach"* and an *"integrated approach"* (Blum & Niss, 1991).

We do believe however that by enacting some of these experiences, children are offered an opportunity to change their beliefs about, and attitudes towards school mathematics. Immersing students in situations more relatable to their direct experience and more consistent with sense-making, provides a means to deepen and broaden their understanding of the scope and usefulness of mathematics as well as learning ways of thinking mathematically that are supported by mathematising situations. Furthermore in this way we can design better opportunities for children to develop mathematical knowledge that is wider than they would develop outside of school, but that also preserves the focus on meaning found in everyday situations. Using appropriate cultural artefacts, which students can understand, analyse, and interpret, we can present mathematics as a means of interpreting and understanding reality. Teaching students to interpret critically the reality they live in, to understand its codes and messages so as not to be excluded or misled should be an important goal for compulsory education.

As we have already had occasion to emphasise, the usefulness and pervasive character of mathematics are merely two of its many facets and can not by themselves capture its very special character, relevance, and cultural value; nonetheless we deem that these two elements can be usefully exploited from the teaching point of view because they can change the common behaviour and attitude held both by teachers and pupils.

For a real possibility to implement this kind of classroom activities, there also needs to be a radical change on the part of teachers. They have to try: i) to modify their attitude to mathematics that is influenced by the way it was learned; ii) to revise their beliefs about the role of everyday knowledge in mathematical problem solving; iii) to see mathematics incorporated into the real world as a starting point for mathematical activities in the classroom, thus revising their current classroom practice, and iv) to investigate the mathematical ideas and practices of the cultural, ethnic, linguistic communities of their pupils. Only in this way can a different classroom culture be attained.

Finally a teacher has to be ready to create and manage open situations, that are continuously transforming and of which he/she cannot foresee the final evolution or result. As a matter of fact, these situations are sensitive to the social interactions that are established, to the students attitudes, reactions, their ability to ask questions, to find links between school and extra-school knowledge; hence the teacher has to be able to modify along the way the content objectives of the lesson. A teacher has to be (and to feel), very strong and qualified both on the mathematical content and on the educational objectives that are potentially contained in these artefacts. A lesson cannot be prepared in advance in all of these aspects; it requires planning for various 'branches' to be drawn together through a process whose management is quite demanding. In accord with Blum & Niss (1991) and Verschaffel et al. (1999), we deem that the effective establishment of a learning environment like the one described here makes very high demands on the teacher, and therefore requires revision and change in teacher training, both initially and through in-service programs.

REFERENCES

Blum, W., & Niss, M. (1991). Applied Mathematical Problem Solving, Modelling, Applications, and Links to Other Subjects – State, Trends and Issues in Mathematics Instruction". *Educational Studies in Mathematics, 22*(1), 37-68.

Bonotto, C. (2001). How to connect school mathematics with students' out-of-school knowledge. *Zentralblatt für Didaktik der Mathematik, 3*, 75-84.

Bonotto , C. (2003). About students' understanding and learning of the concept of surface area. In D. H. Clements, & G. Bright (Eds.), *Learning and Teaching Measurement, Yearbook of the NCTM*, (pp. 157-167). Reston, Va.: NCTM.

Bonotto, C., & Basso M. (2001). Is it possible to change the classroom activities in which we delegate the process of connecting mathematics with reality? *International Journal of Mathematics Education in Science and Technology, 32*(3), 385-399.

Freudenthal, H. (1991). *Revisiting mathematics education. China lectures.* Dordrecht: Kluwer.

Gravemeijer, K. (1999). How emergent models may foster the constitution of formal mathematics. *Mathematical Thinking and Learning. An International Journal, 1*(2), 155-177.

Masingila, J. O., Davidenko, S., & Prus-Wisniowska, E. (1996). Mathematics learning and practices in and out of school: A framework for connecting these experiences. *Educational Studies in Mathematics, 31*, 175-200.

Reusser, K., & Stebler, R. (1997). Every word problem has a solution: The suspension of reality and sense-making in the culture of school mathematics. *Learning and Instruction, 7*, 309-328.

Treffers, A. (1987). *Three dimensions. A model of goal and theory description in mathematics instruction – The Wiscobas Project.* Dordrecht: D. Reidel Publ. Co.

Verschaffel, L., De Corte, E., Lasure, S., Van Vaerenbergh, G., Bogaerts, H., & Ratinckx, E. (1999). Learning to solve mathematical application problems: A design experiment with fifth graders. *Mathematical Thinking and Learning. An International Journal, 1*(3), 195-229.

Verschaffel, L., Greer, B. & De Corte, E. (2000). *Making sense of word problems.* Lisse, The Netherlands: Swets & Zeitlinger.

Chapter 3.2.2

THE RELEVANCE OF MODELLING AND APPLICATIONS: RELEVANT TO WHOM AND FOR WHAT PURPOSE?

Eva Jablonka
Freie Universität Berlin, Email: jablonka@zedat.fu-berlin.de

Abstract: The criteria for relevance of modelling and applications vary with the educational goals and the cultural context. But the competence of judging quantitative information and of evaluating mathematical models is a key competence that is linked to all levels of goals in a diversity of contexts. It is suggested that different mathematical practices should be distinguished and analysed in order to develop a curriculum, which aims at developing this competence.

1. DIFFERENT WORLDS – DIFFERENT MODELS

Mathematical modelling is a purposeful activity; different purposes may result in different mathematical models of the "same" reality. "The relationship between applications and modelling and the world we live in" (cf. the section *Epistemology* in the discussion document) varies, depending on the socio-cultural and economic context of the "we" and on our life-styles, careers and political standpoints. Different perceptions of a problem and thus different criteria for what constitutes an acceptable solution may arise in almost any situation. For example, if the problem of a bank employee, who has to advise a client (aided by a software package), is the comparison of financing offers for a mortgage, for the manager of the bank this is a problem of profitability, and for the customer it is one of planning her personal finances; minimizing production costs need not minimise energy consumption; modelling traffic flow or parking space design will not reduce the amount of cars.

Introducing applications or modelling into mathematics teaching always means introducing social context, no matter whether this aims at teaching formal mathematics via modelling, at dealing with relevant applications, or at teaching modelling.

It is informative to look at the collection of contexts used in the examples from the research studies presented at the International Study Conference (published in the Pre-Conference Volume (Blum & Henn, 2004)). These examples are from personal everyday practices and from professional practices, in which mathematics is used. The progression to professional practices, as reflected in the transition from simple to more advanced modelling tasks, implies a change of perspective from consumer to producer.

Examples from everyday life comprise: filling a swimming pool; light intensity needed for reading; cooling rates of coffee (of green tea or of corn soup); planning bus trips for senior citizens or students; distances given on road signs; shaking hands at birthday parties; buying dishes in restaurants; comparing discount percentages; life spans of batteries; taxi prices; dealing with supermarket bills; railway schedules; bank accounts; savings and loans. Some examples deal with a mathematical analysis of cultural artefacts, such as shapes of ice cream, of hats and umbrellas, of churches, modern bridges and airport buildings, of dress designs by Sonia Delaunay and the surface of a Porsche. It is easy to imagine that these contexts reflect or advocate a lifestyle, which is not that of the majority of the students' families.

Professional practices from which the examples are taken include: designing a 3-question survey, reducing the noise of an aeroplane to a given limit, locating a water reservoir, seismic exploration of oil and gas, data analysis and dynamical systems models in biology, optimising traffic flow, designing a road to limit speeding, selling dishes in restaurants, measuring land, or optimising a relay race.

Two examples presented at the study conference deal with an analysis of fairness (in ranking of Commonwealth games performance and of students' assessment) and one refers to statistics of supposedly rising crimes. These examples differ from all the others in that they do not simulate an out-of-school mathematical practice in the classroom as authentic as possible, but aim at evaluating a practice in which mathematics is used by others. This aim resembles more that of quantitative literacy.

The contexts chosen for application and modelling are not arbitrary, but depend on the educational goals. In mathematics education mathematical practices (e.g. of university mathematicians, of computer scientists, of engineers, biologists, economists, statisticians, of computer users, of skilled manual workers, of consumers) are recontextualised for the purpose of enculturation. Mathematics is a highly specialised activity that consists of a range of practices, some of which employ sophisticated tools and sign systems. The recontextualisation of parts of those practices establishes the school subject mathematics as it is defined in curriculum documents. The process involves decisions about what areas of knowledge are to be selected from which practices, about how these areas are to be related within school mathematics and to other school subjects and what the relation to the everyday knowledge of the students should be. The selection of examples for introducing applications and modelling is a political decision and the resulting curriculum has social implications.

Skovsmose (2004) draws attention to the fact that power can be executed through the mathematics curriculum by establishing relationships to out-of-school practices, both by referring to and by ignoring distinct practices. He distinguishes the practices of constructers (who maintain and develop knowledge), of operators of mathematical technology, of consumers and of the "disposable". In these terms the application tasks in TIMSS and PISA refer to practices of operators and consumers (cf. Jablonka, 2000, Gellert & Jablonka, 2002), the examples from everyday life listed above as well.

2. RECONTEXTUALISATION AND AUTHENTICITY

Bernstein (1996, p.46) argues that pedagogic discourse cannot be identified with any of the discourses it transmits. From this point of view "authentic" examples of mathematical modelling and applications, if enacted in a classroom or included in an assessment, are by definition a simulation; authenticity and large-scale assessment turn out to be an inherent contradiction.

The effects of recontextualisation are well known: many students face a problem when trying to solve typical mathematical "word problems" that are found in textbooks and tests all over the world. These tasks are not easily to be identified as clearly belonging either to the everyday practical context, which would suggest using everyday knowledge, or to the academic context, which would reveal that they are to be solved mathematically. Even though the problem situations are realistic, the questions to be answered often are not questions a person involved in the situation would ask. It is not possible anymore for the learner to evaluate the solution of the problems from a practical point of view; everyday knowledge turns out to be an insufficient base for solving the contextualised problems because the mathematical solution has a structure that differs from any practical solution. Generalisability and formalisation is the achievement of mathematics and not of common sense and practical wisdom. However, the students are expected to believe that behaving mathematically in those situations would be in the elaboration of practical interests. Palm (this volume, Chapter 3.2.3) assumes and provides some empirical evidence that a clear statement of the purpose of a practical task mitigates the difficulties students face.

Recontextualisation also causes a transformation of the unmediated discourses found in out-of-school practices of mathematical modelling, even though a modelling perspective overcomes the philosophy of naïve realism encapsulated in traditional word problems. In many modelling problems that are processed and designed for the classroom, the original practice is not visible anymore. Who are the people modelling a catwalk, a running elephant, or the digestion process of sheep? In teaching materials the examples often are presented as narratives from the perspective of a person acting in a problem situation. So, for example, a problem from traffic engineering is turned into a story of two adolescents (named Kim and Robert), who want to count cars in a street ("James Street") in order to model traffic flow to inform a decision about a pedestrian crossing (Rouncefield, 1993). Such a recontextualisation is based on the assumption that it eases identification with the protagonists on the side of the students. However, in this example, the fact that this question does not resemble the perspective of the pedestrian is concealed in the narrative, as is the fact that the solution method can be generalised and is not valid only for "James Street" (for an analysis of examples see Jablonka, 2001). This implicit pedagogy effects how the students think about mathematical modelling and applications. It would be more sensible to introduce a meta-discourse on the nature of mathematics that helps to differentiate between different practices of using mathematics.

However, *authentic* (i.e. actual, not imitated, not false or adulterated) mathematical modelling takes place, when students and teachers are *bona fide* engaging in a modelling or application activity about an issue relevant to them or to their community.

Gutstein (in press) reports from teaching a seventh-grade mathematics class in K-8 school of 800 students in a working-class, Mexican immigrant community in Chicago. In "real-world projects", which often emerged from students' own questions, mathematics is used as a principal analytical tool to investigate social justice issues. One project, for example, involved comparing the cost of one B-2 bomber (about $2.1 billion, with all development costs) to a four-year scholarship at a prestigious university, and to find out for how many years the money for one bomber would pay for four-year scholarships for the whole graduating class (assuming constant costs and annual number of students). In another project students discussed an article about a report that analysed mortgage rejection rates in relation to the race of the applicants in metropolitan areas in the USA.

Since such activities frequently include judgement of quantitative information and evaluation of underlying mathematical models, the critical analysis of authentic texts forms a part of such genuine modelling activities. This involves not only mathematical (conceptual) understanding, but also contextual knowledge, political awareness and judgements based on values. Engaging in genuine mathematical modelling activities leaves the decision about the problems to be dealt with to the teachers and students and thus goes against standardisation.

3. IDENTIFYING A CORE CURRICULUM

On the whole, a focus on modelling and applications decreases the overlap in curriculum within and across countries. The desirable competencies of distinct groups of people who are engaged in various cultural practices and who live in "different worlds" will be expressed differently (cf. Niss, 2003). Identifying a core curriculum in applications and modelling that aims at contributing to general education (cf. Blum et al., 2002, Issue 5a) would not be desirable if this were done in terms of contexts to be dealt with. Given the differences in cultural context as well as between groups of students, it cannot be assumed that the same problems would be relevant to all, though a list of issues of a world-wide political and social relevance (such as environmental problems, migration, economic inequality, technology of weapons) could be produced. But these can come into conflict with local priorities.

However, as it has been argued in this paper, a key competence is to be able to make judgements about models and quantitative information produced by others (cf. also Blum et al., 2002, Issue 5c). This competence is linked to the goal of identifying and understanding the role that mathematics plays in the world. It is conceivable to approach this goal by identifying and describing types of mathematical models in terms of the different practices in which they emerge. The mathematical models dealt with in the classroom could then be chosen as representative examples. This issue does not address a "small size" of the mathematics curriculum (cf. Usiskin, 1999), such as the problems for one lesson or a single unit or chapter, it is rather about the effect of the school experience until the end of compulsory education.

To achieve this, further epistemological analysis and empirical research of mathematical practices that exist in "the world we live in" and in the worlds others live or lived in, seems to be necessary. An essential assumption here is that mathematics consists of a variety of activities that occur within distinct domains in which mathematics is used and that these activities are associated with specific perceptions of mathematical knowledge. A universal description of the process of mathematical modelling falls short of the varying methodological standards, criteria for validation and evaluation that are relevant in different contexts. The criteria used in selecting fields for applying mathematics and in defining what is considered a solution are external to mathematics and thus reflection should not be restricted to methodological considerations. A problem solution by means of mathematical modelling embodies the interests, aims and associated values of the social and technological practice, in which it is embedded. Perrenet and Morsche (Blum & Henn, 2004, p. 215) give an example of what a "social reflection" of a modelling activity might include. It should take the various interests of the actors (such as technology producers, users, regulators and advisors) into account as well as the evolution of the problem and the intended and concomitant consequences (see also Jablonka, 1999).

Practices in which mathematics is used can be analysed, for example, alongside the dimensions of (i) mathematical methods and styles of reasoning, (ii) the degree of mathematisation, (iii) the tools (such as calculators, software and tables) employed, and (iv) the explicitness and degree of justification of the rules that regulate the application of mathematics.

Different mathematical methods are associated with different cultures concerning the expectations regarding the assumptions, the goals and underlying methodologies. For example, it is obvious that results gained from models based on genetic algorithms differ from those obtained from traditional optimisation methods, or that deterministic and probabilistic approaches are different.

Many research and development practices show a high degree of mathematisation. There is, for example, a substantial development in algorithms and software tools in research fields penetrated by mathematical language, such as mathematical ecology, systems biology or molecular modelling. A problem with including "authentic" examples of these practices in the school curriculum is caused by the fact that the technological transformation of academic mathematical knowledge is a process embedded in a highly specialised division of labour. The transformation is mediated by several disciplines (resulting in software development) and the mathematics involved is in general too sophisticated. There is a lack of studies and descriptive accounts that provide examples and explain the principles of these applications.

In terms of justification of the rules that regulate the application of mathematics one can, for example, distinguish ad-hoc-models and models based on a theory, or distinguish models in terms of consistency, connectivity, complexity and comprehensiveness. Increasing computing power enables more detailed calculations with fewer theoretical assumptions. Ideally, mathematics is conceived as an activity, the practices of which tend to make explicit the principles of their regulation. However, the assumptions on which mathematical models are based often remain implicit. Frequently the ways in which data were gathered and measured cannot be reconstructed. Practices in which the use of mathematics highly depends on mathematical technology tend to show a low degree of clarity of rules. This may be a consequence of the high degree of mathematisation (cf. Keitel, 1993).

To distinguish and further analyse different mathematical practices (including their history) helps clarifying the "relationship between applications and modelling and the world we live in", as it was phrased in the discussion document for the ICMI Study Conference. This could form a starting point for developing a "core curriculum" in applications and modelling that aims at contributing to general education. Mathematical modelling is always embedded in a social practice and thus it is, in the end, not possible to promote goals and a collection of examples of mathematical modelling and applications without, at the same time, (implicitly) promoting a social practice.

Given the amount of available modelling and application problems and materials – for different types and levels of mathematics education and at different levels of authenticity – the purpose and the desirable outcomes of mathematical modelling and application activities should be addressed and the socio-political stakes of mathematical modelling and applications should be taken seriously.

REFERENCES

Bernstein, B. (1996). *Pedagogy, symbolic control and identity: Theory, research, critique. Class, codes and control, vol. 5.* London: Taylor & Francis.

Blum, W. et al. (2002): ICMI-Study 14: Applications and Modelling in Mathematics Education: Discussion Document. *Educational Studies in Mathematics, 51*, 149-171.

Blum, W., & Henn, H.-W. (2004). *ICMI Study 14. Applications and Modelling in Mathematics Education: Pre-Conference Volume.* Dortmund: University of Dortmund.

Gellert, U., & Jablonka, E. (2002). Tasks and Questions Used in International Assessment of Mathematical Literacy. In L. Bazzini & C. Inchley (Eds.), *Mathematical Literacy in the Digital Era, Proceedings of the 53rd Conference of the CIEAEM, Verbania, Italy, July 21-27, 2001* (pp. 114-118). Milano: Ghisettie Corvi Editori.

Gutstein, E. (in press). "So One Question Leads to Another": Using Mathematics to Develop a Pedagogy of Questioning. In N. S. Nasir and P. Cobb (Eds.), *Diversity, Equity, and Access to Mathematical Ideas.* New York: Teachers College Press.

Jablonka, E. (1997). What Makes a Model Effective and Useful (or not)? In S.K. Houston et al (Eds.), *Teaching and Learning Mathematical Modelling: Innovation, Investigation and Applications* (pp. 39-50). Chichester: Albion Publishing.

Jablonka, E. (2000). Perceptions of mathematics and reality in TIMSS mathematics items. In A. Ahmed, J.M. Kraemer, & H. Williams (Eds.), *Cultural Diversity in Mathematics (Education)* (pp. 127-132). Chichester: Ellis Horwood.

Jablonka, E. (2001). Modelling: Good Problems - not only a Question of (Good) Taste. In J.F. Matos, W. Blum, S. Carreira, & K. Houston (Eds.), *Modelling, Applications and Mathematics Education: ICMTA9: Applications in Science and Technology* (pp. 305-316). Chichester: Ellis Horwood.

Keitel, C. (1993). Implicit Mathematical Models in Social Practice and Explicit Mathematics Teaching by Application. In J. de Lange, C. Keitel, I. Huntley, & M. Niss (Eds.), *Innova-*

tion in Maths Education by Modelling and Applications (pp. 19-30) Chichester: Ellis Horwood.

Niss, M. (2003). *Mathematical Competencies and the Learning of Mathematics: The Danish KOM Project*. Paper presented at the National Academy of Sciences' Symposium, Talking it Through: A Cross-National Conversation about Secondary Mathematics Curricula, March 31-April 1, Washington DC.

Rouncefield, M. (1993). Statistical Simulations. In T. Breiteig, I. Huntley, & G. Kaiser-Messmer (Eds.), *Teaching and learning Mathematics in Context* (pp. 81-92). Chichester: Ellis Horwood.

Skovsmose, O. (2004). *Research, Practice and Responsibility*. Discussion Paper of the Survey Team 1: The relation between research and practice in mathematics education at ICME 10, Copenhagen, Denmark, July 4-11, 2004.

Usiskin, Z. (Ed.). (1999). *Developments in School Mathematics Around the World*, vol. 4. Proceedings of the Fourth UCSMP International Conference on Mathematics Education, Aug. 5-7, 1998. Reston, VA: National Council of Teachers of Mathematics

Chapter 3.2.3

FEATURES AND IMPACT OF THE AUTHENTICITY OF APPLIED MATHEMATICAL SCHOOL TASKS

Torulf Palm
Department of Mathematics Umeå University, Sweden, Email: Torulf.Palm@math.umu.se

Abstract: In this paper the issue of the authenticity of applied mathematical school tasks is discussed. The paper includes a description of a framework for reflection, analysis and development of authentic tasks. Its possible uses are exemplified by two studies. One of the studies is an analysis of in what way and to what extent the applied tasks included in the national assessments in Finland and Sweden are, or are not, authentic. In the second example the framework is used in a study of the impact of authenticity on the students' sense making in word problem solving.

1. INTRODUCTION

There are several reasons for the inclusion of applications and modelling in mathematics education (for a review of the arguments presented in the literature on mathematics education see Blum & Niss, 1991). The reasons include the possibilities of the use of applications and modelling to (1) facilitate the learning of necessary skills for being able to use (and critically examine the use of) mathematics outside the mathematics classroom and (2) facilitate the development of an experience of school mathematics as useful and powerful for solving meaningful task situations in real life outside the mathematics classroom, which could provide motivation for and relevance of mathematical studies as well as facilitate the development of a comprehensive picture of mathematics that includes applications and modelling.

To attain the goals inherent in these reasons it is likely that some features of the applications may be crucial. For example, the learning of being able to use mathematics in real life situations beyond school would be facilitated by being confronted with tasks that require the competencies needed for solving problems in life outside school and that respect the features and conditions of out-of-school task situations. To acquire a picture of mathematics that includes applications and modelling, and to develop the experience that mathematics is useful and powerful in out-of-school situations for motivational reasons is probably facilitated by meeting and solving tasks that are experienced as important instances of applications in life beyond school. Thus, for the applications included in the learning environment to function as facilitators for attaining these goals it is likely that a significant proportion of them would have to be close simulations of important real life situations in which the application of mathematics can play an important role. In other words, it is likely that they would have to be *authentic*.

However, the applications in mathematics education, sometimes also called mathematical word problems, have been criticized on the basis that a too large proportion of them lack these important features, which is argued to have contributed to a lack of fulfilment of the described learning goals. Boaler (1994) and Cooper & Dunne (2000) argue that many word problems are pseudo-realistic and show how students are required to think differently than in out-of-school situations when solving such tasks. For example, Cooper & Dunne (2000, p.35) exemplify how students are required to make some considerations of the real situation described in a task, but not too much. Many studies have shown that it is common that students use solution strategies in their word problem solving that do not even include the consideration of the realities of the real world situations described in the word problems (for an overview of such studies see Verschaffel, Greer, & De Corte, 2000). According to Boaler (1994) and Nesher (1980) indeed the students have not formed the belief that their skills learned in school mathematics are useful in life beyond school. In addition, according to Gerofsky (1996) and Sowder (1989) students do not even like word problems.

Based on this introduction it seems that concerning the attainment of important goals of applications and modelling in mathematics education the concept of *authenticity* may play a significant role. But claims aimed at having an impact on mathematics education practice do best accompanied by empirical evidence convincingly showing its importance. It is often not enough to show that goals have not been attained by current education practice. Thus, for widespread acceptance and strong possibilities of influencing practice there is a need for a strong evidence base for claims of the impact (or non-impact) of the authenticity of mathematical applications on important educational goals. It should also be acknowledged that developing authentic mathematical applications, especially for higher levels of mathematics, is not unproblematic and require considerable reflection and time.

2. A FRAMEWORK FOR AUTHENTICITY

It should be noted that the term authenticity with respect to assessment and tasks is being used with very different meanings in the education literature. For a review of different meanings of this concept see (Palm, 2002). In this paper the term *authentic task* refers to a task in which the situation described in the task including a question or assignment, i.e. the *figurative context* (Clarke & Helme, 1998), is a situation from real life outside mathematics itself that has occurred or that might very well happen. In addition, the task situation is truthfully described and the conditions under which the task solving takes place in the real situation are simulated with some reasonable fidelity in the school situation.

For efficiency in the critical examination and development of tasks that mirror this definition, for the purposes of classroom instruction, textbook development and assessments as well as for research purposes, it is useful to operationalize such a definition into a more specified description of its meaning on a more fine-grained level. Such a description is provided in (Palm, 2002). The underlying idea behind this framework for authenticity is the construct of simulation. The constituents of the framework are a set of aspects of real life task situations, which can be simulated with more or less fidelity, and a discussion of the affective dimension in relation to the simulation of real life task situations. The aspects are included in the framework on the basis that a strong argumentation may be offered that their simulations can have an impact on the extent to which students, when dealing with word problems, may engage in the mathematical activities attributed to the real situations that are simulated. The proposed aspects are the Event, Question, Purpose, Information/data, Presentation, Solution strategies, Circumstances and Solution requirements. In addition, there are also subaspects appurtenant to some of the aspects. In the framework, the aspects are described, exemplified and argued for. In the discussion of the affective dimension the affective domain is described and its importance is stressed. It is also argued that significant engagement in the figurative contexts, and consistency between the students' actions in the school situations and in the situations that are simulated, are enhanced when the simulated situations are experienced as familiar and meaningful.

3. THE FRAMEWORK IN USE

3.1 An analysis of tasks

Thus, this framework for authenticity of mathematical applications may be used for the analysis and development of authentic tasks and may serve as a basis for a discussion of the constitution of school tasks that are intended to emulate out-of-school task situations. One example of its possible uses is the analysis of Finnish and Swedish national assessments for upper secondary school in which the framework was used as a tool for describing in what way and to what extent the applications in the assessments could be considered authentic or not (Palm & Burman, 2004).

The analysis showed, for example, that about 50 % of the applied tasks were considered to both describe an *event* that might occur in real life beyond school and include a *question* that really might be posed in that event. About 25 % of the tasks also possessed the quality that the *information /data* given in the task was the same as in some corresponding real life task situation to such a degree that it required the same mathematics for its solution as would have been required to solve the task in the simulated situation. When also the similarity in the *availability of external tools* between the assessment situation and the real-world task situations was considered, 20 % of the tasks were judged to simulate these four aspects with such fidelity that the same mathematics would be required by the assessment tasks and the simulated real world situations.

3.2 A study of the impact of authenticity

The framework may also be used for research on the impact of authenticity, which will be exemplified in this section. The background for the study is the conclusion drawn from a number of studies in the 80's and 90's that students from different countries from different parts of the world have tendencies to neglect an appropriate use of common sense knowledge of the world in their word problem solving and provide solutions that are inconsistent with the realities of the 'real' situations described in the tasks (Verschaffel et al., 2000). The reasons for these tendencies are argued to be the prevalent school culture in which the students' actions are situated.

However, studies have indicated that a difference in the working conditions, such as the requirement of making a telephone call to place an order (DeFranco & Curcio, 1997), having available concrete materials such as planks, a saw and a meter stick (Reusser & Stebler, 1997), or working with the tasks under the heading of another school subject (Säljö & Wyndhamn,

1993) can have a positive influence on the proportion of students providing solutions consistent with the realities of the figurative contexts.

However, such working conditions go beyond many classroom resources and may be difficult to uphold on a regular basis. Therefore, using the framework for authenticity (Palm, 2002) a study was designed to investigate the impact of task authenticity on students' use of real world knowledge in their solutions to word problems, with the restriction that a higher degree of authenticity has to be accomplished within the frames of the practicalities of normal classroom procedures (Palm, 2002).

Guided by the aspects of real life task situations included in the framework a higher degree of authenticity was achieved through a more thorough and true description of the 'real' situations described in the word problems used in earlier studies presented in the research literature (see e.g. Verschaffel et al., 2000). Two versions, one more authentic and one less authentic version, of six word problems divided in three different categories of word problems, were randomly administered to 160 students. The students provided written responses, but in addition, all students were interviewed to gather additional information. The results of the study show that authenticity, even under the restrictive constraints of normal classroom resources, can affect students' tendencies to effectively use their real world knowledge in the solutions to word problems. The students who were faced with the more authentic task variants both provided written solutions that were consistent with the realities of the 'real' situations described in the tasks and activated their knowledge of the 'real' situations, whether or not it affected their written solutions, in a significantly higher proportion of the tasks (it is to be noted however that even if the students working with the more authentic task variants provided a higher proportion of 'realistic' responses some of these more authentic tasks still yielded a large proportion of 'unrealistic' responses). It was also concluded from the interviews that the main reasons for providing solutions that are inconsistent with the situations described in the word problems were the students' frequent use of what may be called superficial solution strategies and their beliefs about mathematical word problem solving. These strategies and beliefs have been developed in an education including many encounters with pseudo-realistic tasks.

An example of a pair of word problems included in the study is Examples 1 and 2 (below). Example 2 is considered to simulate the aspects in the framework to a larger extent than Example 1. For example, the consideration in the task development of the aspect Purpose resulted in a clarification of the purpose of the task solving in Example 2 by making the students order the buses by filling in an ordering form. A clear purpose of the task solving would be known in a corresponding out-of-school situation and can be important since the appropriateness of the answer may be dependent on if the answer to the question is to be used directly to order the buses, or if it is to be used to form the basis of a decision on the number of buses to order, which may involve other considerations as well. In addition, a clear purpose may facilitate a focus on the figurative context instead of on the social context of a school situation. The inclusion of the order form is similar to the requirement of a telephone call in the study by DeFranco & Curcio (1997), but without the difficulties with practical resources.

Example 1:

360 students will go by bus on a school trip. Each bus can hold 48 students.

How many buses are required?

Example 2:

All students in the school will the 15th of May go on a school trip together. You have decided that everyone will go by bus, and that you shall order the buses. You have seen in the student namelists that there are 360 students in the school. Your teacher said that you can order the buses from Swebus, and that each bus can hold 48 students.

Fill in the note below, which you are going to send to Swebus to order the buses.

Swebus – Bus order

Your name:..
School:...
Date of the trip:..
Number of buses to order:...
Other requirements:...
....

Of the students who were faced with the task variant in example 1, 75 % provided the answer 8 buses or gave a realistic comment to why their answer made sense. Of the students who dealt with the more authentic task variant in example 2, 95 % provided the answer 8 buses, which was considered to be consistent with the described situation in the task.

4. SUMMARY AND CONCLUDING REMARKS

To sum up, there are important learning goals in mathematics education which attainments may be facilitated by the use of applications and by students' engagement in modelling activities. In addition, there are reasons to believe that the authenticity of these applications can facilitate the attainment of some of these goals.

However, it is of great significance that claims about the impact (or non-impact) of authenticity can be corroborated by empirical evidence. To assist in the development of authentic tasks, for both instructional and research purposes, a framework specifying the meaning of authentic tasks on a fine-grained level has been suggested. This framework can also be helpful in structuring the studies and the information gathered from them to achieve a more comprehensive picture of the impact of authenticity. The possible uses of the framework in research studies have been exemplified. Together with other research these studies contribute to our understanding of the use and impact of authentic mathematical applications.

However, the available body of research on the impact of authenticity is far from sufficient, which influences the practitioners working in mathematics education. Developing tasks that simulate important aspects of meaningful out-of-school task situations with high fidelity takes a great deal of time, effort and money. Today many teachers, textbook writers and professional assessment developers spend a lot of work trying to develop such tasks, motivated by the belief that their effort is worthwhile. Others do not share this belief and therefore not this direction in their work. Due to the lack of evidence decisions are many times forced to be based on assumptions. A more extensive body of research about the consequences of the authenticity of mathematical applications is needed so practitioners will have better possibilities to base their decisions about task development on empirical evidence grounded in scientific research. This would allow a more efficient use of available resources and better opportunities for developing efficient learning environments.

REFERENCES

Blum, W., & Niss, M. (1991). Applied mathematical problem solving, modelling, applications, and links to other subjects – State, trends and issues in mathematics instruction. *Educational Studies in Mathematics, 22*, 37-68.

Boaler, J. (1994). When Do Girls Prefer Football to Fashion? An Analysis of Female Underachievement in Relation to "Realistic" Mathematics Contexts. *British Educational Research Journal, 20*(5), 551-664.

Clarke, D. J., & Helme, S. (1998). Context as Construction. In O. Björkqvist (Ed.), *Mathematics Teaching from a Constructivist Point of View* (pp. 129-147). Vasa, Finland: Faculty of Education, Abo Akademi University.

Cooper, B., & Dunne, M. (2000). *Social Class, Sex and Problem-solving*. Buckingham Philadelphia: Open University Press.

DeFranco, T. C., & Curcio, F. R. (1997). A division problem with remainder embedded across two contexts: Children's solutions in restrictive vs. real-world settings. *Focus on Learning Problems in Mathematics, 19*(2), 58-72.

Gerofsky, S. (1996). A Linguistic and Narrative View of Word Problems in Mathematics Education. *For the Learning of Mathematics, 16*(2), 36-45.

Nesher, P. (1980). The Stereotyped Nature of School Word Problems. *For the Learning of Mathematics, 1*, 41-48.

Palm, T. (2002). *The realism of mathematical school tasks - Features and consequences.* Umeå University, Umeå, Sweden.

Palm, T., & Burman, L. (2004). Reality in mathematics assessment: An analysis of task-reality concordance in Finnish and Swedish national assessments. *Nordic Studies in Mathematics Education*, 9(3).

Reusser, K., & Stebler, R. (1997). *Realistic mathematical modeling through the solving of performance tasks.* Paper presented at the the Seventh European Conference on Learning and Instruction, Athens, Greece.

Sowder, L. (1989). Searching for Affect in the Solution of Story Problems in Mathematics. In D. B. McLeod & V. M. Adams (Eds.), *Affect and Mathematical Problem Solving: A New Perspective.* New York: Springer-Verlag.

Säljö, R., & Wyndhamn, J. (1993). Solving everyday problems in the formal setting: An empirical study of the school as context for thought. In S. Chaiklin & J. Lave (Eds.), *Understanding practice: Perspectives on activity and context* (pp. 327-342). Cambridge: Cambridge University Press.

Verschaffel, L., Greer, B., & De Corte, E. (2000). *Making sense of word problems.* Lisse, The Netherlands: Swets & Zeitlinger Publishers.

Chapter 3.2.4

ELEMENTARY MODELLING IN MATHEMATICS LESSONS: THE INTERPLAY BETWEEN "REAL-WORLD" KNOWLEDGE AND "MATHEMATICAL STRUCTURES"

Ralph Schwarzkopf
Department of Mathematics, IEEM, University of Dortmund, Germany,
Email: ralph.schwarzkopf@math.uni-dortmund.de

Abstract: Whereas many investigations analyse the strategies of students working on word problems in interviews and tests, empirical investigations on the associated classroom interactions are rarely represented in mathematics education. The goal of the present study is to understand interaction processes in mathematics lessons of early grades, dealing with modelling and solving word problems. This paper assumes theories of symbolic interaction to analyse these interaction processes within the qualitative research paradigm.

1. INTRODUCTION

Solving word problems surely demands the construction of an adequate interplay between the "real–world" and "mathematics" (Blum et al., 2002, p. 265). From a theoretical point of view, this interplay is often described as a "circle of modelling" (De Corte et al., 2000, p. 134), but empirical studies on the students' "problems with problems" (Säljö & Wyndhamn, 1997) have led to the conclusion, that many students do not follow this concept of solving word problems. Whereas early investigations on this topic concluded that the students´ solutions are surprisingly senseless, newer publications focus more and more on the "rationality" of the students' solving strategies, developed within the culture of mathematics classrooms (Reusser & Stebler, 1997; Säljö & Wyndhamn, 1997; Voigt, 1998). In other words: The students do not follow the logic of problem solving but they follow the logic of classroom culture. Hence, the interaction in mathematics lessons becomes an important focus of interest to understand the difficulties in word problem solving.

The present paper is based on first results of an empirical study that follows this discussion. The goal of the study is a better understanding of how "components of modelling" (Blum et al., 2002, p. 270), are involved in processes of solving word problems within everyday classroom interaction. How can we describe from a theoretical point of view the interplay between the real-world and mathematics, realised in everyday mathematics classroom interaction?

2. THEORETICAL FRAMEWORK

The author follows the interpretative paradigm, focusing in this paper on theories of symbolic interactionism and ethnomethodology to analyse the organisation of meanings in interaction processes (Yackel, 2000).

2.1 Aspects of Symbolic Interactionism: Framings

Within classroom interaction, the participants construct their individual sense of the content of the interaction process by participation. They do this within their *framing of the situation* (Krummheuer, 1992). Roughly speaking, the framing of the situation gives a context for the individual to interpret the interaction, and is responsible for his/her decisions about rational acting.

Solving word problems demands an interplay between at least two very different framings, namely "everyday-understandings" of the problem's real-world context and "mathematical" framings. This interplay is complex in nature, because within different framings the participants are acting in different ways concerning the relevance of facts, the meaning of assertions, the acceptance of statements, rules for correct reasoning and many other aspects. Hence, solving word problems demands framings that are not only different, but often contradictory concerning their rules for rational acting. One goal of the present investigation is to reconstruct the changes of rationality within processes of solving word problems in teaching and learning situations. In other words to ask: How do the participants combine their knowledge(s) that are constructed within different framings?

The author will argue, that the wide-spread idea of "translation" between the language of the real-world and the language of mathematics is not adequate to describe the complexity of solving word problems within mathematical teaching and learning processes. For this, it might be helpful to give a short example from the empirical conduct of the author's study.

3. AN EXAMPLE FROM A FOURTH GRADE CLASS

In the following, the question presented shall be illustrated by analysing a process of word problem solving. The data are taken from the regular mathematics lessons of a fourth grade class of primary school (10 year old) students in Germany. Within this lesson, the teacher confronts the students with the following task, taken from their mathematics textbook (see Fig. 3.2.4-1).

On the motorway from Gießen to Dortmund:

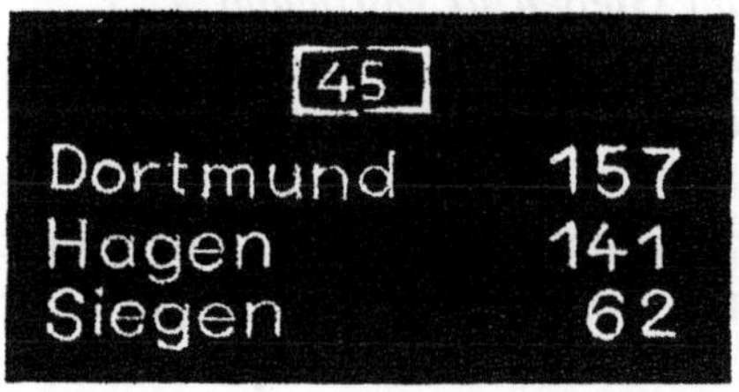

On further signs there are the kilometres information:

Dortmund	151	143	135	132
Hagen	135	127	119	116
Siegen	56	48	40	37

Calculate in all cases the difference

a) Between Dortmund and Hagen
b) Between Dortmund and Siegen.

Figure 3.2.4-1. On the motorway

At the beginning of the lesson, the participants discuss some real-world aspects, giving a context for the task (first episode). Afterwards, the teacher makes a sketch of the motorway and the given signs on the blackboard. The students then do the calculation and observe that they lead to two results only. At the end of the lesson, the teacher wants the children to explain this constancy (second episode).

From the expert's point of view, this task may be very easy. The children would only have to "translate" from mathematics into the rest of the world, understanding the calculated *differences* between the given numbers to be the *distances* between the assigned cities. But, taking processes of teaching and learning in classrooms seriously, it is important that "... social interaction is a process that *forms* human conduct, rather than simply a setting in which human conduct takes place" (Yackel, 2001, p. 11). Hence, from the author's point of view, the complexity of solving word problems originates from the classroom interaction and not from the "internal" structure of the task. Within the discussed lesson for example, the author reconstructed at least five different framings, even if only two of them can be illustrated in the present paper. All of these framings are self-dynamical, i.e. they draw very different and contradictory understandings of the task. The framings involve more than simply different languages, and the organisation of the solution process can't be done by translations. In other words the word problem becomes complex due to the process of solving within the interaction process. Taking the process of teaching and learning itself as the matter of research, the author proposes a theoretical point of view to understand this complexity of solving word problems within classroom interaction.

3.1 First Episode

The teacher asks the students to read the word problem. Afterwards, the children are asked to discuss their first impressions. This is the answer of Werner:

> *W* *They are on the motorway from Gießen to Dortmund. [...] And then they see this blue sign. And then that there are still 157 kilometres left to Dortmund.*

Werner tells a short story about some people ("they") who drive on the motorway and read the given sign of the word task. These people use the sign in a way that is typical for acting in the real-world: they read the distance on the sign, to find out how many kilometres they will still have to drive until they reach Dortmund. Obviously, the required calculation of the word problem is not relevant within this story: Who would subtract in everyday life the numbers given on signs on the motorway?

A few seconds later, the teacher picks up the story of Werner as follows:

> *T* *Okay. Werner found out, where the car drivers are. He already told us, but some of you were talking so I would like to hear it again. Where are they driving, or if you are in the car, where are you driving? [...]*

By her reaction, the teacher shows herself satisfied with the story of Werner as an adequate rationality for interpreting the task. Furthermore, she wants the children to imagine that *they* are the drivers on the motorway. Hence, the context-giving story of Werner becomes an officially required framing through the reaction of the teacher. In the ongoing lesson, the teacher tries several times to re-activate this "story-bound-framing".

3.2 Second Episode

This episode deals with an explanation for the observed constant calculation results. In advance, a sketch was made on the blackboard and the children solved the calculations. It is already clear that there are only two results for the calculation, and this observation is marked within the sketch by vertical lines in some of the signs (see Fig. 3.2.4-2).

The sketch shows a part of the motorway and all of the signs given in the word problem. Some arrows painted below the motorway illustrate the direction of "the" car on its way from Gießen to Dortmund.

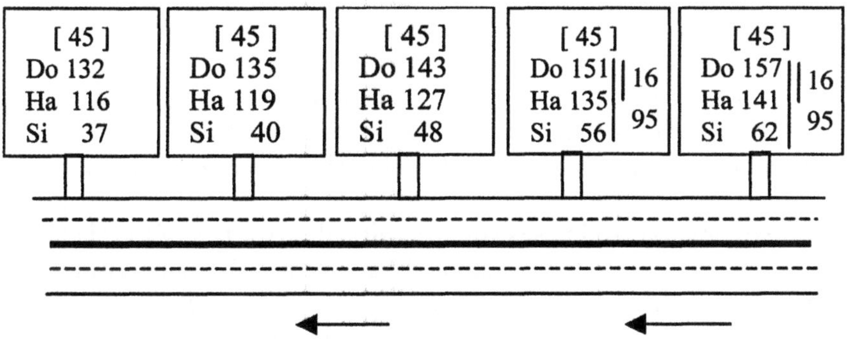

Figure 3.2.4-2. The way from Gießen to Dortmund

This sketch is compatible with the story-bound-framing of the word problem: one can imagine cars passing the signs and the drivers reading them to inform themselves about the distance left to reach Dortmund. But regarding the sketch concerning the observed phenomenon, it is rarely helpful, for the positions of the cities are not illustrated. Hence, the real-world meaning of the results, namely the distances between the cities, cannot be found in the sketch. Nevertheless, the sketch becomes important for the ongoing lesson.

At the beginning of this episode, the teacher demands reasons for the constant results:

T *[...] Is there any reason for this? We are driving on the motorway and see the blue signs. It has to be like this. Just look: You are driving by car (points along the motorway on the blackboard) read the signs and if you calculate this then the difference between these cities (points at "Do" and "Ha" on the most right sign) noted on the signs is equal, why?*

On the one hand, she stresses the story-bound-framing as an adequate point of view to produce arguments. Pointing at the sketch is compatible with the rationality of the story – one can imagine passing and reading the signs while driving on the motorway. On the other hand, the teacher stresses the calculations and their constant results. This seems to be strange within the required framing, because the story – probably bound to the real-world experiences of the children – doesn't give any motivation for the drivers to calculate the differences. Briefly speaking, the rationality of the real-world has changed, and the students have to create a new framing, with new, yet unknown rules for rational acting, that may provide rationality for the calculations.

However, the student Anja is able to construct a relation between the calculated results and the given sketch in the following way:

> *A [...] Well I think the signs are all standing in the same distance to each other and if they are standing in this same distance then the numbers are of that kind that it is exactly equal.*

In her explanation, Anja uses some empirical properties of the sketch – in fact, the signs painted on the blackboard have nearly the same distances, so she fulfils the teacher's demand concerning the logic of classroom culture. Anja uses the sketch to build an alternative sense for the numbers given in the signs on the motorway. According to this interpretation, one could name her underlying framing a "sketch-bound" one. The rationality within this framing is different from the children's experiences within the real-world, but it is compatible with the children's social experiences of demands of the teacher – if the teacher stresses the importance of the sketch, the sketch must show the required reason.

The lesson ends with an adequate reasoning, based on more interventions by the teacher. Due to space restrictions of this paper, these episodes can't be discussed here.

4. THE INTERPLAY BETWEEN THE REAL-WORLD AND MATHEMATICS

In the above episodes, two framings, leading to different horizons of understanding of the word problem, can be reconstructed:

The *story-bound-framing* leads to a realistic understanding of the given information, but not to an adequate interpretation of the calculation task.

Within the *sketch-bound-framing* it is possible to interpret the mathematical differences as distances in the real-world. But the empirical properties of the sketch are taken as reality and the interpretation leads to distances between the signs.

Within both framings, the real-world concept "distance" is the main concept needed to understand the context of the word problem. But the framings differ in their rational acting with this concept. Within the story-bound-framing, there is no need to calculate distances – one can read the interesting distances from the signs. Within the sketch-bound-framing, the only important distances are given by the empirical properties of the sketch on the blackboard. The interaction process becomes complex, because both rationalities are very strong on the one hand, but on the other hand they are not compatible with the mathematical background of the word problem.

Although the reconstructed framings lead to different understandings of the word problem, they also have something in common: Within the framings, the word problem is somehow *empirically enriched*. For example, the story is about people driving on the motorway and the sketch allows us to imagine the motorway itself – this context is constructed within the interaction process and is not given by the task. But this empirical enrichment doesn't lead to a structural understanding of the task, because neither the story nor the sketch focuses on the distances that are important for understanding the calculations – there is no rationality to focus on the distances between the cities. In other words the empirical enrichment *restricts the mathematical structure* of the word problem.

From the author's point of view, the contrary would be necessary for an adequate interplay between mathematics and the real-world: The main demand to solve a word problem is to find a balance between *empirically restricting* the "visible" real-world to mathematically relevant aspects, and at the same time to *structurally enrich* the real-world by "invisible" mathematical structures. For this structural enrichment, problem solvers often have to change their understanding of the word problem's real-world context. In the example given above, the students' real-world experience (driving on the motorway and reading the signs) is not helpful, because it leads to the *changing distances* between the car and the cities, assigned by the motorway-signs. But from a mathematical point of view, explaining the constant differences, the numbers on the signs divide the motorway into the *static distances* between the signs and the cities. This meaning of the signs differs from their meaning in the children's real-world experiences. Hence, the participants can't find an adequate reason for the constant results by translating their knowledge from the real-world to mathematics because, roughly speaking, the intention of a translation is to replace words of one language by words from another language, while trying to reproduce the original meaning. That is a translator searches for different words with the same meaning, but the demand for the children is contrary to that intention, for they would have to change the meanings of the numbers on the signs. Using language of theories of symbolic interaction: the students have to *modulate their framings* (Krummheuer, 1992) in order to find a new structural meaning for the numbers on the signs, changing their empirical, story-bound rationality.

Briefly speaking, the ongoing project works on these modulations of framings that allow the construction of an adequate interplay between the real-world and mathematics. Do teachers and students manage to find a balance between empirically restricting and structurally enriching the context of word problems, and if they do, how can we understand these processes from a theoretical point of view?

REFERENCES:

Blum, W. et al. (2002). ICMI Study 14: Application and Modelling in Mathematics Education – Discussion Document. *Educational Studies in Mathematics, 51*, 149-171.

De Corte, E., Greer, B., & Verschaffel, L. (2000). *Making Sense of Word Problems.* Lisse: Swets and Zeitlinger.

Krummheuer, G. (1992). *Lernen mit „Format".* Weinheim: Deutscher Studien Verlag.

Reusser, K., & Stebler, R. (1997). Every Word Problem has a Solution – the Social Rationality of Mathematical Modeling in Schools. *Learning and Instruction, 7*(4), 309-327.

Säljö, R., & Wyndhamn, J. (1997). Word Problems and Mathematical Reasoning – A Study of Children's Mastery of References and Meaning in Textual Realities. *Learning and Instruction, 7*(4), 361-382.

Voigt, J. (1998). Negotiating the Mathematical Meaning of Empirical Phenomena. In F. Seeger, J. Voigt, & U. Waschescio (Eds.), *The Culture of the Mathematics Classroom* (pp. 191-220). Cambridge: University Press.

Yackel, E. (2000): Explanation, Justification and Argumentation in Mathematics Classrooms. In M. van den Heuvel-Panhuizen (Ed.), *Proceedings of the 25th International Conference for the Psychology of Mathematics Education* (Vol. I, pp. 9-24). Utrecht, The Netherlands: Freudenthal Institute.

Section 3.3

MODELLING COMPETENCIES

Edited by
Brian Greer and Lieven Verschaffel

Chapter 3.3.0

MODELLING COMPETENCIES – OVERVIEW

Brian Greer and Lieven Verschaffel[1]
Graduate School of Education, Portland State University, Email: brian1060ne@yahoo.com / University of Leuven, Belgium, Email: lieven.verschaffel@ped.kuleuven.be

Abstract: Descriptions of levels of modelling activity ranging from basic technical skills to philosophical, and even ethical, considerations have been developed. These analytical efforts serve to provide guidelines for the teaching/learning and assessment of modelling.

In this overview, we consider how modelling competencies may be characterized, using a framework of three levels of modelling activity that we label implicit modelling (in which the student is essentially modelling without being aware of it), explicit modelling (in which attention is drawn to the modelling process), and critical modelling (whereby the roles of modelling within mathematics and science, and within society, are critically examined).

1. IMPLICIT MODELLING

As pointed out by Usiskin (this volume, Chapter 3.3.5), much of what is done in standard mathematics curricula, even at the elementary stage, can be characterized as modelling even though it is not acknowledged as such.

1.1 Arithmetical operations as models

The core of the modelling process comprises setting up a correspondence between some aspect of a real-world situation and a mathematical structure, carrying out appropriately motivated operations within that structure, and interpreting the results of those operations back in the real-world context. This frame is applicable, in principle, to the solution of the simplest of word problems involving a single operation. A judgment is required whether or not the operation provides an appropriate model for the situation described. This fundamental insight is masked in typical traditional teaching which omits instructive counterexamples where the superficially appropriate operation turns out, on deeper reflection, to be inappropriate (for several examples, see Usiskin, this volume, Chapter 3.3.5). As a result, when researchers have posed such counterexamples to students, the dominant response is to apply the superficially indicated operation as an implicit model that is inappropriate (see Verschaffel et al., 2000 for a comprehensive review of this research).

The damaging effects of concealing from students the nature of what they are doing, by (a) ignoring complexity without motivating this simplification, and (b) not using counterexamples, can be seen particularly strikingly in the documentation of the "illusion of linearity" (De Bock et al., this volume, Chapter 3.3.3). Similar remarks apply throughout the curriculum (Usiskin, this volume, Chapter 3.3.5). At the most extreme, the teaching of probability is particularly ill-served by pretending that the world organizes itself cleanly in alignment with the elements of probabilistic modelling, such as equiprobable events and statistical independence.

In general, modelling demands a balance between idealized simplification and precision (Singer, this volume, Chapter 3.3.2), which can be seen very clearly, for example, in the application of geometry to real-life problems. There is a fundamental difference, of course, between the student who ignores the complexity of real-world situations through blind conformity with the rules of a didactical contract, and the mathematician or scientist who mindfully simplifies in the course of modelling.

1.2 Competencies for implicit modelling

The set of competencies expected within the view of the curriculum that leaves modelling implicit, and how it is typically implemented, corresponds to what Hatano has labelled "routine expertise", defined as "simply being able to complete school mathematics exercises quickly and accurately without (much) understanding" (Hatano, 2003, p. xi). It is, of course, desirable that students should be able to associate a wide range of situations with the arithmetical schemes relating to the basic operations and proportionality (Usiskin, this volume, Chapter 3.3.5). However, such a program clearly does not go far enough. The pervasive "illusion of linearity" (De Bock et al., this volume, Chapter 3.3.3) may be interpreted as the result of students achieving routine expertise in the solution of stereotyped problems involving proportionality. Their "expertise" is routine to the extreme extent that it is clear that even their apparent understanding reflected in correct answers on such problems is illusory, since it cannot withstand slight perturbation.

2. EXPLICIT MODELLING

The deliberate introduction of examples of models, together with the concepts and terminology of modelling as a generic process, have become part of mainstream curricula in many countries (see the series of books emanating from the biennial International Conferences on the Teaching of Mathematical Modelling and Applications (ICTMA), Part 6 of this volume). The core description of the modelling process, mentioned above, has been elaborated in a variety of versions such as that presented by Blomhøj and Jensen (this volume, Chapter 2.2) (and see Part 1 of this volume).

2.1 Competencies for explicit modelling

Houston (this volume, Chapter 3.3.4) describes a number of very detailed and explicit analyses of the phases of modelling that provide useful frameworks for teaching and assessment. He argues for the power of such schemes for providing detailed profiles of students' strengths and weaknesses across the many subprocesses of the modelling process. Clearly these schemes are useful for constructing targeted assessment items.

As is the case generally, it is easy to assess aspects of modelling that can be reduced to routine expertise. However, as pointed out by De Bock et al. (this volume, Chapter 3.3.3) modelling, by its nature, implies adaptive expertise. Moreover, as acknowledged by Houston, modelling – as an authentic part of mathematical practice – is typically a group activity. In general, we may suggest that the elements of modelling that are inherently social, such as the goals of the modelling, the resources available in terms of other people, tools, and distributed information, debate about alternative models, and communication of an interpretation of the model to a target audience lie outside the scope of standard forms of assessment which are closed "in terms of time, in terms of information, in terms of activity, in terms of social interaction, in terms of communication" (Verschaffel et al., 2000, p. 72).

By contrast, there have been notable examples of attempts to assess and teach modelling expertise in ways that transcend those limitations – typically, with assessment closely integrated with the teaching. As a pioneering example dating from the 1980s, in a small-scale and rather short-lived project in Northern Ireland (in which both Houston and Greer were involved), students were able to opt for an alternative A-level course in mathematics which included a year-long project under the supervision of their teacher that could be either an investigation in pure mathematics or a modelling exercise (Greer and McCartney, 1989). Assessment was carried out holistically on the basis of the report written by the student. Other exemplary endeavors include the series of modules developed by the Shell Centre at Nottingham (Swan et al., 1987 – 9/2000), the very extensive work of Lesh and others associated with him (e.g. Lesh & Doerr, 2003), and some important design experiments (e.g. Verschaffel et al., 1999, and see Verschaffel et al., 2000, Chapter 6 for other examples).

3. CRITICAL MODELLING

3.1 The place of modelling within the mathematics curriculum

As strongly argued by Blomhøj and Jensen (this volume, Chapter 2.2), any adequate description of what it means to be mathematically competent must include modelling competency. Efforts to include modelling as a core part of mathematical curricula have succeeded in some countries, not without problems and resistance (De Lange, 1996; Keitel, 1993). From our point of view, modelling is the link between the "two faces" of mathematics, namely as a means of describing aspects of the physical and social worlds, and as a set of autonomous formal structures. Understanding this relationship is fundamental to the understanding of mathematics, unless it is to be reduced to a formal game like chess.

Singer (this volume, Chapter 3.3.2) discusses how developments of the twentieth century have forced reconsideration of the nature of mathematical descriptions of our physical and mental worlds. As she comments (p. 234) "All we can do ... is model". Clearly, it would be unrealistic to introduce students too early to the philosophical complexity of these arguments. Nevertheless, it is not unrealistic (as Usiskin (this volume, Chapter 3.3.5) argues) to lay the groundwork early for understanding fundamental principles of modelling and development of a modelling disposition.

3.2 Teaching about the role of modelling in society

If modelling of social, as well as physical, phenomena, is accepted as a core part of learning mathematics, it has major benefits and ramifications. In particular, as argued in our plenary talk (Verschaffel, Mukhopadhyay, and Greer, this volume, Chapter 2.6) it potentiates authentic links between school mathematics and the lived experience of students. Growing recognition of the societal importance of mathematics education is reflected in recent policy statements such as that of the Programme for International Student Assessment (PISA) (Henning and Keune, this volume, Chapter 3.3.1, and see Part 1).

There are fundamental issues in the use of modelling for social phenomena. As Jablonka (2003) points out, any such use must be qualified by consideration of the diversity of people's lives, their practices, experiences, goals – in general, culture. Skovsmose uses the term "the formatting power of mathematics" to refer to the subtle, and mostly undetected, ways in which mathematics does not just reflect our view of the world, but also helps to shape it, so that "when part of reality becomes modelled and remodelled, then this process also influences reality itself" (Skovsmose, 2000, p. 5). Most people uncritically "consume" the products of mathematical modelling, without having any understanding of the models used, the assumptions on which they are based, or the general concept of modelling itself.

3.3 Competencies for critical modelling

Blomhøj and Jensen (this volume, Chapter 2.2) provide a very comprehensive view of what mathematical competency entails, in which modelling competency is one component (others particularly relevant to modelling are communicating and tool use). They emphasize the importance of developing a critical attitude towards all parts of the modelling process. Moreover, as they point out, mathematical competency further entails a judgmental conception of the application of mathematics, its historical development, and the nature of mathematics as a subject area.

Henning and Keune (this volume, Chapter 3.3.1) discuss some recent PISA examples in which attempts are made to probe students' ability to critique the models under discussion. Arguably, a major failing of mathematics education has been that people in general are unaware of the nature and assumptions of models that affect their lives.

4. CONCLUSIONS

Modelling should be developed through a coherent curricular strand, beginning in the earliest years, and recognizing its crucial role in the development of an appropriate mathematical disposition (Verschaffel, 2002). As such, the development of modelling competencies is too complex to be fitted within a simple model of sequential stages. Several of the themes visited above testify to this complexity, and to the dialectical nature of tool use in linking experience to formal mathematics through a variety of activities within designed environments.

By its nature, modelling demands adaptive expertise and is typically a social activity, and its vulnerability to impoverished forms of teaching and assessment that do not reflect these core aspects should be resisted. Moreover, modelling is often situated in social and political contexts and learning to model should go beyond the merely technical aspects to address its human purposes (Blomhøj and Jensen, this volume, Chapter 2.2; Mukhopadhyay and Greer, 2001; Verschaffel et al., this volume, Chapter 2.6).

REFERENCES

De Lange, J. (1996). Using and applying mathematics in education. In A.J. Bishop, K. Clements, C. Keitel, J. Kilpatrick, & C. Laborde (Eds.), *International handbook of mathematics education*, Vol. 1 (pp. 49-97). Dordrecht, The Netherlands: Kluwer.

Greer, B., & McCartney, R. (1989). The Further Maths project: A response to Cockcroft at the sixth-form level. In B. Greer, & G. Mulhern (Eds.), *New directions in mathematics education* (pp. 131-148). London: Routledge.

Hatano, G. (2003). Foreword. In A.J. Baroody, & A. Dowker (Eds.), *The development of arithmetic concepts and skills* (pp. xi-xii). Mahwah, NJ: Lawrence Erlbaum Associates.

Jablonka, E. (2003). Mathematical literacy. In A.J. Bishop, M.A. Clements, C. Keitel, J. Kilpatrick, & F.K.S. Leung (Eds.), *Second international handbook of mathematics education* (pp. 77-104). Dordrecht, The Netherlands: Kluwer.

Keitel, C. (1993). Implicit mathematical models in social practice and explicit mathematics teaching by applications. In J. de Lange, C. Keitel, I. Huntley, & M. Niss (Eds.), *Innovation in mathematics education by modelling and applications* (pp. 19-31). Chichester, England: Ellis Horwood.

Lesh, R., & Doerr, H.M. (2003). *Beyond constructivism. Models and modelling perspectives on mathematics problem solving, learning, and teaching*. Mahwah, NJ Lawrence: Erlbaum Associates.

Mukhopadhyay, S., & Greer, B. (2001). Modeling with purpose: Mathematics as a critical tool. In B. Atweh, H. Forgasz, & B. Nebres (Eds.), *Socio-cultural research on mathematics education* (pp. 295-311). Mahwah, NJ: Lawrence Erlbaum Associates.

Skovsmose, O. (2000). Aporism and critical mathematics education. *For the Learning of Mathematics*, *20*(1), 2-8.

Swan, M., Binns, B., Gillespie, J., & Burkhardt, H. (1987-9/2000). *Numeracy through problem solving: Five modules for curriculum and assessment in quantitative literacy*. Nottingham, London: Longman, and Shell Centre Publications (www.mathshell.com).

Verschaffel, L. (2002). Taking the modelling perspective seriously at the elementary school level: Promises and pitfalls (Plenary Lecture). In A.D. Cockburn, & E. Nardi (Eds.), *Proceedings of the 26th Annual Meeting of the International Group for the Psychology of Mathematics Education*, Vol. 1 (pp. 64-80). Norwich, England: University of East Anglia.

Verschaffel, L., De Corte, E., Lasure, S., Van Vaerenbergh, G., Bogaerts, H., & Ratinckx, E. (1999). Design and evaluation of a learning environment for mathematical modelling and problem solving in upper elementary school children. *Mathematical Thinking and Learning*, 1, 195-229.

Verschaffel, L., Greer, B., & De Corte, E. (2000). *Making sense of word problems*. Lisse, The Netherlands: Swets and Zeitlinger.

---

[1] We would like to thank the members of the Working Group on Modelling Competencies for their contributions: Werner Blum, Cinzia Bonotto, Rita Borromeo Ferri, Rosalind Crouch, Dirk De Bock, Helen Doerr, Solomon Garfunkel, Ken Houston, Toshikazu Ikeda, Mike Keune, Jerry Legé, Vimolan Mudaly, Swapna Mukhopadhyay, Jacob Perrenet, Mihaela Singer, Gloria Stillman, Zal Usiskin. Particular thanks are due to Swapna Mukhopadhyay for acting as recorder.

Chapter 3.3.1

LEVELS OF MODELLING COMPETENCIES

Herbert Henning and Mike Keune
Department of Mathematics, Otto-von-Guericke Universität Magdeburg, Germany, Email: herbert.henning@mathematik.uni-magdeburg.de / Ökumenisches Domgymnasium Magdeburg, Germany, Email: keune@domgymnasium-magdeburg.de

Abstract: The concept of a competence-oriented approach towards modelling is examined and a level-oriented description of modelling competencies introduced. The characteristic abilities associated with each level are listed and some illustrative examples are provided. The level-oriented description is related to the concept of mathematical literacy and briefly compared with other descriptions of modelling competence.

1. COMPETENCE-ORIENTED APPROACH

In this paper we follow Weinert's (2001) definition of competence as the sum of available or learnable abilities and skills together with willingness to solve upcoming problems and to act responsibly and critically concerning the solution. Klieme et al. (2003), reporting on the development of national educational standards in the Federal Republic of Germany, suggest that we expect education, through the learning processes involved, to provide individuals with the abilities necessary to act independently and responsibly in society.

If you look at the teaching and learning of modelling there are at least two possible approaches. One approach aims at describing necessary abilities, skills and attitudes of students. Results gained from this approach are called *component descriptions*. On the other hand, the examination of competencies in terms of complexity of modelling processes results in *level descriptions*. Klieme et al. (2003, p. 61) call these two types of description "Komponentenmodelle" and "Stufenmodelle". Here we follow these distinctions between a list of abilities, skills and attitudes (components) and the examination of different levels of these abilities, skills and attitudes, considering these two perspectives as complementary means for the description of modelling competencies.

2. COMPONENT-ORIENT DESCRIPTIONS OF MODELLING COMPETENCIES

Following a definition of the term modelling competences by Maaß (2004), we include in the term modelling competences those abilities, skills, attitudes that are important for the modelling process and the willingness of students to deploy them. Similarly, Blum (2002) defined modelling competence as the ability to structure, mathematize, interpret and solve problems and, in addition, the ability to work with mathematical models, validate the models, analyze them critically and assess models and their results, communicate the models and observe and self-adjustingly control the modelling process.

3. LEVEL-ORIENTED DESCRIPTION OF MODELLING COMPETENCIES

Here we introduce a level-oriented description of the development of modelling competencies, characterized in three levels:

Level 1: Recognition and understanding of modelling
Level 2: Independent modelling
Level 3: Meta-reflection on modelling

Competence, as a theoretical construct, cannot be observed directly. One can only observe students' behaviour and actions as they solve problems, for example. Competence is understood here as a measurable variable, in the sense that level of competence can be inferred by observing the behaviour of students.

In a pilot study (Henning & Keune, 2004; Henning et al., 2004; Keune et al., 2004) students' behaviour was observed as they worked on modelling problems, with the goal of reaching conclusions concerning their levels of modelling competencies. The theoretical assumption here was that at the first level procedures and methods can be recognized and understood, as a prerequisite to being able to independently solve problems at the second level.

Conscious solving of problems in the sense of this paper requires, accordingly, knowledge of the procedure. Furthermore, the authors make the assumption that meta-reflection on modelling would at the very least require both familiarity with modelling and personal experience.

Within this perspective the levels of modelling competencies could be considered as one dimension of at least three dimensions in which a modelling activity takes place, the other two being level of complexity (contexts, methods, technical skills), and educational level.

4. CHARACTERISTIC ABILITIES

Level 1 – Recognize and understand modelling

Characterized by the abilities to recognize and describe the modelling process, and to characterize, distinguish, and localize phases of the modelling process.

Level 2 – Independent modelling

Characterized by the abilities to analyze and structure problems, abstract quantities, adopt different perspectives, set up mathematical models, work on models, interpret results and statements of models, and validate models and the whole process.

Pupils who have reached this second level are able to solve a problem independently. Whenever the context or scope of the problem changes, then pupils must be able to adapt their model or to develop new solution procedures in order to accommodate the new set of circumstances that they are facing.

Level 3 – Meta-reflection on modelling

Characterized by the abilities to critically analyze modelling, formulate the criteria of model evaluation, reflect on the purposes of modelling, and reflect on the application of mathematics.

At this third level of competency, the overall concept of modelling is well understood. Furthermore, the ability to critically judge and recognize significant relationships has been developed. Consideration concerning the part played by models within various scientific areas of endeavour as well as their utilization in science in general is present. This implies that finished models are examined and any inferences drawn from them evaluated (Jablonka, 1996), while at the same time criteria for model evaluation are scrutinized (Henning & Keune, 2002).

5. MATHEMATICAL LITERACY AND MODELLING COMPETENCIES

The Programme for International Student Assessment (PISA) gives a precise definition of the term mathematical literacy as "an individual's capacity to identify and understand the role that mathematics plays in the world, to make well-founded mathematical judgements and to engage in mathematics, in ways that meet the needs of that individual's current and future life as a constructive, concerned and reflective citizen." (Organization for Economic Co-operation and Development (OECD), 1999, p. 41). The concept of mathematical literacy connects the development of mathematical structures with the treatment of realistic tasks. This connection can be considered as analyzing, assimilating, interpreting and validating a problem – in short, modelling. Within this perspective modelling competencies form a part of mathematical literacy and the examination of modelling competencies are helpful in clarifying the mathematical literacy of students.

For example, in the work of Haines et al. (2001) a component-oriented approach to modelling skills is applied. They distinguish between modelling competences and skills based on the phases of the modelling process, which also affords a framework for assessment (Houston, this volume, Chapter 3.1.4).

Based on the work of Niss (1999, 2003), Blomhøj & Jensen (this volume, Chapter 2.2) characterize modelling competences within three dimensions. According to them, the competences acquired by students can vary in terms of "technical level", "radius of action" and "degree of coverage".

Our level-oriented description of modelling can be considered as another perspective on modelling competencies. It can be used as a descriptive, normative and meta-cognitive aid when assessing student performance, planning lessons, and selecting teaching contents.

6. EXAMPLES

The following three examples for assessing level of modelling competencies are based on PISA study examples (OECD, 2003) which have been reformulated.

WATERTANK

During a math class students are asked to describe a watertank as it is filled. The tank is one meter wide, empty at the beginning and is filled with one liter of water per second. The students receive further informations from the teacher as to the shape and measurements of the tank.

Here you see one student's results. He sketched the tank of water and depicted in a graph how the water-level changed over time.

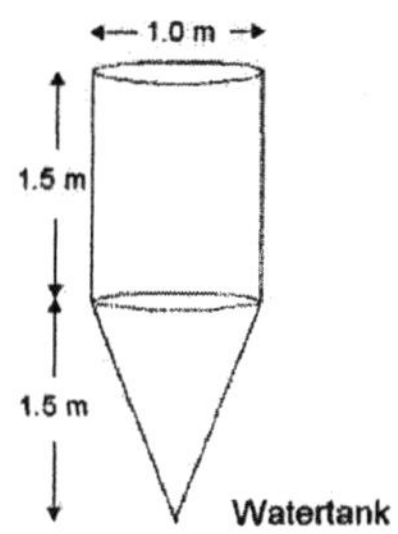

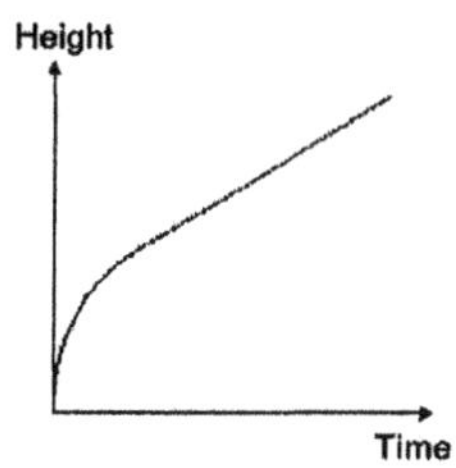

A1) How could the student have established the course of the graph?

A2) Are there other informations which the student did not use?

The teacher judges that the results so far are good and encourages the student to find a formula for calculating the water-level.

A3) What steps would the student have to take in order to set up a formula for calculating the water-level?

Figure 3.3.1-1. Watertank

While solving the water tank problem the students have to demonstrate their ability to recognize that the water tank as depicted is a compound object, that material thickness does not play a role in the solution of the problem, that a qualitative graphical model is used and that the quantitative data given are not used in the model. These are abilities situated at level one in terms of our description.

The second example is appropriate for assessing competencies from the second level (set up and work with models).

SCHOOL PARTY

It has been announced that a famous band is going to play in the gym at a school party in our school. Almost all the students from your school and many students from neighboring schools would like to come to the concert. From the organizers of the party you receive the task of calculating the maximium possible number of spectators for the gym.

B1) Plan how you will proceed with solving the problem and write out the steps needed for the solution.

B2) Complete the task which the organizers gave you. If any details are missing, figure them out by estimating.

The organizers would like you to show your work to the heads of the school in a short presentation.

B3) Make up a sheet of key points which you would like to tell the heads of the school.

Figure 3.3.1-2. School party

The third example is based on the PISA study problem entitled: "Rising Crimes" and has been reformulated to assess competencies at level three.

ALARM SYSTEMS

Every year the police record statistics of the number of house-burglaries in their city. From these statistics a manufacturer of alarm systems has picked out the following years.

year	1960	1965	1970	1975	1980	1984
number of crimes	110	200	330	480	590	550

The manufacturer has used this data to make the following statement in his advertisements: *Every 10 years the number of burglaries doubles or tribles! Buy an alarm system now before your house is robbed too!*

C1) Is the first sentence of the advertising slogan correct? Support your answer.

C2) Why could the manufacturer have specifically chosen this data

Imagine that your parents work for the police and tell you that the police aren't going to record these statistics in the future.

C4) Explaine briefly the advantages/disadvantages of this type of statistics.

Figure 3.3.1-3. Alarm systems

In this problem pupils are asked to demonstrate their ability to reflect critically on the modelling process and its use in a real world application. Furthermore, they are tested on their ability to evaluate the use of models in general. When considering models and the modelling process, one must be incessantly aware of the possible misuse of mathematics, as well as the social relevance of models, their interpretations, and the predictions that they can make.

7. CONCLUSION

A level-oriented description of modelling competencies has been presented and compared with other descriptions of modelling competencies, and it has been put into the framework of mathematical literacy. Important issues for further research are the examination of the level-orientated description at different educational levels and the role of the context of the modelling tasks.

REFERENCES

Blomhøj, M., & Jensen, T. H. (2006). What's all the fuss about Competencies. This volume, Chapter 2.2.

Blum, W. et al. (2002). ICMI Study 14: Applications and modelling in mathematics education – Discussion document. *Educational Studies in Mathematics 51*(1-2), 149-171.

Haines, C.R., Crouch, R.M., & Davis, J. (2001). Understanding students' modelling skills. In J.F. Matos, W. Blum, K. Houston, & S.P. Carreira (Eds.), *Modelling and Mathematics Education: ICTMA9 Applications in Science and Technology* (pp. 366-381). Chichester: Ellis Horwood.

Henning, H., Hartfeld, C., Kubitza, T., Hammermeister, S., & Keune, M. (2004). *Anwendungen – Modelle –Lösungen*. Technical Report Nr. 2. Magdeburg, Germany: Otto-von-Guericke-Universität.

Henning, H., & Keune, M. (2002). Modelling and Spreadsheet Calculation. In Vakalis, I., Hallett, D. H., Kourouniotis, C., Quinney, D., & Tzanakis, C. (Eds.), *Proceedings of the Second International Conference on the Teaching of Mathematics (at the undergraduate level).* ID114 CD-Rom, Hersonissos: Wiley, from http://www.math.uoc.gr/~ictm2/Proceedings/ICTM2_Proceedings_Table_of_Contents.html.

Henning, H., & Keune, M. (2004) Levels of Modelling Competencies. In W. Blum, & H.-W. Henn (Eds.), *ICMI Study 14: Application and Modelling in Mathematical Education. Pre-Conference Volume* (pp. 115-120). Dortmund: University of Dortmund.

Jablonka, E. (1996). *Meta-Analyse von Zugängen zur Mathematischen Modellbildung und Konsequenzen für den Unterricht.* Berlin: Transparent.

Keune, M., Henning, H., & Hartfeld, C. (2004). *Niveaustufenorientierte Herausbildung von Modellbildungskompetenzen im Mathematikunterricht.* Technical Report Nr. 1. Magdeburg: Otto-von-Guericke-Universität.

Klieme, E. et al. (Eds.). (2003). *Zur Entwicklung nationaler Bildungsstandards – Eine Expertise.* Frankfurt a. M.: Deutsches Institut für Internationale Pädagogische Forschung (DIPF), also from www.dipf.de/aktuelles/expertise_bildungsstandards.pdf.

Maaß, K. (2004) *Mathematisches Modellieren im Unterricht – Ergebnisse einer empirischen Studie.* Hildesheim: Franzbecker.

Niss, M. (1999). Kompetencer og uddannelsesbeskrivelse. *Uddannelse 9*, Undervisningsministeriet, also from http://udd.uvm.dk.

Niss, M. (2003). *Mathematical Competencies and the Learning of Mathematics: The Danish KOM Project.* From www7.nationalacademies.org/mseb/Mathematical-Competencies_and_the_Learning_of_Mathematics.pdf.

Organisation for the Economic Co-operation and Development (1999). *Measuring student knowledge and skills: A new framework for assessment.* Paris: OECD.

Organisation for the Economic Co-operation and Development (2003). *The PISA 2003 assessment framework: mathematics, readings, science and problem solving knowledge and skills.* Paris: OECD.

Weinert, F. E. (2001). Vergleichende Leistungsmessung in Schulen – eine umstrittene Selbstverständlichkeit. In F. E. Weinert (Ed.), *Leistungsmessung in Schulen* (pp. 17-31). Weinheim, Basel: Beltz Verlag.

Chapter 3.3.2

MODELLING BOTH COMPLEXITY AND ABSTRACTION: A PARADOX?

Mihaela Singer
Institute for Educational Sciences, Romania, Email: mihaela_singer@yahoo.com

Abstract: Do we need models in explaining the outer world and the self? What types of models might be helpful in school to explain both complexity and abstraction? What level of representation is appropriate? What dimensions of training should be focused on in constructing an inquiry-based learning? How could these dimensions be reflected in developing students' competencies? Analysing the dual relationship between complexity and abstraction, the study proposes some strategies to enhance learning in a model-building environment.

1. DO WE NEED MODELS? BETWEEN UNCERTAINTY AND INCOMPLETENESS

In 1927, Werner Heisenberg formulated the Uncertainty Principle: The more precisely the position of a particle is determined, the less precisely its momentum is known at this instant, and vice versa.

In 1931, Kurt Gödel proved the Incompleteness Theorem: Any logical deductive system, within which arithmetic can be developed, is *essentially incomplete.*

With these two claims, our wonderful logical world broke down into infinitely many pieces. The Uncertainty Principle says that we cannot make a measurement with enough precision and this is not happening because of the imperfection of our instruments – that would be too simple. Rather, it is happening because, given the quantum structure of matter, the measurement itself is affecting the state of the system we are trying to measure. The Incompleteness Theorem says that within any given branch of mathematics there would always be some propositions that could not be proven either true or false using the axioms of that mathematical branch itself. We might be able to prove every conceivable statement about numbers within a system by going *outside* the system in order to come up with new rules and axioms but, in doing so, we will only create a larger system with its own unprovable statements. The implication is that *all* logical systems of any complexity are incomplete.

Metaphorically speaking, the Uncertainty Principle is saying that we cannot measure our physical world; the Incompleteness Theorem is saying that we cannot totally understand our mental world. All what we can do is ... to model. We think in models. Models lead us to build representations of reality in order to explain that reality. Models are necessary for many reasons: they help us concentrate more closely on some specific aspects, simplifying reality; they offer various perspectives in analysing an entity; they permit us to emphasise specific aspects, or, conversely, to have a general overview of a given phenomenon; they afford abstraction by neglecting or reducing dimensions of entities, or by including entities in various classes and categories. Some models are preferred because they better explain a phenomenon, others because they reflect a dominant way of thinking at a specific time and place, others because they permit a number of applications that are otherwise difficult or impossible. Models are the *vehicles* through which we build the understanding of our material and immaterial worlds.

2. WHAT TYPES OF MODELS? BETWEEN COMPLEXITY AND ABSTRACTION

Mathematics learning is difficult because it aims to achieve an increase in *complexity*. At the same time, it is difficult because it aims to achieve an increase in *abstractness*. These two tendencies are paradoxically contradictory. The following example from geometry will explain this assertion. Consider a geometrical point: it has no dimension; it is the simplest geometrical figure. A more complex figure, having one dimension, is a line. The next stage of complexity is represented by geometrical plane figures, which have two dimensions, and the fourth, by the solids – the three-dimensional shapes. There is no doubt: passing from the geometrical point to the 3-D space, we are facing an increase in complexity, and the construction of the axiomatic geometries carefully followed this way of development. But let's change the perspective: the 3-D space is the space we live in, it is the most familiar one; in this space we can move from right to left (and vice-versa), back and forth, or up and down – it has three dimensions. Imagine a space whose beings could only move from right to left (and vice-versa), or back and forth. These beings, moving on a plane, might look like ∠, or like □, or like ○, etc. (e.g. Abbot, 1884). Prisoners of their space, these beings could not move up and down – their space has only two dimensions. Imagine another space in which the motion is possible in only one direction, namely a line – this space has only one dimension. In building these representations, we are gradually increasing the distance from the concrete level; we are building more complex ... abstractions. The point is the most abstract representation we can imagine; it has no dimension, no motion inside. How can we build both complexity and abstraction in the child's mind, while avoiding this construction becoming paradoxically contradictory? Through developing patterns, the human mind is able to work with both abstraction and complexity. To make this ability functional in children, it is necessary to be aware of the critical aspects, to make them explicit, then to incorporate them in adequate training procedures, making it possible to internalise the contextual coherence. Neuroscience findings show that the brain works by encoding environmental information into specific representations in the cortex, followed by decoding the representations into specific commands for the sensory-motor system (e.g. Quartz & Sejnowski, 1997). An efficient learning should help to develop the representational power of the mind, and this representational power – fundamental for developing competencies – is increased when pathways of building abstraction interact with pathways of developing complexity.

While complexity implies an increase in dimensionality, abstraction implies neglecting or grouping dimensions in order to concentrate on a higher level of generality. Moreover, complexity supposes an analysis on a small scale – with a lot of details interacting with one each other (e.g. Bar-Yam, 1997; Singer, 2003), while abstraction supposes an analysis on a large scale – with a mostly focussed perspective.

As I previously stressed, models are vehicles for building both abstraction and complexity. In training students, we could make more explicit these two aspects of modelling by differentiating between models and simulations. Models seek to explain complex systems through abstracting characteristics, while simulations seek to describe situations by offering rich multi-sensory stimuli that try to recreate the complexity of the situation. As Schwartz (1998) underlines, given this difference in purpose, the makers of models seek to limit the complexity of their products so as to make the underlying causal and/or structural mechanisms more salient. In contrast, designers of simulations tend to incorporate as much of the richness and complexity of the referent as possible to make the simulation a rich perceptual experience. In the educational context, this significant difference in goals points to quite different roles for models and simulations. In the elementary grades, the emphasis is on having the students exposed to a variety of phenomena. For example, simulations are helpful in checking a large number of particular cases, with the purpose of training the ability to discover patterns. Secondary students, on the other hand, need to learn more about correlations and underlying causes, without the overlay of complexity that can so readily make them unable to select the essential. Both models and simulations could be used as interactive tools to provide evidence of how changing a parameter might influence the components of a system.

3. WHAT LEVEL OF REPRESENTATION? BETWEEN CONCRETE AND ABSTRACT

Lesh and Doerr (2003) define models as conceptual systems that are expressed using external notation systems and are used to construct, describe, or explain the behaviours of other systems. As conceptual systems, models could have a physical concrete representation or could be abstract. What is the relationship between these two categories of systems during the period of schooling in terms of levels of abstraction? Abstracting is a developmental process supposing a gradual progression from operating with concrete objects to operating with symbols and symbol systems. Constructing abstraction implies reorganising previous knowledge by incorporating it within new systems, which are hierarchically structured. The progression in abstraction of the mathematical concepts learned during school and the progression in abstraction of the models for those could be seen as non-linear functions of time. Fig. 3.3.2-1 represents two generic curves for these functions and their interaction.

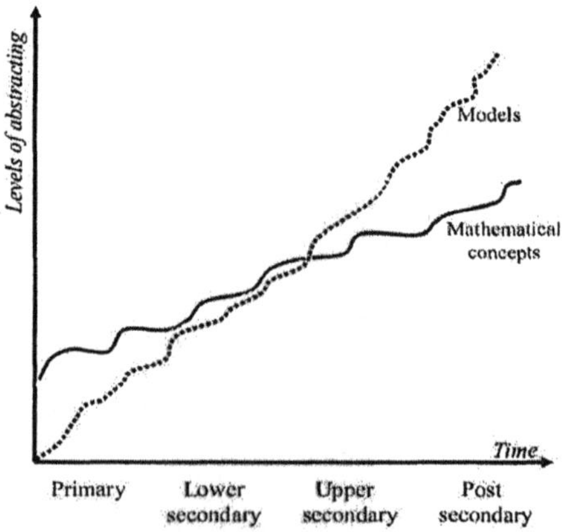

Figure 3.3.2-1. The progression in abstraction for systems of knowledge and for the models used to facilitate their understanding

The mathematical conceptual systems are "slowly" progressing in abstraction over time, but are "strongly" progressing in complexity. As a consequence, in primary grades, the models used by the teacher and the models developed by the students have to be concrete, they have to allow and stimulate manipulative activities. Later on, in order to tackle the increasing complexity of the concepts to be learned, more abstract models that simplify some characteristics of the phenomena under study might be brought into the teaching and learning practice.

The human mind uses models to make sense of other systems. Models are basic instruments for understanding the world we live in. Consequently, the training for building models has to start very early in school (e.g. Usiskin, 2005). It supposes two main components: offering powerful models for learning (powerful representations for the concepts the child has to internalise) and coaching the child in modelling, starting with the simplest problems. Even more, in consensus with the previous sections, to optimise learning, the teacher should develop *exploratory model-building environments*. From this perspective, it is significant to identify those models – powerful representational devices – that are able to foster learning. To enhance their efficiency, the models for teaching and learning should track the development of skills from concrete sensory-motor actions to abstractions. Instruction in a model-building environment that follows a gradual progression of abstracting creates conditions for transferring knowledge and understanding within the purpose of building new knowledge.

4. WHAT IS THE GOAL? AN APPLICATION: MODELLING THE PRACTICE OF COMPREHENSIVE LEARNING

A correct building up of abstraction in students' minds supposes automatised access to fundamental basic elements of that abstraction – a kind of a proto-history of abstraction as a dynamic process. The dynamics of this process are not a simple progression; on the contrary, there are pitfalls, gaps, and discontinuities, as well as spurts, jumps and smooth transitions. Studies focused on the way skills are developed in children and adults show discontinuity, rather than a simple cumulative or progressive process. Spurts and regressions are frequently recorded, until the acquisition of a skill moves to a steady state. Even in the steady state, there are fluctuations (e.g. Fisher & Yan, 2002). With optimal support, the pattern of individual evolution of a skill shows a non-linear progressive variation. Such a variation is generically represented in Fig. 3.3.2-2.a.

Following classroom observations and other research (Singer, 2001, 2003), it seems that, to optimise the process of mastering a specific skill, the training should offer a variation of tasks that mirror the skill development and cover a range around the hypothetical optimal non-linear evolution. To be more specific, it is necessary to systematically reinforce the lower levels of a skill and, at the same time, tasks belonging to higher levels might be introduced. An adequate dosage of these anticipatory tasks challenges the development, contributing to the emergence of an over-learning phenomenon. Fig. 3.3.2-2.b offers a representation for the range covered by an adequate training aiming to optimise the development of a specific skill.

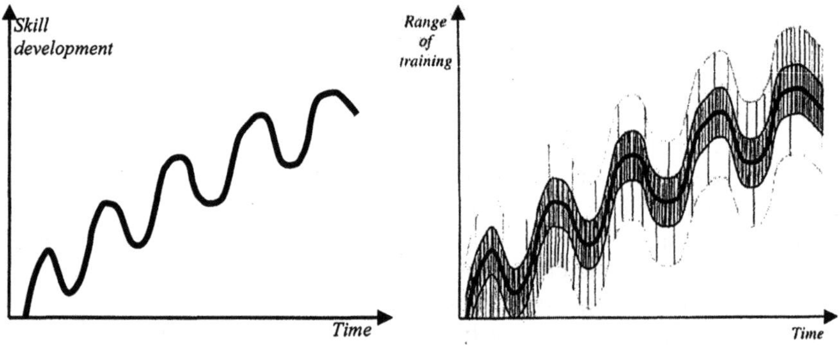

Figure 3.3.2-2.: a) Pattern of variation in developing a skill;
b) Pattern of variation in organising the content of training

An example might give a better idea about the range of training while learning a new concept. The example is excerpted from a lesson on solving linear inequations in grade 9 (15 – 16 years old). The starting point was the following problem:

Two taxis companies have the following offers:
QUICK TAXI Initial cost 6 € plus 1.3 € per 1km.
SPEED TAXI Initial cost 3 € plus 1.9 € per 1km.
What is the best choice for a journey of: a) 3 km? b) 10 km?

Snapshots into the classroom activities show the next sequences:

The teacher invited the students to work in groups with the following purposes: to discuss how the problem could be solved (brainstorming, without any suggestions from the teacher); to identify the mathematical objects involved in the task; to explore ways to solve the problem by examining particular cases and analogous situations; to express the data using variables. In this phase, *the focus is on manipulating and evoking.*

In the next sequence, the students adopted various ways to solve the problem. Some found an appropriate model for the problem (as a linear inequation), some guessed and checked the solutions by trial and error on particular cases, some used graphic representation for the functions discovered during the study. Based on the students' observations, the teacher discussed the degree of generality of each solution. In this phase, *the focus is on the passages between concrete and abstract.*

Next, the students checked the results obtained in the previous phases, analysed and described the behaviours of the two functions brought to attention during the problem-solving process, and expressed their connections in mathematical terms. In this phase, *the focus is on the complexity of the mathematical phenomena involved.*

The context was then extended by varying some parameters. In this idea, the teacher used the same problem to devise other tasks connected with analysing the variation of a linear function, such as "What would happen if the two companies double the starting costs/ increase the costs per kilometre by 0.5 €?". The students were stimulated to vary some data and to analyse the results, and then, to devise their own new problems that could be solved using similar patterns. In this phase, *the focus is on anticipating further developments.*

In the lesson described above, while the majority of the tasks were concentrated at the standard level, there were some in the proximal vicinity, some grounded in previous levels – supposed already internalised – and others, anticipatory, belonging to the future levels intended to be reached by the students. As with progressing in developing a skill, the core of the tasks, in number and structure, moves from the previous basic levels to the anticipated levels. The process of abstracting followed two interacting pathways, one from the *external* environment to *internal* representation and the other from *concrete* objects to *standard symbolic* representation, passing through various *unconventional symbolic* representations. The complexity of information acquired through learning is added to this developmental pattern that underlies the progression in abstracting.

5. WHAT ARE THE BENEFITS? OPENING THE WAY TO BUILD COMPETENCIES

It might be argued that one of the most important outcomes of a model-building learning environment is the self-development of competencies. Competencies can be defined as *structured sets of knowledge and skills* acquired through learning, which allow the identifying and solving of problems that are characteristic for a certain field of activity, in a variety of contexts

(Singer, 1999). In the attempt to model ill-structured real-life problems, the students might develop the following general competencies in mathematics:

- Identifying relationships among the mathematical concepts
- Interpreting quantitative, qualitative, structural and contextual data included in mathematical statements
- Using algorithms and mathematical concepts to characterise a given situation locally or globally
- Expressing the quantitative or qualitative mathematical features of a contextual situation
- Analysing problem situations to discover strategies to optimise solutions
- Generalising properties by modifying the original context or by improving or generalising algorithms.

In order to be developed in students, these competencies might be particularized at each specific grade and level. What is to unravel, from the perspective of this article, is that complexity and abstraction interact in developing each competence because they are deeply involved in building the mathematical knowledge. On the one hand, this knowledge embodies both complexity and abstraction and, on the other hand, acquiring them implies modelling both complexity and abstraction. A model-building learning environment is the way to solve the paradox.

REFERENCES

Abbot, E. (1884). *Flatland: A Romance of Many Dimensions.* Mineola, NY: Dover Thrift Editions.

Bar-Yam, Y. (1997). *Dynamics of complex systems.* Boston: Addison-Wesley Longman Inc.

Doerr, H., & Lesh, R. A. (2003). *Beyond constructivism: Models and modelling perspectives on mathematics teaching, learning, and problem solving.* New York: LEA.

Fischer, K.W., & Yan, Z. (2002). Always under construction: Dynamic variations in adult cognitive microdevelopment. *Human Development, 45*(3), 141-160.

Quartz, S., & Sejnowsky, T. J. (1997). The neural basis of cognitive development: A constructivist manifesto. *Behavioral and Brain Sciences 20*(4), 537-596.

Schwartz, J. L. (1998). *Models, simulations and exploratory environments: A tentative taxonomy.* From http://www.gse.harvard.edu~faculty/schwartz/modeltaxonomy.

Singer, F. M. (Ed.). (1999). *The new National Curriculum.* Bucharest: Prognosis P. H.

Singer, F. M. (2001). Structuring the information – a new way of perceiving the content of learning. *Zentralblatt für Didaktik der Mathematik /IRME* 6/, 204-217.

Singer, F. M. (2003). From cognitive science to school practice: Building the bridge. In N.A. Pateman, B.J. Dougherty, & J. Zilliox (Eds.), *Proceedings of the 27th PME,* CRDG (pp. 207-214).

Usiskin, Z. (2006). The arithmetic operations as mathematical models. This Volume, Chapter 3.3.5.

Chapter 3.3.3

STUDYING AND REMEDYING STUDENTS' MODELLING COMPETENCIES: ROUTINE BEHAVIOUR OR ADAPTIVE EXPERTISE

Dirk De Bock[1+2], Wim Van Dooren[1+3] and Dirk Janssens[4]
[1] *Center for Instructional Psychology and Technology, University of Leuven, Belgium*
[2] *EHSAL, European University College Brussels, Belgium*
[3] *Postdoctoral Fellow of the National Fund for Scientific Research - Flanders, Belgium*
[4] *Department of Mathematics, University of Leuven, Belgium*
Email: dirk.debock@avl.kuleuven.be; wim.vandooren@ped.kuleuven.be; dirk.janssens@wis.kuleuven.be

Abstract: First, we summarise some studies on students' overuse of the linear model when solving problems in various domains of mathematics, showing to what extent they are led by routine behaviour in mathematical modelling. Second, we discuss a teaching experiment that aimed at enabling 8th graders to adaptively choose between a linear, a quadratic or a cubic model while solving geometry problems. The results show that, after the experiment, the students applied the linear model less automatically, but tended to switch back and forth between applying it either "everywhere" or "nowhere", indiscriminately.

1. INTRODUCTION

Contemporary reform documents and curricula in most countries more or less explicitly assume that one of the most important goals of mathematics education is that students gain the competence to make sense of everyday-life situations and complex systems stemming from our modern society, which can be called "modelling competencies". In this chapter, we argue why such modelling competences necessarily imply that students have adaptive rather than routine expertise. Hatano (2003, p. xi) describes adaptive expertise as "the ability to apply meaningfully learned procedures flexibly and creatively" and opposes it to routine expertise, i.e. "simply being able to complete school mathematics exercises quickly and accurately without (much) understanding".

The adaptive level of competency refers to a modelling process wherein the mathematical interpretation of a problem, the selection of an appropriate model, and/or the interpretation of the result is not straightforward or trivial and wherein solutions may involve several modelling cycles in which descriptions, explanations and predictions are gradually refined, revised or rejected (Lesh & Doerr, 2003; Niss, 2001; Verschaffel et al., 2000). A non-adaptive problem-solving process goes directly from superficial elements in the text to computations, without passing through a "situation model" of the problem context (Nesher, 1980; Verschaffel et al., 2000). It involves only routine computational expertise combined with cue-spotting, and thus does not constitute modelling in any real or deep sense. For instance, an elementary school pupil who immediately solves the following problem "*Pete lives at a distance of 9 km from school and Ann lives at a distance of 5 km from school, how far do they live from each other?*" by either adding (9 + 5 = 14 km) or subtracting (9 – 5 = 4 km) the two given numbers (without acknowledging the possibility of other alternatives), demonstrates routine expertise, whereas a pupil who, after having generated, explored and compared different mathematical models, answers that the solution can be any number between (9 + 5 =) 14 km and (9 – 5 =) 4 km demonstrates adaptive expertise.

This chapter focuses on one of the clearest examples of non-adaptive modelling behaviour, namely students' tendency to over-rely on the proportional or linear model when solving mathematical problems. Numerous documents and research reports on a wide variety of mathematical domains, and dealing with students of diverse ages, mention this tendency to routinely apply the proportional model irrespective of the mathematical model(s) appropriate to the problem situation. For example, many upper secondary students routinely answer the following unfamiliar probabilistic problem "*The probability of getting a six in 1 roll with a die is 1/6. What is the probability of getting at least one six in 2 rolls?*" by applying the direct proportionality model (2 × 1/6 = 2/6). Adaptive expertise for such a problem could consist of writing down all possibilities that can occur in 2 die rolls and counting all cases where there is at least a six. Doing this would unmask the inappropriateness of the direct proportionality model and lead the student to a correct answer. Empirical evidence for students' overgeneralisation of the linear model in the domain of probability can be found in Van Dooren et al. (2003).

Recent research (Van Dooren et al., 2005; Verschaffel et al., 2000) has shown that current mathematics instruction practices can encourage students to acquire (at least initially) a routine expertise instead of an adaptive one. Accordingly, students tend to overgeneralize the validity and relevance of some well-trained schemes, and begin to transfer them to settings to which they are neither relevant nor valid. In this respect, the linear model seems to have a special status because it is prominently present in students' minds. Van Dooren et al. (2005) analyzed 2nd to 8th graders' solutions of linear and non-linear arithmetic problems (e.g., "*Ellen and Kim are running around a track. They run equally fast but Ellen started later. When Ellen has run 5 rounds, Kim has run 15 rounds. When Ellen has run 30 rounds, how many has Kim run?*"), in order to determine *when* the tendency to routinely apply the linear model originates and *how* it develops with students' increasing age and educational experience. It was found that the skills to correctly solve linear problems considerably increased with age, from 53% correct answers on this type of problems in 3rd grade to 93% in 8th grade. Most learning gains were made between 3rd and 5th grade. In addition, it was shown that the tendency to overrely on linear methods in non-linear situations developed remarkably parallel with these emerging proportional reasoning skills. In 3rd grade, 30% of a series of non-linear problems were erroneously solved linearly, and this tendency increased considerably until 51% in 5th grade (and decreased afterwards to 22% in 8th grade). This inverted U-shaped evolution of the unwarranted linear answers was found for all non-linear problems, but there were some minor differences according to the specific form of non-linearity in the mathematical structure implicit in the problems. It can be concluded that students – while they acquire linear reasoning skills by being trained in solving "typical'" linearity problems – start to overgeneralise linear models and learn to apply them on the basis of superficial problem characteristics. This tendency was already present in the 2nd grade, increased up to Grade 5, before slightly decreasing from Grade 6 to 8.

In the next paragraphs, we focus on the routine application of proportionality in one specific case. Students, and even adults, tend to think that if the lengths of a figure are multiplied by factor r to produce a similar figure, the area and volume will also be multiplied by factor r, whereas areas and volumes are, respectively, multiplied by r^2 and r^3 (National Council of Teachers of Mathematics, 1989; Tierney et al., 1990).

2. IN SEARCH OF THE ROOTS OF STUDENTS' IMPROPER PROPORTIONAL REASONING IN GEOMETRY

In recent years, the geometrical misconception that the area and volume of a figure enlarge r times when a figure is enlarged r times, has been extensively studied. In a series of experimental studies by De Bock et al. (1998, 2002b), large groups of 12 – 16-year old students were administered paper-and-pencil tests with proportional and non-proportional word problems about lengths, perimeters, areas and volumes of different types of figures. For example, they used the following non-proportional item about the area of a square: "Farmer Carl needs approximately 8 hours to manure a square piece of land with a side of 200 m. How many hours would he need to manure a square piece of land with a side of 600 m?" The vast majority of students (i.e. more than 90% of the 12-year olds and more than 80% of the 16-year olds) failed on this type of problem because of their alarmingly strong tendency to routinely apply linear methods. Even with considerable support (such as the request to make a drawing of the problem situation before solving the problem or the provision of a ready-made drawing on plain or squared paper), only very few students made the shift to correct, non-linear responses. It was found that the request to make a drawing was often neglected and the ready-made drawings were seldom effectively used. In cases when students made a drawing, this was often of a low representational quality and therefore not helpful for finding the correct solution. The only experimental manipulation that had some meaningful impact on students' responses was rephrasing the missing-value problems to a so-called comparison format (e.g., for the earlier mentioned item: "Today, farmer Carl manured a square piece of land. Tomorrow, he has to manure a square piece of land *with a side being three times as big. How much more time* would he approximately need to manure this piece of land?"). In this study, the number of correct answers increased from 23% in the group that received missing-value problems to 41% in the group that received comparison problems. When, as a consequence of the provided help in these studies, students discovered that some of the problems are not linear, remarkably they sometimes started to apply non-linear methods to linear problems too. These students replaced one type of routine behaviour ("proportionality anywhere") by another one ("proportionality nowhere"), but didn't show any sign of adaptive expertise.

While the previous studies showed to what extent students' routine use of linearity is affected by various characteristics of the task, they did not provide adequate information about the thinking processes and modelling competency underlying students' improper proportional reasoning. Therefore, De Bock et al. (2002a) made a shift in their research methodology from collectively testing large groups of 12–16-year old students to individual in-depth interviews with a limited number of students. During these interviews, students' solution processes were revealed through a number of well-specified questions by the interviewer with respect to one single non-linear application problem, as well as through their reactions to subsequent (increasingly stronger) forms of help. This research revealed both students' inclination towards routine problem solving as well as conceptual shortcomings and misconceptions:

- The majority of the students used a proportional model in a spontaneous, almost intuitive way (in the sense of Fischbein, 1987) being unaware of their choice for a proportional model, while others really were convinced that linear functions are applicable "everywhere" or for this situation in particular;
- Many students showed particular shortcomings in their geometrical knowledge (e.g., the misbelief that the concept of area only applies to regular figures, or that a similarly enlarged figure is not necessarily enlarged to the same extent in all dimensions);
- Many students had inadequate habits, beliefs and attitudes towards solving mathematical word problems in a school context (for example, the belief that drawings are not helpful, or that the first solution is always the best), which proved to be a fertile soil for a superficial or deficient modelling process. These factors often prevented students from unmasking their proportional solution as inadequate, and discovering the correct solution to the problem.

3. TEACHING FOR ADAPTIVE EXPERTISE

The next stage of the research program involved the design, implementation, and evaluation of a learning environment aimed at enabling students to adaptively choose between a linear, quadratic and cubic model in the context of enlargements and reductions of geometrical figures.

A series of 10 one-hour experimental lessons was created for use with 13 – 14-year old students. In the development of the lesson series, the results and the conclusions of the earlier studies discussed above were taken into account. Moreover, the development of the learning environment was strongly inspired by (1) the principles of realistic mathematics education (e.g., Gravemeijer, 1994): using realistic problem situations aimed at challenging students' mathematical (mis)conceptions and beliefs, rediscovery of mathematical notions; (2) building on students' informal knowledge, instructional techniques that enhance higher-order thinking (e.g., articulation and reflection); and (3) using multiple representations of the learning contents. The following topics were successively addressed: recognizing and constructing similar figures/objects, proportional relations and their properties, linear growth of the lengths and perimeter in similar figures, quadratic growth of the area and cubic growth of the volume. The lesson series ended with an integrative project about the "Life and Work of the Gnomes" (Poortvliet & Huygen, 1976), in which all learnt contents could be applied in an attractive, challenging and authentic context. Examples of learning activities from the experimental lessons (as well as a more extensive description of the results of this study) can be found in Van Dooren et al. (2004). The learning gains were assessed by means of a word problem test consisting of proportional and non-proportional items. This test was administered before the intervention (pretest), after the intervention (posttest), and three months afterwards (retention test).

A group of eighteen 13–14-year old students followed the experimental lessons. A significant improvement in students' performances on the non-proportional items was observed from pretest (29.2%) to posttest (61.1%). This was followed by a non-significant decrease in the performances from posttest to retention test (to 50.0% correct answers). Contrary to the results for the non-proportional items, the score of the students on the proportional items decreased from 83.3% correct answers on the pretest to 58.3% on the posttest, and went further down from posttest to retention test (although not significantly) to 52.8%. An additional qualitative analysis of the answers revealed first of all that on the pretest, about 70% of all the solutions on the non-proportional items could indeed be characterized as linear. This number of unwarranted linear answers strongly decreased in the posttest to about 18%, while in the retention test, the percentage raised again to about 30%. But students who no longer applied linear solutions to solve non-linear problems, did not always perform better than before. In the posttest and retention test they made errors in applying non-linear solutions on these non-linear problems (such as confusing area and volume, just taking the square of one of the given numbers). The qualitative analysis also confirmed the overgeneralisation effect: while on the pretest only 13% of all the solutions to linear items could be characterised as an application of non-linear strategies, this number raised to 36% on the posttest and retention test. A careful analysis of the videotapes of the experimental lessons supported these conclusions. Certain observations indicated that non-linear relations and the effect of enlargements on area and volume remained intrinsically difficult and counterintuitive for many students. For example, there were students who at the same time understood that the area of a square increases 4 times if the sides are doubled in length (since the enlargement of the area goes "in two dimensions"), while they had difficulty in understanding why this does *not* hold for the perimeter (which also increases in two "directions").

We can hardly argue that the lesson series achieved its goal. After the lessons, the students still experienced serious difficulties in knowing which model they should use in which situation. Although the dominance of the linear model was broken, many students still continued to rely on superficial cues (such as key words or phrases in the problem statement) to decide which mathematical operations to apply. Such a process is far removed from the envisaged adaptive modeling process, in which every phase of the modeling cycle is adequately and thoroughly worked through. The experimental lessons were unable to develop in the students a deeper understanding of proportionality and non-proportionality, and a disposition to switch adaptively between proportional and non-proportional models in accordance with the problem situation.

4. CONCLUSION

The different studies that were briefly reported in this chapter have empirically demonstrated that students routinely rely on linear models and that this routine behaviour leads them to fail on problems that cannot be appropriately modelled linearly. The question arises why the linear model is so predominant in students' minds. Both psychological and educational factors seem to be at the roots of this phenomenon. First, linear models have a strongly intuitive nature and are – from early childhood on – omnipresent in our everyday-life experience. Second, the effect is doubtlessly exacerbated by usual traditional teaching, giving no attention to ideas of modelling but grouping problems together with similar solution methods, seldom using counterexamples for contrast, assuming an unnaturally "cleaned-up" world, sticking to "easy" numbers, etc.

By its very nature, mathematical modelling requires adaptive expertise. It involves a fluent and flexible selection of mathematical models related to different aspects of a real situation and not the automatic application of particular well-trained computational schemes in a straightforward way. Conversely, the modelling approach – as a key part of a real mathematical disposition – can promote adaptive expertise in a diversity of learners. Once more, the results of the reported studies made clear that the application of a modelling perspective can hardly be reached by means of an experimental manipulation or a short-term intervention, but needs a long-term and maintaining strategy (Blum & Niss, 1991; Lesh & Doerr, 2003; Verschaffel et al., 2000). In this respect, it might be appropriate to introduce the modelling perspective much earlier in the child's education (see, e.g., Usiskin, this volume, Chapter 3.3.5) in order to prevent – rather than remedy – routine behaviour and to continue this preventive effort throughout the mathematics curriculum.

REFERENCES

Blum, W., & Niss, M. (1991). Applied mathematical problem solving, modelling applications, and link to other subjects – State, trends and issues in mathematics education. *Educational Studies in Mathematics* 22, 37–68.

De Bock, D., Van Dooren, W., Verschaffel, L., & Janssens, D. (2002a). Improper use of linear reasoning: An in-depth study of the nature and irresistibility of secondary school students' errors. *Educational Studies in Mathematics 50*, 311-334.

De Bock, D., Verschaffel, L., & Janssens, D. (1998). The predominance of the linear model in secondary school pupils' solutions of word problems involving length and area of similar plane figures. *Educational Studies in Mathematics 35*, 65-83.

De Bock, D., Verschaffel, L., & Janssens, D. (2002b). The effects of different problem presentations and formulations on the illusion of linearity in secondary school students. *Mathematical Thinking and Learning 4*(1), 65-89.

Fischbein, E. (1987). *Intuition in Science and Mathematics: An Educational Approach.* Dordrecht: D. Reidel.

Gravemeijer, K. (1994). *Developing Realistic Mathematics Education.* Utrecht, The Netherlands: Freudenthal Institute.

Hatano, G. (2003). Foreword. In A.J. Baroody, & A. Dowker (Eds.), *The Development of Arithmetic Concepts and Skills* (pp. xi–xiii). Mahwah, NJ: Lawrence Erlbaum Associates.

Lesh, R., & Doerr, H.M. (Eds.). (2003). *Beyond Constructivism. Models and Modeling Perspectives on Mathematical Problem Solving, Learning and Teaching*. Mahwah, NJ: Lawrence Erlbaum Associates.

National Council of Teachers of Mathematics (1989). *Curriculum and Evaluation Standards for School Mathematics*. Reston, VA: Author.

Nesher, P. (1980). The stereotyped nature of school word problems. *For the Learning of Mathematics 1*(1), 41-48.

Niss, M. (2001). Issues and problems of research on the teaching and learning of applications and modelling. In J.F. Matos, W. Blum, S.K. Houston, & S.P. Carreira (Eds.), *Modelling and Mathematics Education. ICTMA 9: Applications in Science and Technology* (pp. 72-89). Chichester: Ellis Horwood.

Poortvliet, R., & Huygen, W. (1976). *Leven en Werken van de Kabouter.* [Life and Work of the Gnomes]. Houten: Holkema & Warendorf.

Tierney, C., Boyd, C., & Davis, G. (1990). Prospective primary teachers' conceptions of area. *Proceedings of the 14th Conference of the International Group for the Psychology of Mathematics Education*, Vol. 2 (pp. 307-314). Oaxtepex, Mexico.

Van Dooren, W., De Bock, D., Depaepe, F., Janssens, D., & Verschaffel, L. (2003). The illusion of linearity: Expanding the evidence towards probabilistic reasoning. *Educational Studies in Mathematics 53*, 113-138.

Van Dooren, W., De Bock, D., Hessels, A., Janssens, D., & Verschaffel, L. (2004). Remedying secondary school students' illusion of linearity: A teaching experiment aiming at conceptual change. *Learning and Instruction 14*(5), 485-501.

Van Dooren, W., De Bock, D., Hessels, A., Janssens, D., & Verschaffel, L. (2005). Not everything is proportional: Effects of problem type and age on propensities for overgeneralization. *Cognition and Instruction 23*(1), 57-86.

Verschaffel, L., Greer, B., & De Corte, E. (2000). *Making Sense of Word Problems.* Lisse, The Netherlands: Swets & Zeitlinger.

Chapter 3.3.4

ASSESSING THE "PHASES" OF MATHEMATICAL MODELLING

Ken Houston
University of Ulster, UK, Email: sk.houston@ulster.ac.uk

Abstract: The seven-phase modelling cycle is well known and has been used successfully for over twenty years as a heuristic for teaching modelling. Assessment methods used over this period are reviewed and current research that investigates the use of multiple choice questions to assess the individual phases is reported.

1. INTRODUCTION

The seven phases or stages of the modelling cycle are often given as:

Table 3.3.4-1. Phases of the modelling cycle (Penrose, 1978)

1. Specify the real problem	2. Create a mathematical model
3. Specify the math problem	4. Solve the math problem
5. Interpret the math solution	6. Validate the model
7a. Revise	7b. Report

In the ten-year period centred on 1980, many UK institutions of higher education were beginning to include mathematical modelling in their undergraduate programmes. The biennial series of International Conferences on the Teaching of Mathematical Modelling and Applications (ICTMA) was getting under way and lecturers were telling war stories and writing about their new courses. In Section 2, two of these early courses are described and their assessment criteria are mapped to the seven phases of the modelling cycle. By the early 1990s, people were becoming more concerned to establish the reliability and validity of their assessment instruments and work by a

UK Assessment Research Group (ARG) is reported in Section 3. Ten years later some members of ARG were developing and testing methods of "micro-assessment" - the use of multiple choice questions in assessing modelling abilities. This work and some advantages are described in Section 4.

2. HOLISTIC ASSESSMENT

In this section two of the papers presented at the first ICTMA conference held in 1983 at Exeter and published in Berry, Burghes, Huntley, James and Moscardini (1984) are reviewed. The mathematical modelling movement was well under way and the organisers of ICTMA-1 believed it was time for proponents to get together for an international discussion. This review will give the reader some insights into the state of the art of assessing mathematical modelling at that time. The emphasis was on "assessing the whole thing" although most authors did recognise that the different phases of the modelling cycle, taken in groups, should be apportioned different fractions of the total mark, with the details of the apportionment reflecting the importance attributed to each group by the particular author.

Hall (1984), in his paper *The Assessment of Modelling Projects*, suggests that the modelling skills of students should be assessed in three groups – *Content*, *Presentation* and *Drive*. The details of Hall's three groups are given in Table 3.3.4-2, with their mapping to the phases of the cycle, which are given in Table 3.3.4-1.

Hall (1984, p. 145) writes, "The first group collects together the more technical aspects of modelling, the second is concerned with the written project itself while the third allows for originality and management."
Item 6 implies that "desired objectives" are specified so that the modeller knows what is to be achieved by manipulating "the mathematical expressions", but it is interesting that Hall did not mention explicitly phase 3: "Specify the mathematical problem". And items 10 and 11, Hall's "ability" items that allow for originality and for research, must be present to enable the enterprise to make any headway at all.

Table 3.3.4-2. Hall's items and their mapping to the Phases

Hall's items	Phases
Content	
1. Ability to handle and make sense of natural or experimental data	1
2. Determination of variables and parameters which describe observation	2
3. Recognition of patterns in data and in processes	2
4. Generation of mathematical expressions to summarize observations	2

Hall's items	Phases
5. Ability to set up a model representing the system and relating its significant variables	2
6. Technical ability to manipulate the mathematical expressions of the model to achieve desired objectives	3 and 4
Presentation	
7. Representation and interpretation of data	7b
8. Translation of information into and out of pictorial form	7b
9. Ability to communicate clearly, especially in writing	7b
Drive	
10. Ability to identify situation and to formulate problems	1 and 3
11. Ability to consult books etc. for additional techniques or information	1 and 4
12. Understanding of when to change a model, method or objective in discussing a problem	7a
13. Recognition of what constitutes a solution – evaluation of success of models	5 and 6
14. Ability to work effectively in a group	1 thru 7

The number of items that map to phase 2 is an indicator of the complex nature of the task: "create a mathematical model". It has long been recognised that this is the most difficult phase of the cycle, involving making assumptions to simplify the situation, identifying variables and parameters and determining relationships between them.

The *Presentation* items make it clear that "communicating results" has always been an essential component of modelling, while item 14, "ability to work effectively in a group", indicated that modelling was usually thought of as a group activity. This is still the case, but the difficulties associated with assessing the working of a group and attributing individual marks to the members of a group have led many teachers to abandon group projects as high stakes assessments such as are found in the final year of a programme. Group working is a feature of "the way of life" of a professional mathematician and so should be part of a student's learning experience, and many teachers include this in the early years. The skills of good communication and effective group work are desirable graduate attributes and therefore help to make modelling an excellent curriculum device for preparing students for employment. This idea is explored fully by Challis and Houston (2000) and by Challis, Gretton, Neill and Houston (2002). The assessment of these skills has been the subject of much reflection and research, some of which is described below.

Hall also suggests a *product* model for combining the marks awarded to the three groups, rather than an *addition* model. This was an interesting innovation that did not catch on, probably because a zero score for one group would result in a zero score overall, thus negating the (possibly) good work in the other groups. A similar idea has been proposed again more recently by Gibbs (2001). In his suggestion, in a *product* model for combining marks certain assessment tasks are scored "1" if they are satisfactory or "0" if they are not. The idea here is that these tasks are "course requirement" tasks and the omission of just one task results in failure to meet all the learning outcomes. They are included (a) to ensure that the student tackles them and (b) to provide a vehicle for detailed formative assessment comments. Gibbs suggests elsewhere (Gibbs and Simpson, 2002) that "feedback without grades [on tasks that are 'course requirements'] has a more positive impact on subsequent performance than grades, or even than feedback and grades combined."

The paper by Berry and Le Masurier (1984) is the second to be reviewed. These authors describe a module called *Mathematical Models and Methods* offered to students by the Open University in the UK. Despite this title, students were indeed required to engage in *modelling* in addition to studying *models* and *methods*. Modelling was assessed using two written assignments, the first of which, by concentrating only on phases 1 to 3, was to ensure that the student had started work and was progressing in a direction likely to achieve success. The second assignment contributed 80% and required all of the phases. The published marking scheme is arranged in six groups of items and probably relates more closely to the seven phases than Hall's items. This is not unexpected since it was the Open University that first proposed the seven-phase cycle (Penrose, 1978). The items used are given in Table 3.3.4-3 with the mapping to the phases.

Table 3.3.4-3. Berry and Le Masurier's items and their mapping to the Phases

Berry and Le Masurier's items	Phases
Abstract	
1. Statement of the problem to include both the starting point and the actual conclusion reached	1
2. Significance of problem	1
3. Sources of data	4 and/or 6
Formulation	
4. Assumptions	2
5. Simplifications	2
6. Important features	2
Initial model	
7. Variables defined	2
8. Model (following on from assumptions) and its solution	2 and 4
9. Interpretation of solution and criticism of initial model	5 and 6

Berry and Le Masurier's items	Phases
Data	
10. How collected	4 and/or 6
11. Relevance of data	4 and/or 6
12. Presentation of data (e.g. diagrams, graphs, etc.)	7b
Revisions to the model	
13. Revised models based on criticism	7a
14. Interpretation and criticism of revised models	5 and 6
15. Criticism of final model	6
Conclusions	
16. Brief summary of the main results of the modelling	7b

Items 3, 10, and 11, which relate to the source, collection and relevance of data could map to either or both of the "solve" and "validate" phases, depending on the nature of the problem and the models developed. Each of items 9 and 14 include both the "interpretation" and "validation" phases. Later developments, described below, indicate that it is more desirable to separate these into two items.

3. DEVELOPING ROBUST ASSESSMENT CRITERIA FOR PROJECTS

About a decade later a number of university teacher/researchers formed an ad hoc group, the UK Assessment Research Group (ARG), to review the assessment schemes currently in use for the assessment of undergraduate project work in mathematics. A product of this work was a set of robust criteria-based assessment procedures. A by-product was the development of an extremely effective means of peer-assisted professional development in assessment. The ARG story is told by Haines and Houston (2001) and this paper contains references to the four reports written by ARG and published in limited print runs, and other pertinent publications by the Group. Much of the thinking and products of ARG are included in the resource pack *Mathematics Teaching and Learning - Sharing Innovative Practices* (Haines and Dunthorne, 1996).

The assessment criteria developed for the "modeling" aspects of modelling projects comprise 11 items which can be mapped very closely to phases 1 to 6 and 7a of the modelling cycle, and 9 items dedicated to written reports (phase 7b). Since many courses also required students to report in other media such as an oral presentation or a poster, items suitable for these activities were also developed and tested. The great importance attached to good communication skills is emphasised by the attention given to phase 7b. Assessment criteria were also developed for projects in pure mathematics, statistics, and historical/educational investigations. Table 3.3.4-4 gives the ARG items for modelling and written communication, with the mapping to the 7 phases.

Each should be prefixed by "The student..."

Table 3.3.4-4. ARG's items and their mapping to the Phases

ARG Items	Phases
Modelling	
1. States objectives of task	1
2. Identifies the main features of the task	2
3. Makes simplifying assumptions	2
4. Identifies possible variables of interest	2
5. Explores relationships and develops a mathematical model	2
6. States the mathematical problem	3
7. Finds solution	4
8. Interprets solution	5
9. Validates solution	6
10. Shows evidence of research	2
11. Demonstrates initiative, determination, flair	1 to 6 and 7a
Communication skills (written)	
12. Gives a free standing abstract or summary of the report	7b
13. Gives an introduction to the report	7b
14. Structures the report logically	7b
15. Makes the structure of the report verbally explicit	7b
16. Demonstrates a command of the appropriate written language	7b
17. Complements logical structure with visual presentation and layout	7b
18. Makes appropriate use of references and appendices	7b
19. Gives a concluding section in the main report	7b
20. Gives a well reasoned evaluation	7b

In the ARG publications each item has a short descriptor which effectively describes a "good performance"; these are omitted here. In use, each item would be scored from "0" (not shown) to "4" (high) on an integer scale and a teacher's ability to make a reliable judgement of a score is enhanced through professional development and experience. These item scores can then be combined or used in a suitable way to give a holistic score for the project. "A suitable way" is one that reflects the intended learning outcomes of the module, and different teachers may wish to give more or less weight to particular items at different times or in different contexts.

4. MICRO ASSESSMENT

Recently Haines, Crouch and Davis (2000) published a seminal paper on developing and testing multiple choice questions (MCQs) that examine most of the phases individually. This initial work was extended by Haines, Crouch and Davis (2001), Haines, Crouch and Fitzharris (2003), Haines and Crouch (2001), Houston and Neill, (2003a, 2003b) and Izard, Haines, Crouch, Houston and Neill (2003).

MCQs were written to test eight modelling skills of a student, which map to the earlier phases. Details are given in Table 3.3.4-5.

Table 3.3.4-5. MCQ-tested abilities and their mapping to the Phases

Modelling abilities	Phases
1. Making simplifying assumptions	1 and 2
2. Clarifying the goal	1 and 2
3. Formulating the problem	2 and 3
4. Assigning variables, parameters and constants	2 and 3
5. Formulating mathematical statements	3
6. Selecting a model	2
7. Interpreting graphical representations	5
8. Relating back to the real situation	5

The evidence of this research suggests very strongly that it would be beneficial, when teaching modelling, to spend some time practising these skills one at a time. Thus a number of exercises concentrating only on, for example, "making simplifying assumptions" could be given to students, and so on. Students tended to score least well on "clarifying the goal" an on "selecting a model" from a list of possible equations. Matching an equation to a graph and vice versa proved difficult.

5. CONCLUSIONS

Assessment criteria used over a twenty-year period have been reviewed and mapped to the seven phases of the modelling cycle. The recent work on micro-assessment of individual phases suggests that it is beneficial to teach modelling, not only holistically but also through a detailed study of the different phases, giving students critical feedback at all times.

REFERENCES

Berry, J.S., Burghes, D.N., Huntley, I.D., James, D.J.G., & Moscardini, A.O. (Eds.) (1984). *Teaching and Applying Mathematical Modelling*. Chichester: Ellis Horwood.

Berry, J.S., & Le Masurier, D. (1984). Open University Students do it by themselves. In J.S. Berry, D.N. Burghes, I.D. Huntley, D.J.G. James, & A.O. Moscardini (Eds.), *Teaching and Applying Mathematical Modelling* (pp. 48-85). Chichester: Ellis Horwood.

Challis, N., & Houston, S.K. (2000). Embedding Key Skills in the Mathematics Curriculum. *New Capability, 4*(3), 7-10 and 50.

Challis, N., Gretton, H., Neill, N., & Houston, S.K. (2002). Developing Transferable Skills - Preparation for Employment. In P. Khan, & J. Kyle (Eds.), *Effective Learning and Teaching in Mathematics and its Applications* (pp.79-91). London: Kogan Page.

Gibbs, G. (2001). *Learning and Teaching Strategies: What lessons can we learn about assessment?* Keynote Lecture at a National Assessment Conference in Birmingham, UK, organised by the LTSN Generic Centre.

Gibbs, G., & Simpson, C. (2002). *How assessment influences student learning – a conceptual overview.* From http://cehep.open.ac.uk/ssrg/reports/documents/42_02.pdf (Accessed 23 July 2004).

Haines, C.R., & Dunthorne, S. (Eds.) (1996). *Mathematics Teaching and Learning – Sharing Innovative Practices*. London: Arnold.

Haines, C.R., Crouch, R.M., & Davis, J. (2000). *Mathematical modelling skills: A research instrument.* Hatfield: University of Hertfordshire, Department of Mathematics Technical Report No. 55.

Haines, C.R., Crouch, R.M., & Davis, J. (2001). Recognising students' modelling skills. In J.F. Matos, W. Blum, S.K. Houston, & S.P. Carreira (Eds.), *Modelling and mathematics education ICTMA 9: Applications in science and technology* (pp. 366-380). Chichester: Ellis Horwood.

Haines, C.R., Crouch, R.M., & Fitzharris, A. (2003). Deconstructing Mathematical Modelling: Approaches to Problem Solving. In Q.-X. Ye, W. Blum, K. Houston, & Q.-Y. Jiang (Eds.), *Mathematical modelling in education and culture* (pp. 41-53). Chichester: Ellis Horwood.

Haines, C., & Houston, K. (2001). Assessing Student Project Work. In D. Holton (Ed.), *Teaching and Learning Mathematics at Undergraduate Level* (pp. 431 - 442). Dordrecht: Kluwer.

Haines, C., & Crouch, R. (2001). Recognizing constructs within mathematical modeling. *Teaching mathematics and its applications 20*(3), 129-138.

Hall, G. (1984). The Assessment of Modelling Projects. In J.S. Berry, D.N. Burghes, I.D. Huntley, D.J.G. James, & A.O. Moscardini (Eds.), *Teaching and Applying Mathematical Modelling* (pp. 143 –148). Chichester: Ellis Horwood.

Houston, K., & Neill, N. (2003a) Investigating students' modelling skills. In Q.-X. Ye, W. Blum, K. Houston, & Q.-Y. Jiang (Eds.), *Mathematical modelling in education and culture* (pp. 54-66). Chichester: Ellis Horwood.

Houston, K., & Neill, N. (2003b) Assessing modelling skills. In S.J. Lamon, W.A. Parker, & S.K. Houston (Eds.), *Mathematical modelling: a Way of Life – ICTMA 11* (pp. 155-164). Chichester: Ellis Horwood.

Izard, J., Haines, C., Crouch, R., Houston, K., & Neill, N. (2003). Assessing the Impact of Teaching Mathematical Modelling: Some Implications. In S.J. Lamon, W.A. Parker, & S.K. Houston (Eds.), *Mathematical modelling: a Way of Life – ICTMA 11* (pp. 165-178). Chichester: Ellis Horwood.

Penrose, O. (1978). How can we teach Mathematical Modelling? *J Mathematical Modelling for Teachers 1*(2), 31-42.

Chapter 3.3.5

THE ARITHMETIC OPERATIONS AS MATHEMATICAL MODELS

Zalman Usiskin
The University of Chicago, USA, Email: z-usiskin@uchicago.edu

Abstract: Mathematical modelling begins in the early primary grades even though the language and ideas of mathematical modelling are not employed. The common arithmetic operations are mathematical models for various counting and measure situations found in the real world. These models parallel the theoretical properties of the operations and provide the basis for more sophisticated mathematical models found in algebra, geometry, analysis, and statistics. The advantages of drawing attention early in instruction to modelling acts involving arithmetic operations are outlined.

1. INTRODUCTION

As illustrated in this volume, mathematical modelling is predominantly treated as an advanced topic to be introduced in courses at the tertiary or, to an increasing extent, secondary level of instruction. These courses typically assume that the learner has little or no experience in mathematical modelling, and start from scratch, with definitions of what is meant by a mathematical model and some broad description of the process (e.g., Meerschaert, 1993). The definitions of mathematical model and the modelling process, however, do not require mathematics at the secondary or tertiary level. Indeed, the process is often described in such a way that makes mathematical modelling synonymous with what might be termed real-world problem solving in the sense of a problem being a situation for which the solver has no algorithm (e.g., Dossey et al., 2002). This similarity suggests that modelling might begin as early as other mathematical problem solving, namely at the primary level, but in the literature one finds very little discussion of the broad principles of modelling in the context of primary school. For example, Blum (1991), in a clear and concise overview of the subject, notes a number of real-world examples of modelling using arithmetic: genetics, rates of interest, price index, income tax, elections, and musical scales. In each case, however, other mathematical topics are listed along with arithmetic, and when the aims of mathematical modelling are summarized in a table (p. 16), it seems as if there is no place for mathematical modelling in the curriculum before the lower secondary level.

With these practices, we make modelling seem as if it is an advanced and relatively obscure idea. Yet mathematical models are often implicitly used in primary and secondary mathematics classrooms, though the concepts and language of modelling are absent and though scarcely any attention is given to discrimination training to enable students to distinguish cases in which operations furnish appropriate (at some level of precision) or inappropriate models for described situations. I argue here that applications of the four basic operations of arithmetic can be re-conceptualised as simple modelling exercises, with considerable benefits.

2. MODELS INVOLVING ADDITION

If you have 3 cookies and I have 5 cookies, then together we have 8 cookies, a result students find first by counting. This type of situation is so common that we give it and its generalization to x cookies and y cookies through addition little thought. If A and B are discrete finite sets with $N(A) = a$ and $N(B) = b$, then $N(A \cup B) = a + b$ (where $A \cup B$ is the union of the sets). This is as fundamental as the commutative or any other property of the operation. Throughout the world, this real-world model is used to teach students basic addition facts. Later, because of the ubiquity of such situations, students are asked to memorize answers when x and y are small whole numbers, and to learn algorithms for obtaining answers when x and y are large whole numbers.

After students apply the mathematical model of addition to answer certain counting problems involving small whole numbers, they are asked to apply the same model to situations such as populations in which the numbers are larger (and the answer cannot be found quickly by counting), to financial situations where the numbers are often written as decimals, and to recipes or probabilities where the numbers are written as fractions. But the language and the limitations of the model are seldom described. Students become so accustomed to the model that they apply it where it does not apply, to situations like the following:

1. Carl has 5 friends and Georges has 6 friends. Carl and Georges decide to give a party together. They invite all their friends. All friends are present. How many friends are there at the party? (Verschaffel et al., 2000, p. 19).
2. The price of a chair is reduced 20% on sale, and then its sale price is reduced by 10%. What is the total reduction?
3. A cup of milk is added to a cup of popcorn. How many cups of the mixture will result? (Davis & Hersh, 1981, p. 71)
4. What will be the temperature of water in a container if you pour 1 jug of water at 80° F and 1 jug of water at 40° F into it? (Nesher, 1980, p. 46)

The resolution of the incorrect application is different in each of these situations but is similar to the resolution of more advanced models. To resolve (1), we refine the model to encompass situations in which there is overlap: $N(A \cup B) = N(A) + N(B) - N(A \cap B)$ (where $A \cap B$ is the intersection of the sets). For (2), though we could use a generalization of this refined model (the percent reduction is 10% + 20% – 10%·20%), the situation is more insightfully viewed as involving multiplicative models. A reduction of 20% is equivalent to a size change or scale factor of 80%. Then the situation calls for the application of scale factors of 80% and 90%, for a total multiplicative change of 80%·90%, or 72%, a reduction of 28%. The resolutions of (3) and (4) are more complex, and must take into account the chemistry and physics, respectively, of the situations.

In geometry, the model is generalized to determine the total length of segments placed end to end, as in calculating perimeter. It is applied to determine the measure of the angle formed by the outer rays of two adjacent angles, the area of the union of disjoint planar regions, and the volume of the union of disjoint 3-dimensional solids. In these situations, the property may be called Angle Addition or an Additive Property of Area or an Additive Property of Volume. In combinatorics or probability, the model is typically identified as a Fundamental Counting Principle. Researchers in mathematics education have called this model the Putting-Together Model for Addition or Part-Part-Whole.

Thus, through all of schooling in mathematics, this single core model appears, but its appearance is disguised in various forms and quite different settings. For this reason, to most students, these applications do not share a commonality, so the student misses an extraordinarily important point: A fundamental reason for all students to learn addition is the ubiquity of important applications of the Putting-Together Model.

The Putting-Together Model by no means encompasses all of the applications in which addition of numbers is involved. Indeed, Davis and Hersh (1981, p. 74) declare "There is and there can be no comprehensive systematization of all the situations in which it is appropriate to add". Suppose a temperature of –4° C were to increase by 15° C. We find the answer by the addition –4 + 15 = 11. This addition can be interpreted as a putting-together situation only if one stretches the idea of putting together. It is easier to think of this as the same mathematical model applied to a different set of situations, those involving slides or shifts. In fact, in most textbooks, the geometric idea of slides is used to reinforce or to determine the rules for addition of positive and negative numbers. Students see the geometry as a device or rule to obtain sums and do not realize that, through this process, addition is again a model for a set of real situations. We call it the Slide Model of Addition: If a slide x is followed by a slide y, the result is a slide x+y. This model accounts for applications of complex number and vector addition but its study begins in late primary or early secondary school.

3. MODELS INVOLVING SUBTRACTION

Two models for subtraction have long been in the literature, namely take-away (if a quantity y is taken away from an original quantity x, the quantity left is x – y) and comparison (the quantity x – y tells how much y is less than the quantity x). In English, the two most common names for the answer ("remainder" for a take-away situation, "difference" for a comparison) reflect the different feels that these models have to the user. These models are first encountered in small whole-number situations but later extended to all positive numbers and, later, real numbers, and to the geometry of length, area, and volume. Comparison has its own special cases: change and directed error, and (with the help of absolute value) undirected error and distance on the number line. Thus, as with the addition models, these models appear in different forms and settings, so that the learner does not usually realize the common features.

Some books treat the Putting-Together Model for Addition and the Take-Away Model for Subtraction within a single model scheme: Part-Part-Whole. In the same way, the Slide Model for Addition can be related to the Comparison Model for Subtraction within a single model scheme: Start-Shift-Finish. This amalgamation of models parallels the usual relationship between addition and subtraction in mathematical theory (a – b = c if and only if a = c + b), where subtraction is defined in terms of addition and is not treated by itself. The other way of defining subtraction in terms of addition (a – b = a + -b), which students encounter when subtracting positive and negative numbers, also is interpretable in terms of models for the operations. A situation in which a temperature of 10° goes down 7° can be viewed as Addition Shift 10 + (–7) or as a Subtraction Shift (a new model) 10 – 7. Thus we see that there is a structure to the common models of addition and subtraction that parallels the (pure) mathematical theory of these operations.

4. MODELS INVOLVING MULTIPLICATION

Whereas the counting of real objects is almost universally present in the learning of addition facts, multiplication tends to be defined initially as repeated addition and multiplication "facts" are often memorized by students with no real world situation to check them against, even though multiplication models rich and important situations. The lack of connection with the real world results in many children lacking the development of multiplicative reasoning patterns (e.g., see Harel & Confrey, 1994).

One way of categorizing the classes of situations which model, or are modeled by, multiplication is by the units of the quantities being multiplied. Multiplication by a scalar covers a class of applications grouped as the Size Change Model of Multiplication: When a quantity x is multiplied by a scalar k, $k > 0$, then the product kx is k times the size of the original. Scalar multiplication includes among its applications the "part of" situations resulting from wanting to find a fraction or a percent of a quantity, the "times as many" situations resulting from wanting to enlarge a quantity by a certain factor, and the size change transformations in geometry that result in similar figures. Discounts, taxes, simple interest, scale models, expansions, and contractions all fall under this framework. By viewing multiplication by -1 as changing direction through 180°, the geometry can be extended to explain multiplication by negative numbers. This further extends to the view of multiplication by the complex number z as combining a size change of $|z|$ with a rotation of Arg(z). Repeatedly multiplying by different scalars explains the multiple discount problem mentioned earlier as an application of putting-together addition. Repeatedly multiplying by the same scalar leads to discrete models of exponentiation such as are used in the calculation of compound interest.

Multiplication by a quantity with a unit also covers a broad class of applications. One type in this class fits the Area Model of Multiplication: The area of a rectangle with length x units and width y units is xy square units. The discrete version is sometimes used as a check of an answer to a multiplication fact: The number of elements in a rectangular array with x rows and y columns is xy. The volume of a rectangular solid extends this model to three dimensions.

In calculating the area of a rectangle, we may think of its length as acting across its width. By summing many rectangles and taking a limit, the area model generalizes to give the area interpretation of direct integrals in calculus. When one factor is a rate (as in the formula distance = rate × time, or total cost = number of items × unit cost), then the rate acts across the other quantity. This describes the Acting Across or Rate Factor Model of Multiplication: When a rate x of unit 1 per unit 2 acts across a quantity y of unit 2, the total is xy unit 1.

A student equipped with these multiplicative models is far more likely to understand why, when y = kx, y varies directly as x, or why the total number of students in a school can be found by multiplying the average number of students per grade by the number of grades. Unfortunately, many students often know multiplication only as repeated addition, and they consequently find it difficult to find any real world explanation or use of multiplication of fractions, decimals, or positive and negative numbers. Yet multiplication is taught, and knowledge of multiplication is required of all students, because of the ubiquity of important applications of various models of multiplication.

5. MODELS INVOLVING DIVISION

The mathematics education literature has long distinguished two models for division: partitive (or partition) and quotitive (or measurement). Sutherland (1947) divided these into six different categories. Two of these categories, ratio (partition) and rate (measurement), are found in an analysis of operations by Usiskin and Bell (1983) and, with the other models mentioned here, applied in materials for early secondary school students (University of Chicago School Mathematics Project, 2002). The ratio model applies when x and y are quantities with the same units, where x/y tells how many y's are in x (or what part of y that x is). The rate model applies when x and y are quantities with different units, where x/y is the amount of quantity x per quantity y. In this conception, ratios are scalars, while rates are unitized quantities.

In the measurement model described in the literature, the divisor is most often a rate, as when there are 18 cookies and 2 cookies per child are to be served, and we wish to know how many children can be served, or when one has $18 to spend on items that are $2 each and one wants to know how many items can be bought. But most people mentally change the rate to a simple quantity; 2 cookies per child becomes 2 cookies, and $2 each becomes $2, and then the problem is reduced to a ratio division.

Multiplication and division are theoretically related as addition and subtraction are, so, correspondingly, their models can be grouped together. The Rate Model of Division and the Rate Factor Model for Multiplication are related by the definition of division $a \div b = c$ if and only if $a = c \cdot b$, as are the Ratio Model of Division and the Size Change Model of Multiplication.

Like addition and subtraction, many other classes of situations are modeled by multiplication and division (Greer, 1992).

6. THE IMPORTANCE OF EARLY EXPOSURE TO MODELLING IDEAS

When no language of modelling and models to describe the applications of arithmetic is present, it becomes more difficult for children to apply arithmetic than it ought to be. Given a word problem with certain given numbers, operations are performed on those numbers without a solid base underlying the selection of the operation. Moreover, the lack of systemic exposure to counterexamples renders students vulnerable to problems in which linguistic or other cues suggest an inappropriate operation, as thoroughly documented and analysed in Verschaffel et al. (2000) and illustrated for illusory applications of proportionality by De Bock et al. (this volume, Chapter 3.3.3).

Models of the operations give a basis for applying mathematics; they are the postulates that connect mathematics to the world of real and fanciful problems. The properties of (applied) models of the operations should be treated as we do the (theoretical) properties of the operations. When models for arithmetic operations act as postulates, the theorems are statements deducible from the models. A very large number of examples of theorems of this sort exist, within which models involving single operations are combined within more complex models. For example, we can combine models for addition and multiplication to explain why some situations lead to linear models of the form $y = mx + b$ or $Ax + By = C$. We can use the Rate Model for Division and the Comparison Model for Subtraction to explain why the formula for slope involves two subtractions and a division. With models for exponentiation in arithmetic, we can derive exponential models. And, just as importantly, we can examine why models in some circumstances are descriptive and not necessarily causal, as is often the case when linear regression is used to obtain equations of lines to fit data. And on and on.

It seems worthy of a study of applications and modelling in mathematics education to consider the contributions that discussions of models of arithmetic operations and other primary school content do make, could make, and should make in the development of the skills and concepts necessary to be a competent user of mathematics. While there may be pitfalls in taking the modelling perspective seriously in early schooling (Verschaffel, 2002), they are heavily outweighed by the advantages.

REFERENCES

Blum, W. (1991). Applications and modelling in mathematics teaching: A review of arguments and instructional aspects. In M. Niss, W. Blum, & I. Huntley (Eds.), *Teaching of mathematical modelling and applications* (pp. 10-29). Chichester: Ellis Horwood.

Davis, P.J., & Hersh, R. (1981). *The mathematical experience*. Boston: Birkhauser.

De Bock, D., Van Dooren, W., & Janssens, D. (2005). Studying and remedying students' modelling competencies: Routine versus adaptive expertise. This volume, Chapter 3.3.3.

Dossey, J.A., McCrone, S., Giordano, F.R., & Weir, M.D. (2002). *Mathematics methods and modelling for today's mathematics classroom.* Pacific Grove, CA: Brooks/Cole.

Greer, B. (1992). Multiplication and division as models of situations. In D. Grouws (Ed.), *Handbook of research on mathematics teaching and learning* (pp. 276-295). New York: Macmillan.

Harel, G., & Confrey, J. (Eds.). (1994). *The development of multiplicative reasoning in the learning of mathematics*. Albany, NY: State University of New York Press.

Meerschaert, M. M. (1993). *Mathematical modeling.* New York: Academic Press.

Nesher, P. (1980). The stereotyped nature of school word problems. *For the Learning of Mathematics 1*(1), 41-48.

Sutherland, E. (1947). One-step problem patterns and their relation to problem solving in arithmetic. *Contributions to Education, No 925.* New York: Teachers College, Columbia University.

University of Chicago School Mathematics Project (2002): *Transition mathematics.* Glenview, IL: Prentice Hall.

Usiskin, Z., & Bell, M. (1983). *Applying arithmetic: A handbook of applications of arithmetic. Volume II: Operations.* Arithmetic and Its Applications Project, ERIC document ED 264 088, Chicago, IL.

Verschaffel, L. (2002). Taking the modelling perspective seriously at the elementary school level: Promises and pitfalls [Plenary talk]. In A.D. Cockburn, & E. Nardi (Eds.), *Proceedings of the 26th Annual Conference of the International Group for the Psychology of Mathematics Educatio.* Norwich, England: University of East Anglia.

Verschaffel, L., Greer, B., & De Corte, E. (2000). *Making sense of word problems.* Lisse, The Netherlands: Swets and Zeitlinger.

Section 3.4

APPLICATIONS & MODELLING FOR MATHEMATICS

Edited by
Hugh Burkhardt and Eric Muller

Chapter 3.4.0

APPLICATIONS AND MODELLING FOR MATHEMATICS – OVERVIEW

Eric Muller and Hugh Burkhardt

Brock University, St. Catharines, Ontario, Canada, Email: emuller@brocku.ca / Shell Centre for Mathematical Education, Nottingham, UK, Email: Hugh.Burkhardt@nottingham.ac.uk

Abstract: These five chapters address important issues on mathematics teaching and learning. They include, amongst others, how Applications and Modelling help students learn mathematics in ways that result in a deep and holistic understanding; are central to the development of Mathematical Literacy, and; are enriched by the creative use of technology.

1. INTRODUCTION

Mathematics has long had a large slice of curriculum time in every country, mainly because of its perceived utility[1] in solving problems that people face in some other school subjects, and in life and work. We teach and we learn mathematics to develop:

- a powerful toolkit of mathematical strategies, concepts and skills, and
- competency in using it to tackle problems from the "real world" or as Henry Pollak (1979), a pioneer in this area, has suggested, the "rest of the world".

Modelling competence is essential for such problem solving and much of this book is about how it develops with appropriate teaching. In these five chapters we focus on the benefits that flow the other way – the contributions of modelling activities to the development of other mathematical competencies.

What are these other mathematical competencies?

As with other complex activities, there are many descriptions of mathematical performance; they have much in common but each has a different emphasis, particularly on the centrality they give to modelling. The Danish KOM model (KOM, 2002), described in Chapter 2.2 of this volume, identifies eight competencies of two types: (a) asking and answering mathematical questions (which requires mathematical thinking, problem tackling, modeling, and reasoning) and (b) dealing with mathematical language (through representation, symbols and formalism, communicating, and tools). The US National Council of Teachers of Mathematics (NCTM, 2000) also identifies two types of mathematical competencies: content (number and operations, algebra, geometry, measurement, data analysis and probability) and processes (problem solving, reasoning and proof, communication, connections, representation). The U.S. National Research Council (NRC, 2001) identifies five "interwoven and interdependent" strands of mathematical proficiency: conceptual understanding, procedural fluency, strategic competence, adaptive reasoning, and productive disposition. The complementary UK Tomlinson (2004) and Smith (2004) reports stress respectively the importance of "functional mathematics" and of "mathematics: for its own sake; for the knowledge economy; for science, technology and engineering; for the workplace; and for the citizen". These analyses all envisage the same range of mathematical practices that, developed in classrooms, will help students to learn mathematics in ways that result in deep understanding and the ability to use mathematics where it matters.

Characteristic of all these schemes of classification is a much broader view of "doing mathematics" than is shown in most traditional curricula. Each places emphasis on mathematical processes. In sharp and deliberate contrast, most school programs, most teachers and parents, and most official "high-stakes" school examinations treat mathematics as only the grammar and syntax of mathematical language, and often only a small procedural subset of these. The broader view requires that students engage with mathematics as a connected whole, not just as a succession of separate topics, chapters, and formulas. Time, of course, makes teaching sequential; in contrast, the multiple connections that are essential to a robust understanding of mathematics do not arise naturally – they require learning activities specifically designed to develop them.

Mathematical models of authentic situations do this well. They reveal more readily than do artificial textbook problems that, to be effective, mathematics must be approached holistically rather than as an accumulation of bits and pieces of de-contextualized knowledge. Although the development of mathematical expertise has traditionally been approached by decomposing problems into component skills that are taught separately, evidence from many sources (e.g. Schoenfeld, 1992, de Corte et al., 1996) shows that this is not an effective way to build expertise. In "doing mathematics" the whole is much more than the sum of the parts. Neither is it easier to learn in fragments. "It is harder, not easier, to understand something broken down into all the precise little rules than to grasp it as a whole" (Thurston, 1990).

In considering applications and modelling in relation to other mathematical competencies, it is important to distinguish two different types of application, shown in Fig. 3.4.0-1 (due to Malcolm Swan, see Chapter 3.4.1).

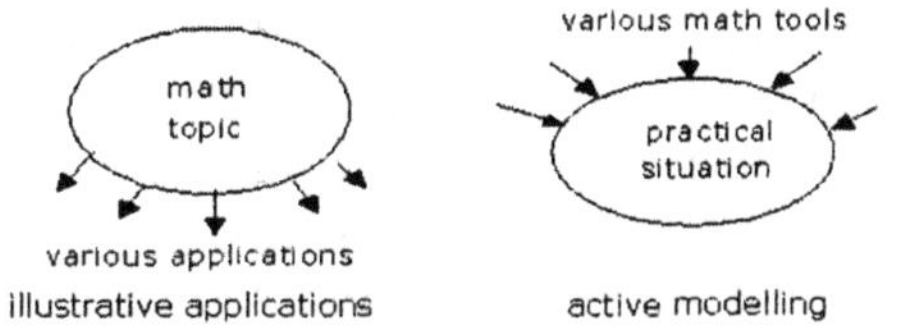

Figure 3.4.0-1. Goal types for applications

Most curricula offer *illustrative applications*; there the focus is on a specific mathematical topic, showing the various practical domains where it can be useful and practising its use in those contexts. The student has no doubt as to the mathematics to be used – it is the topic just taught. In contrast, in *active modelling* the focus is on the practical situation and understanding it better. Usually, a variety of mathematical tools will be useful for different aspects of the analysis. (This is a good indication as to the real goal.) Choosing and using tools appropriately is a major part of the challenge to the student. Both types of activity are important in learning mathematics. Both provide connections between mathematics and practical situations. However only active modelling, as opposed to learning models, involves the full range of mathematical competencies. Modelling is all about *applied power.*

Thus context-based mathematical modelling provides ideal settings to blend content and process so as to produce flexible mathematical competence. The iterative self-correcting cycle of asking questions, using tools, producing answers, and then asking new questions helps *students develop the cognitive connections* required to understand mathematics as a discipline. Concrete, contextualized models can be especially effective as a glue that binds together in the minds of students the many abstract and otherwise disconnected facets of mathematics.

Another complementary role played by modelling in developing mathematical competence is *enhancing student motivation.* Students confronted with appealing applications and models will learn, from direct experience, convincing answers to the universal question that plagues mathematics teachers everywhere: "Where am I going to use this?"

Although connecting mathematics to authentic contexts helps make mathematics meaningful, it demands delicate balance. On the one hand, contextual details camouflage broad patterns that are the essence of mathematics; on the other hand, these same details offer associations that are critically important for many students' long-term learning. Few can doubt that the tradition of de-contextualized mathematics instruction has failed the many students who leave high school with neither the usable mathematical skills nor the quantitative confidence required for today's society. The tradition of formal mathematics, used mainly as a 'gatekeeper' to future academic study, leaves many able students both innumerate and undereducated. However, when the traditional symbol-intensive curriculum is anchored in authentic applications and modelling by the students, many will reveal aptitude for mathematics that was previously undeveloped. A diverse curriculum featuring both abstract and applied mathematics can help break the rigidity of traditional expectations and enable more students to achieve higher levels of mathematical competence. In this chapter we discuss how, with appropriate teaching, modelling competence can support the development of other mathematical competencies in the learning process.

2. PAPER SUMMARIES

Applications and Modelling for Mathematics is structured in five chapters. After this introduction, Swan, Turner and Yoon describe, analyse, and provide examples on ways modelling encourages the asking and the answering of mathematical questions, and how it promotes the use of mathematical language. They highlight the fact that, in modelling situations, students develop mathematical expertise based on an integrated field of knowledge, make multiple connections both within and outside mathematics, and not only reinforce their mathematical understanding but also develop new mathematical knowledge. The third chapter is on *Mathematical Literacy*. Steen and Turner describe what ML means, the kinds of problem it involves and how it is developed, along with an outline of some contentious issues. In the fourth chapter, Antonius, Haines, Jensen and Niss discuss the pattern of classroom activities needed, and the roles of the teacher, in supporting the learning of other mathematical competencies through modelling and applications. The fifth chapter explores uses and possibilities of various technologies in mathematical modelling while it focuses on the development of other mathematical competencies. Pead provides multiple examples from the secondary school level while Ralph describes a university modelling program that is technology-centred.

We conclude this overview by drawing attention to a few of the key questions that need further research in depth, and associated development:

- In what ways do concrete applications and active modelling build understanding of mathematical concepts?
- How far do the extended chains of reasoning involved in modelling real

situations encourage students to improve the reliability of their technical skills in mathematics?

- How far does modelling improve the performance of students in pure mathematical problem solving, and in which aspects?
- How can typical mathematics teachers be enabled to effectively help their students in their classroom to learn to handle real world problems?
- In what ways can technology help enough with all these goals, so as to encourage teachers and schools to make the necessary investments
- How can mathematical literacy be 'sold' to teachers and to school systems, as a prime goal of mathematics education?

Reliable answers to these questions will need warrants for their generality from replication in many parallel but diverse projects and school systems (Burkhardt & Schoenfeld, 2003).

REFERENCES (FOR ALL CHAPTERS IN SECTION 3.4)

Alrø, H., & Skovsmose, O. (2003). *Dialogue and Learning in Mathematics Education. Intention, Reflection, Critique*. Dordrecht: Kluwer.

ARISE. (1999). *Applications based reform in secondary education*. Freudenthal Institute, The Netherlands.(see http://www.onderzoekinformatie.nl/en/oi/nod/onderzoek/OND1286563/).

Berry, J. S., & Davies, A. (1996). In C.R. Haines, & S. Dunthorne (Eds), *Mathematics Learning and assessment: Sharing Innovative Practices*, (pp. 3.3-3.11). London: Arnold.

Berry, J., & Houston, K. (1995). Students using posters for communication and assessment. *Educational Studies in Mathematics, 29*(1), 21-27.

Berry, J., & Houston, K. (2004). Investigating students working styles in mathematical modeling activities. In H-W. Henn, & W. Blum (Eds.), *ICMI Study 14: Applications and Modelling in Mathematics Education, Pre-Conference Volume*, (pp. 35-40). Dortmund: University of Dortmund.

Blomhøj, H., & Jensen, T. H. (2003). Developing mathematical modelling competence: Conceptual clarification and educational planning. *Teaching Mathematics and its Applications, 22*(3), 123-139.

Brock-a (n.d.). (see http://www.brocku.ca/mathematics/resources/learningtools/index.php).

Brock-b (n.d.). (see http://www.brocku.ca/ctl/pdf/Brock_Teaching_1.pdf).

Brousseau, G. (2003). *Glossaire de quelques concepts de la théorie des situation, p. 24.* (see http://perso.wanadoo.fr/daest/guy-brousseau/textes/Glossaire_Brousseau.pdf).

Burkhardt, H. (1981). *The Real World and Mathematics.* Glasgow: Blackie. Reprinted 2000, Shell Centre Publications, Nottingham. (see http://www.mathshell .com /scp/index.htm).

Burkhardt, H., & Schoenfeld, A. H. (2003). Improving Educational Research: towards a more useful, more influential and better funded enterprise. *Educational Researcher 32,* 3-14.

Burkhardt, H. (2004). Assessing Mathematical Proficiency: What is important? In A. H. Schoenfeld (Ed.) *Assessing Students' Mathematics Learning: Issues, Costs and Benefits.* Volume XXX. Mathematical Sciences Research Institute Publications. Cambridge: Cambridge University Press.

Burkhardt, H. (2006). From design research to large-scale impact: Engineering research in education. In J. Van den Akker, K. Gravemeijer, S. McKenney, & N. Nieveen (Eds.), *Educational design research.* London: Routledge.

C++ programming language (n.d.). (see http://www.research.att.com /~bs/C++.html).

Cabri, Cabri Géomètre (n.d.). (see http://www-cabri.imag.fr/).

Coypu (n.d.). (see http://www.mathshell.com/).

Crouch, R., M., & Le Masurier, D. (1996). In C.R. Haines, & S. Dunthorne (Eds.), M*athematics Learning and assessment: Sharing Innovative Practices,* (pp. 3.20-3.30). London: Arnold.

Crowther Report 15-18 (1959). *A Report of the Central Advisory Council for Education,* HMSO, London.

De Bock, D., & Roelens, M. (1993). The journey of the drilling rig Yatzy: Today on television, tomorrow as a large scale modelling exercise. In J. de Lange, I. Huntley, C. Keitel, & M. Niss (Eds.), *Innovation in Maths Education by Modelling and Applications,* (pp. 211-224). Chichester: Ellis Horwood.

de Corte, E., Verschaffel, L., & Greer, B. (1996). Mathematics teaching and learning. In D. Berliner, & R. Calfee (Eds.), *Handbook of Educational Psychology,* (pp. 491-549). New York: MacMillan.

Dunn, R.S., & Dunn, K.J. (1998). *The Complete Guide to the Learning Styles Inservice System.* Boston: Pearson, Allyn & Bacon.

Fathom (n.d.). (see http://www.keypress.com/fathom).

Ferri, R.B. (2004). Mathematical Thinking Styles and Word problems. In Henn, H.-W. & Blum, W. (Eds.), *ICMI Study 14: Applications and Modelling in Mathematics Education, Pre-Conference Volume,* (pp. 47-52). Dortmund: University of Dortmund.

FI, Freudenthal Institute (n.d.). Utrecht, Netherlands. (see http://www.onderzoekinformatie.nl/en/oi/nod/organisatie/ORG1235806/).

Giordano, F. & Weir, M. (1991). *Differential Equations, A Modelling Approach.* Menlo Park: Addison-Wesley.

GSP, Geometer's Sketchpad (n.d.). (see http://www.keypress.com/sketchpad/).

GS, Geometric Supposer (n.d.). Schwartz, J., Yerushalmy, M., Shternberg, B. (see http://cet.ac.il/math-international/software5.htm).

GT, Graphing tools (n.d). (see http://www.freedownloadscenter.com/Multimedia_and_Graphics/Mathematical_Graphing_Tools/).

Haines, C. R., & Dunthorne, S. (Eds.). (1996). *Mathematics Learning and Assessment.* London: Arnold.

Haines, C.R., & Le Masurier, D.W. (1986). Model validation – interactive lecture demonstrations in Newtonian Mechanics. In J.S. Berry et al. (Eds.), *Mathematical Modelling Methodology, Models and Micros,* (pp. 115-129). Chichester: Ellis Horwood.

ICTMA (n.d.). A complete listing of ICTMA Conference Proceedings. (see http://www.infj.ulst.ac.uk/ictma/books.html).

Ikeda, T., & Stephens, M. (2001). The effects of students' discussion in mathematical modeling. In J.F.Matos, W.Blum, K.Houston, & S.P.Carreira (Eds.), *Modelling and Mathematics Education: ICTMA9 Applications in Science and Technology,* (pp. 381-400). Chichester: Ellis Horwood.

JAVA programming language (n.d.). (see http://java.sun.com/).

JavaMath (n.d.). (see http://smard.cqu.edu.au/Database/Teaching /JavaMath.html).

Jensen, T. H. (to appear). *Udvikling af matematisk modelleringskompetence som matematikundervisningens omdrejningspunkt – hvorfor ikke?* Unpublished doctoral dissertation, Roskilde University, Denmark.

KOM (2002). *Competencies and mathematical learning.* Niss, M., (Chair), Roskilde University, Denmark.

Laborde, C. (1994). Working in small groups: A learning situation? In Biehler et al. (Eds.), *Didactics of Mathematics as a Scientific Discipline.* Dordrecht: Kluwer.

Lesh, R. A., & Doerr, H. M. (Eds.). (2002). *Beyond constructivism: Models and modeling perspectives on mathematics teaching, learning, and problem solving.* Mahwah, New Jersey: Lawrence Erlbaum Associates.

Lesh, R., Post, T., & Behr, M. (1987). Representations and translations among representations in mathematics learning and problem solving. In C. Janvier (Ed.), *Problems of representation in teaching and learning mathematics.* Hillsdale, New Jersey: Lawrence Erlbaum.

Lesh, R., Yoon, C., & Zawojewski, J. (in press). John Dewey Revisited – Making Mathematics Practical VERSUS Making Practice Mathematical. In R. Lesh, E. Hamilton, & J. Kaput (Eds.), *Foundations for the Future in Mathematics Education.* Hillsdale, New Jersey: Lawrence Erlbaum.

Maple (n.d.). (see http://www.maplesoft.com/).

MARS (2002-2004). Swan, M., Crust, R., & Pead, D., with the Shell Centre team for the Qualifications and Curriculum Authority. *World Class Tests of Problem Solving in Mathematics Science and Technology.* London: nferNelson.

MARS (2004) Swan, M., Crust, R., & Pead, D., with the Shell Centre team. *Developing Problem Solving,* 6 teaching and learning modules. London: nferNelson.

Mason, J., Burton, L., & Stacey, K. (1982) *Thinking Mathematically.* Boston: Addison Wesley.

Mathematics Archives – Topics in Mathematics (n.d.). (see http://archives.math.utk.edu/cgibin/intertive.html).

Matsuzaki, A. (2004). 'Rub' and 'Stray' of Mathematical Modelling In H-W. Henn & W. Blum (Eds.), *ICMI Study 14: Applications and Modelling in Mathematics Education, Pre-Conference Volume,* (pp. 181-186). Dortmund: University of Dortmund.

MDM4U (2002). *Mathematics for Data Management*, Ontario Grade 12 Course. (see http://www.brocku.ca/cmt/mdm4u).

Merlot, Multimedia Educational Resource for Learning and Online Teaching (n.d.). (see http://www.merlot.org).

Muller, E. R. (2001). Reflections on the sustained use of technology in undergraduate Mathematics education. In D. Holton (Ed), *The Teaching and Learning of Mathematics at the University Level: An ICMI Study*, (pp. 381-394). Dordrecht: Kluwer.

NCTM, National Council of Teachers of Mathematics (2000). *Principles and Standards for School Mathematics*, Reston. (see http://www.nctm.org /standards/).

NRC, National Research Council (2001). *Adding it Up: Helping Children Learn Mathematics.* Washington, DC: National Academy Press.

NSDL, National Science, Mathematics, Engineering and Technology Education Digital Library (n.d.). (see http://nsdl.org/).

NTPS, Numeracy Through Problem Solving (n.d.). Swan, M., Binns, B., & Gillespie, J., and Burkhardt, H, with the Shell Centre team. (1987-89). Harlow: Longman. (see http://www.mathshell.com/scp/index.htm).

OECD (2001). *Knowledge and Skills for Life - First Results from PISA 2000.* Paris: OECD. (see http://www.pisa.oecd.org/dataoecd/44/53/33691596.pdf).

OECD (2003). *The PISA 2003 Assessment Framework: Mathematics, Reading, Science and Problem Solving Knowledge and Skills*, Paris: OECD. (see https://www.pisa.oecd.org/dataoecd/38/51/33707192.pdf).

OECD (2004). *Learning for Tomorrow's World – First Results from PISA 2003.* Paris: OECD. (see http://www.pisa.oecd.org/dataoecd/1/60/ 34002216.pdf).

Ottesen, J., T. (2001). Do not ask what mathematics can do for modelling: Ask what modelling can do for mathematics. In D.Holton (Ed.), *The Teaching and Learning of Mathematics at the University Level: An ICMI Study*, (pp. 335-346). Dordrecht: Kluwer.

PISA, Program for International Student Assessment (n.d.). (see http://www.pisa.oecd.org/pages/0,2987,en_32252351_32235731_1_1_1_1_1,00.html).
Pollak, H.,O. (1979). The Interaction between Mathematics and other School Subjects. In UNESCO (Ed.), *New Trends in Mathematics Teaching IV*, (pp. 232-248). Paris.
Ralph, B. (1999). *Journey Through Calculus.* Brookes/Cole.
RME, Realistic Mathematics Education: (see http://www.fi.uu.nl/en/rme/TOURdef+ref.pdf).
Schoenfeld, A. H. (1992). Learning to think mathematically: Problem solving, metacognition, and sense-making in mathematics. In D. Grouws (Ed.), *Handbook for Research on Mathematics Teaching and Learning,* (pp. 334-370). New Yori: MacMillan.
Shell Centre (1985). Swan, M., Pitts, J., Fraser, R., and Burkhardt, H, with the Shell Centre team. *Problems with Patterns and Numbers,* Joint Matriculation Board, Manchester. (see http://www.mathshell.com/scp/index.htm).
SMILE (n.d.). (see http://www.smilemathematics.co.uk/
Smith Report. (2004). *Making Mathematics Count*, Department For Education and Skills, HMSO, London. (see http://www.mathsinquiry.org.uk/report/index.html).
Steen, L. A. (Ed). (2002). *Mathematics and Democracy: the case for quantitative literacy,* the National Council on Education and the Disciplines (NCED), USA. (see http://www.maa.org/ql/mathanddemocracy.html).
Steen. L. A., & Foreman, L. S. (2001). *Why Math? Applications in Science, Engineering, and Technological Programs*, Research Brief, American Association of Community Colleges.
Swan, M., Pitts, J., Fraser, R., & Burkhardt, H, with the Shell Centre team (1985). *The Language of Functions and Graphs,* Joint Matriculation Board, Manchester. (see http://www.mathshell.com/scp/index.htm).
Thurston, P. (1990). Mathematics Education. *Notices of the American Mathematical Society, 37*(7), 844-850.
TIMSS (n.d.). Trends in (formerly Third) International Mathematics and Science Study. (see http://nces.ed.gov/timss/).
Tomlinson Report (2004). *14-19 Curriculum and Qualifications Reform.* Department For Education and Skills, HMSO, London. (see http://www.14-19reform.gov.uk/).
Treilibs, V., Burkhardt, H., & Low, B. (1980). *Formulation processes in mathematical modelling.* Nottingham: Shell Centre Publications.
UCSMP (1999). University of Chicago School Mathematics Project. Chicago. (see http://socialsciences.uchicago.edu/ucsmp/).
USMES (n.d.). Unified Sciences and Mathematics for Elementary Schools: (see http://books.nap.edu/openbook/0309052939/html/129.html).
Verschaffel, L., Greer, B., & De Corte, E. (2000). *Making sense of word problems.* The Netherlands: Swets and Zeitlinger.
VB.net (n.d.). Visual Basic.net, programming language. (see http://msdn.microsoft.com /vbasic).
Wisweb (n.d.). (see http://www.wisweb.nl).

---

[1] Other less "useful" subjects with comparable cultural importance and intellectual challenge, such as music, get far less time in schools.

Chapter 3.4.1

THE ROLES OF MODELLING IN LEARNING MATHEMATICS

Malcolm Swan, Ross Turner and Caroline Yoon, with Eric Muller
Shell Centre for Mathematical Education, University of Nottingham, UK, Email: Malcolm.Swan@nottingham.ac.uk / Australian Council for Educational Research, Camberwell, Victoria, Australia, Email: turner@acer.edu.au / Indiana University, Bloomington, Indiana, USA, Email: csyoon@indiana.edu

Abstract: This chapter illustrates how Applications and Modelling promote the learning of mathematics by developing the student's mathematical language and her use of tools, and by developing the learner's capacity to ask and answer questions in, with, and about mathematics.

1. INTRODUCTION

Modelling activities offer such powerful means of developing mathematical understanding and retaining mathematical knowledge that a number of projects have placed modelling at the centre of the mathematics curriculum. In the US, for example, the Applications Based Reform in Secondary Education project (ARISE, 1999) developed a high school mathematics curriculum based entirely on models and modelling. The University of Chicago School Mathematics Project (UCSMP, 1999) is built on the belief that modelling helps to develop ability both in pure and applied mathematics. In these projects mathematics educators have developed a vision of the teaching and learning of mathematics, where the teacher and the student travel freely between pure and applied mathematics – so much so that there is little distinction between the two. Around the world there are other examples of this approach, notably *Realistic Mathematics Education* (RME) from the Freudenthal Institute.

In this chapter we outline and illustrate how modelling promotes the learning of mathematics. Some examples from the Shell Centre's *Numeracy through Problem Solving* materials (NTPS, 1987-89) are used to show how core mathematical tools, both concepts and skills, can be developed through modelling activities. From the work of Lesh et al. (2002), we draw examples to illustrate how modelling promotes the development of mathematical abilities that are traditionally ignored in elementary texts, and yet are a core component of the mathematical activity done in both mathematics research in math-heavy fields like engineering and business, and in tackling everyday life problems.

It is our theme here that an effective way to promote the growth of mathematical competencies is to forge strong and explicit connections between mathematical knowledge on the one hand, and the contexts within which that knowledge can be used on the other. We must recognise the importance of giving students the opportunity to think about and make use of the mathematical features of their surroundings. Such a goal can be powerfully approached through a mixture (see Chapter 3.4.0) of *active modelling* and *illustrative applications* – or *mathematising reality* and *realising mathematics* respectively.

We will discuss and illustrate the two groups of the KOM framework: *dealing with mathematical language and tools* (this includes representing, using symbols, formalism and communicating) and *'asking and answering questions in, with, and about mathematics'* (this includes modelling as well as mathematical thinking, tackling problems and reasoning).

2. MODELLING DEVELOPS MATHEMATICAL LANGUAGE AND TOOLS

In the 1980's, the Shell Centre for Mathematical Education and a UK examination board developed the curriculum and assessment scheme on *Numeracy through Problem Solving* (NTPS, 1987-89), built around five modules: *Design a Board Game, Produce a Quiz Show, Plan a Trip, Be a Paper Engineer* and *Be a Shrewd Chooser.* Though primarily aimed at developing mathematical literacy (see Chapter 3.4.2), they also show how modelling activities can promote representing, symbols and formalism, and tools competences.

Example 3.4.1-1: Learning to be pro-active with algebra
In *Be a Paper Engineer*, students design pop-up cards in various ways, as in the example illustrated in Fig. 3.4.1-1. In this type of design, the fold lines are parallel and one piece is glued to the base of the card. The design func-

tions correctly if the card pops up when it is opened flat on the table and if the interior piece folds only along the designated lines and becomes completely concealed within the base of the card as it is closed.

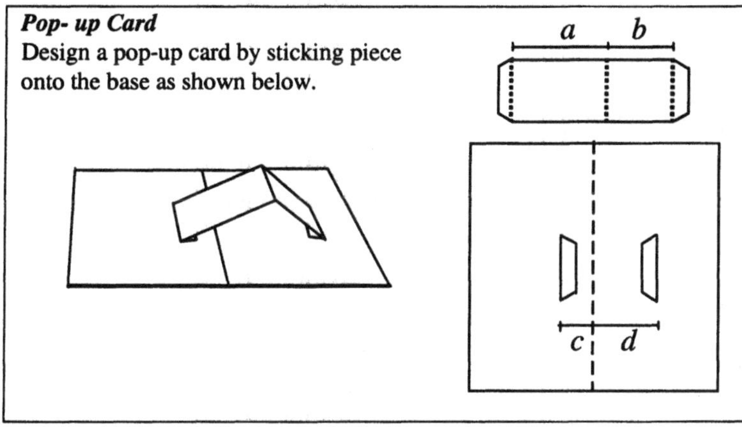

Figure 3.4.1-1. Pop-up card instructions

In exploring such designs, students formulate general laws such as:

- for the card to shut properly: $a-b = d-c$,
- for the card to open to 180°: $a+b \geq c+d$,
- for the interior not to 'stick out': *base length* $\geq 2(a+c)$ *or* $\geq 2(b+d)$.

Students generate equations and inequations in different forms and explore whether or not these are equivalent. A necessary and sufficient set is derived. More complex constructions (such as when the fold lines of the piece stuck on are not parallel) result in more advanced relationships.

Example 3.4.1-2: Understanding of the binomial distribution

In the *Great Horse Race* (Fig. 3.4.1-2) every student can make progress, including many who would normally have great difficulty with the binomial distribution of probabilities. All quickly recognise that Horse 1 is not a good bet.

A few have the misconception that because there are two dice, higher numbered horses will move more rapidly. Even they, through playing the game, soon realise that horses in the middle will move faster because "there are more ways of making 7 than 11 or 12". Most will enumerate these "1+6, 2+5, 3+4..."; some initially miss that 4+3 etc is different from 3+4 until the teacher suggests two different coloured dice. More advanced students consider the effect the length of the track on the likely outcome. This game thus proves an effective (and quick) stimulus to concept formation. Furthermore if every student colours the squares covered by each horse, they obtain a frequency distribution. When these are displayed around the classroom, students have an immediate visual of the variability of sample data. If the teacher enters these data in a spreadsheet, the frequency distribution of the totals will provide a better fit to the theoretical probabilities than will those of individual students.

The Great Horse Race Rules

1. Put the horses on their starting positions, 1 to 12.
2. Each player chooses a different horse.
   If there are only a few players, then each can choose two or three horses. The remaining horses are still in the race but no one owns them.
3. Roll two dice and add the scores.
4. The horse with that number moves one square forward.
5. The first horse past the finish wins.

Figure 3.4.1-2. Instructions for the Great Horse Race

When students work on modelling tasks, they express their thoughts in a variety of representations, such as words, diagrams, tables, spreadsheets, equations and graphs. Indeed, many "real world" mathematics problems begin with information provided in a variety of representations. Mathematisers utilise different representational media throughout the solution process to highlight structural nuances in the mathematical system being developed. Since any representational medium will highlight some aspects of the structure, while obscuring others, students who engage in modelling activities develop the ability to switch back and forth between different representational media – referred to as "translation skills" by Burkhardt (1981) and "representational fluency" by Lesh et al. (1987).

In contrast, most textbook problems rarely require students to use more than one representation at a time. Students tend to work almost exclusively within a symbolic representation. Since these textbook problems do not call for students to examine the same mathematical structure in different representations, students are less likely to gain the insight that is obtained through seeing different aspects of the structure in this way. For example, they may correctly work an equation algebraically by applying rules of symbolic manipulation without ever seeing the same equation being expressed graphically, and may consequently not realise that solving an equation corresponds to finding the intersection point of two function graphs.

Modelling thus provides the opportunity for (or even demands) the exploration of *alternative representations* of problem situations. Representational fluency is developed in that part of the modelling cycle where students forge links between the real world context and the mathematical expression of the kernel of the problem. This process develops a deeper connected understanding of the representations themselves, and of their interrelationships. A *realising mathematics* approach illustrates how a particular mathematical tool or technique represents features of some real world situation. As part of a *mathematising reality* modelling approach, the student can propose and explore alternative representations, or the teacher can present a variety of possible representations for students to use. Working with different representations facilitates reflection on, and evaluation of, alternative solutions with respect to the problem context.

Example 3.4.1-3: Using a spatial representation in a solution strategy.

In the process of designing a league tournament for 22 teams, from *Problems with Patterns and Numbers* (Shell Centre, 1984), some students wanted to know how many matches would be involved. Some began listing the pairings (AvB, AvC, AvD, ...) in an organised list. The student shown (Figure 3.4.1-3), however, used the link between a spatial and algebraic representation to rapidly generate and generalise the solution

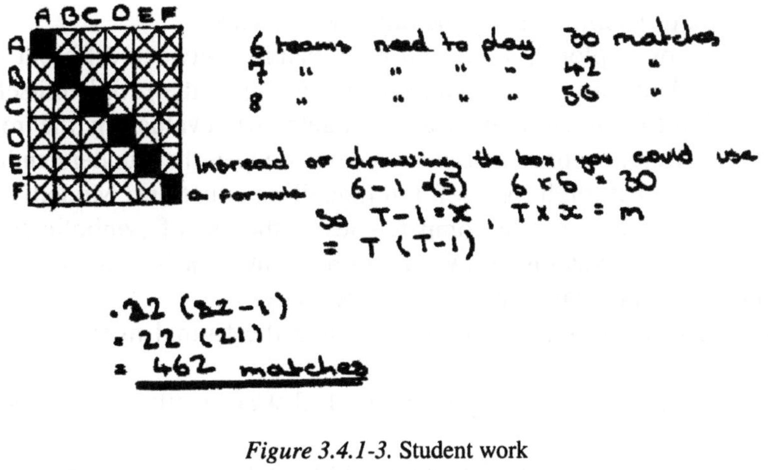

Figure 3.4.1-3. Student work

Example 3.4.1-4: Using graphical representation in a solution strategy

The Ffestiniog Railway task, from *The Language of Functions and Graphs* (Swan et al., 1985), is an example of a substantial real-life problem that requires simple arithmetic, reliably sustained through long chains of reasoning. Fig. 3.4.1-4 provides the instructions for this task.

The Ffestiniog Railway

This railway is one of the most famous in Wales.
Your task is to devise a workable timetable.

- There are six stations along the track.
- The line is single track, with two passing places.
- Three steam trains travel back and forth along the line.
- The journey from Porthmadog to Blaenau Ffestiniog takes 65 minutes (including stops).
- There is a ten-minute stop at each end of the line.
- Intermediate stops are very short and may be neglected when timetabling.
- Trains should depart at regular intervals.
- All trains start and end the day at Porthmadog.
- The first train leaves Porthmadog at 9.00am.
- The last train returns to Porthmadog by 5.00pm.

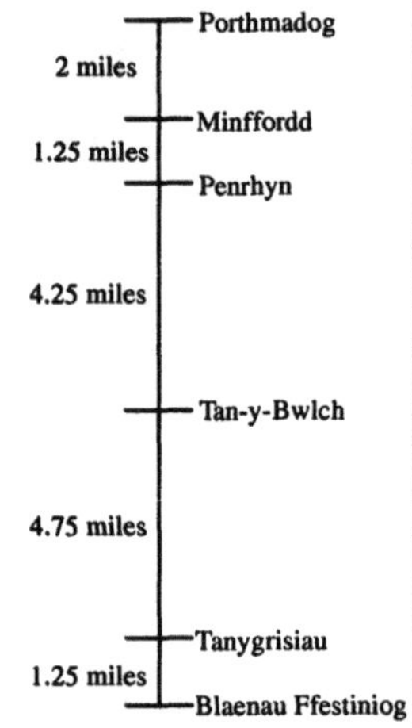

Figure 3.4.1-4. Instructions for devising a timetable

Though various approaches can succeed, students found that the most elegant and effective approach used a graphical solution process. Students constructed a distance-time graph showing the three trains, deduced from this the location of necessary passing places and read off times that the train stops at various stations, thus constructing a timetable. This was then compared with the timetable used by the railway company.

In general, by exploring links and relationships between different representations, students develop a better understanding of the *problem* itself and develop and refine potentially useful models. By evaluating the relative strengths of different representations, students deepen their understanding of the *mathematics* that might be useful in modelling situations. Modelling thus facilitates the development of competencies in the use of symbolic and formal mathematical systems. Powerful opportunities arise for students to strengthen their understanding of such systems by:

- forging connections between contexts and the formal mathematical expressions related to those contexts;
- motivating the study of applications of abstract mathematical formulations.

Once again, both the *realising mathematics* and the *mathematising reality* approaches provide these opportunities.

Modelling requires good communication, both with others and with oneself. As students interact, they refine their own thinking. In struggling to sort out what they need to say, they begin to sort out what they understand.

Each phase of the modelling process is helped by group discussions. As students formulate problems, discussion can enable them to distinguish the relevant and the irrelevant and to construct relationships between variables. They may then 'brainstorm' alternative solution ideas together. and may help each other interpret, critique and validate solutions. Throughout, talk clarifies the connections between the shared mathematical knowledge and the external context, and activates cognition.

The ability to communicate mathematically includes:

- the capacity to hear, understand and interpret the communications of others;
- the capacity to formulate mathematical descriptions and to articulate these in writing or verbally in ways that may be understood by others.

These communication abilities develop with practice, and students need to be given opportunities to interpret the mathematical reasoning of others, and to express their own thinking, in speech and writing. Modelling activities provide these opportunities.

3. MODELLING PROMOTES THE ASKING AND ANSWERING OF MATHEMATICAL QUESTIONS

A rich modelling situation provides opportunities for both asking and answering questions. In a previous ICMI Study on The Teaching and Learning of Mathematics at the University Level, Ottesen (2001, p. 344) writes: "..., students learn to ask certain types of questions that can only be answered by means of mathematics, as well as types of questions that can only be posed by means of mathematics". Examples of student modelling work (MDM4U, 2002) in an Ontario mathematics course, that Suurtamm and Roulet (this volume, Chapter 5.2) describe, demonstrate that students are motivated by asking questions and searching for answers.

Mason, Burton and Stacey (1982) in their book *Thinking Mathematically* provide innumerable concrete applications which prompt the reader to raise mathematical questions. The experimental nature of the situations prompts the reader to make conjectures that can be explored further and proved or disproved.

Modelling is a powerful promoter of meaning and understanding in mathematics. When presented with problems set in some real world context, students formulate questions about the context and think about the usefulness of their mathematical knowledge to investigate the questions. They are immediately encouraged to connect their mathematical knowledge with the external context. Mathematical thinking is promoted, and reasoning skills are exercised, as students seek to make those connections.

Modelling encourages reasoning through the complementary processes of simplification and elaboration. *Simplification* involves: analysing the elements of a problem situation; identifying features that are more or less important; making assumptions that might assist in making the problem more amenable to analysis; identifying sub-problems; breaking the problem down to its essential components; expanding components and looking for suitable representations to help clarify and explore the selected components and to work towards a useful mathematisation of the problem; and defining a clear way of approaching the problems to be solved. *Elaboration* works through reviewing and refining the initial outcomes of modelling, to enable progress towards further development of a more complete model and more generally applicable solution to the original problem. These processes engage the student in long chains of reasoning.

The defining of appropriate measures is central to mathematics. The following task (Burkhardt, 2004) provides great opportunities for discussing the merits and weaknesses of alternative measures.

Example 3.4.1-5: Comparing alternative measures

Our school has to select a girl for the long jump at the regional championship. Three girls are in contention. We have a school jump-off. Their results, in metres, are given below:

Elsa	Ilse	Olga
3.25	3.55	3.67
3.95	3.88	3.78
4.28	3.61	3.92
2.95	3.97	3.62
3.66	3.75	3.85
3.81	3.59	3.73

Hans says "Olga has the longest average. She should go to the championship".
Do you think Hans is right? Explain your reasoning.

Figure 3.4.1-5. Comparing alternative measures

In the TIMSS video lesson (TIMSS, n.d.) on which this task is based, the students are prompted to calculate the mean length of jump for each girl and to use that for selection. (Olga wins, despite having shorter longest jumps than either of the others) The teacher moves on without comment! There is no discussion of other measures, their strengths and weaknesses. Is this good mathematics? What does this divorce from reality do for the image of mathematics among students?

Example 3.4.1-6: Learning to aggregate unlike quantities and to define complex phenomena operationally

In many practical situations the variables represent unlike quantities and the student is challenged to look for ways to handle them.

In one problem (Lesh et al., in press) students are required to sort eighteen players into three equal volleyball teams. They are given the following information to incorporate into their model:

- The players' heights in feet and inches (a one-dimensional spatial quantity).
- The players' vertical leap in inches (a measure of displacement).
- The players' times for 40 metre dash trials (this is the only set of quantities where a lower value is more desirable) .
- The players' serve results (a success rate out of ten attempts).
- The players' spike results (the outcome of five spike attempts).
- The coach's comments for each player (qualitative information).

Such a variety of quantities is rarely encountered in most school mathematical textbooks problems. Instead, typical mathematics exercises will usually require students to operate on only one kind of quantity. However, the Volleyball problem is representative of many "real world" problems that do not involve only pre-mathematised quantities that have already been converted into a uniform type of element, but also include an assortment of unlike quantities that need to be integrated into a common model. To sort the players into three equal teams, students need to operationally define what "equal teams" means. One of the typical ways that students do this is to make the sum total of each teams' players abilities rank scores as similar as possible. However, using this measure, they often find that one team does not have any good servers, and is thus disadvantaged. To deal with this situation students revise their initial operational definition of "equal teams" from being "equal composite rank scores", to one that each team contains a similar assortment of players of different strengths. Students are exploring different definitions of "optimal". The practice of operationally defining phenomena is not only called for in "real world" problems. "Pure" mathematical research depends heavily on defining measures of concepts like "smoothness", "connectedness", and "denseness" of numbers, functions, and other mathematical objects. However, most elementary textbook problems only require students to deal with clearly defined mathematical phenomena, thus disallowing students the opportunity to engage in the mathematical practice of defining them themselves.

4. CONCLUDING REMARKS

In modelling, students experience concrete embodiments of new mathematical concepts. Because of the interdependence of various parts of the model, students quickly realise that careless mathematical work that remains unchecked can result in a lot of additional work. Students practice working with mathematical skills and tools to generate mathematical representations, to handle estimations, approximations, error analysis, to work reliably through the long chains of reasoning that the modelling process involves, to check the consistency of their solutions, to communicate using mathematical language, etc..

In modelling situations, a student develops mathematical expertise based on an integrated field of knowledge. The student exercises multiple connections both within and outside mathematics. In mathematics the connections overcome the topic classifications that are introduced by the curriculum. The National Council of Teachers of Mathematics in its *Principles and Standards for School Mathematics* (NCTM, 2000) lists connections as one of its process standards: "They (students) see mathematical connections in the rich interplay among mathematical topics, in contexts that relate mathematics and other subjects, and in their own interests and experience".

Although modelling is one of the mathematical competencies that students must develop, it in turn promotes and supports the development of other mathematical competencies. Thus by providing mathematical modelling experiences, students will not only reinforce their acquired mathematical knowledge but will also develop new mathematical knowledge. The iterative self-correcting cycle of asking questions, using mathematical tools, producing answers, and then asking new questions helps students develop the cognitive connections between product and process required to understand mathematics as a discipline.

Chapter 3.4.2

DEVELOPING MATHEMATICAL LITERACY

Lynn Arthur Steen and Ross Turner with Hugh Burkhardt
St. Olaf College, Northfield, USA, Email: steen@stolaf.edu / Australian Council for Educational Research, Camberwell, Victoria, Australia, Email: turner@acer.edu.au

Abstract: *Mathematical literacy* has received increasing attention in many countries over the last few years. This is partly driven by concerns of employers that too many students leave school unable to function mathematically at the level needed in the modern world of work. Further, it is increasingly recognised that people can only tackle many of the challenges of modern life effectively if they are mathematically literate in key areas. Planning in personal finance, assessment of risk, design in the home or on the computer screen, and critical appraisal of the flood of statistical information from advertising, politicians and the press – these are just a few of the domains where mathematics is an essential tool in sensible decision making, not just an exotic luxury. Mathematical literacy, like literacy in language, is empowering.

1. WHAT DO WE MEAN BY MATHEMATICAL LITERACY?

The term and its variants are used in a variety of ways[1] alongside other terms with overlapping meanings – *quantitative literacy, numeracy, functional mathematics, quantitative reasoning* and more, are all used. Here we shall mainly call it *mathematical literacy* (ML), focussing on the core idea:

Mathematical literacy is the capacity to make effective use of mathematical knowledge and understanding in meeting challenges in everyday life.

In contrast, mathematics in school largely continues the legacy of Euclid, Newton, and Euler – a school-based, "scholastic" discipline of major importance conveying the basic ideas of geometry, algebra, and calculus. It also equips a minority of students well for their chosen specialised professions –

in mathematics, physics, and traditional engineering. Valuable as this may be, and it is of immense and irreplaceable value, a *central* goal of schooling in the modern age must be to prepare *all* students for life in an increasingly technological society. That is what ML is all about: *mathematics acting in the daily lives of citizens.*

In this respect, school mathematics does not look so good. Most adults use little of the mathematics they first learned in secondary school in their every day lives. Much of that mathematics could be useful for mathematical literacy but the additional modelling skills that would make it so are not covered in most curricula.

Here we shall summarise key developments in recent years, including recent thinking in the US, described in *Mathematics and Democracy: the case for quantitative literacy* (Steen, 2002), and the suggestions in the UK Government Tomlinson report (2004) that *functional mathematics* should be at the core of learning for all, with additional *specialist mathematics* for students who want more. Of particular note is the OECD's *Program for International Student Assessment* (PISA, n.d.). This is important because it represents an international consensus on mathematical literacy. The PISA test instrument, while its exclusively short items are not "cutting edge", is a big step forward. It complements the narrower view of mathematics embodied in the Third International Mathematics and Science Study (TIMSS, n.d.), which devotes little serious attention to applications, and none to modelling or other non-routine problem solving. In PISA's definition (OECD, 2003, p. 24):

> *Mathematics literacy is an individual's capacity to identify and understand the role that mathematics plays in the world, to make well-founded mathematical judgments and to use and engage with mathematics in ways that meet the needs of that individual's life as a constructive, concerned and reflective citizen*

This broader conception looks at the life circumstances, contexts and needs of an individual, and considers the importance of their capacity to engage with and use mathematics in those life contexts. It involves recognising mathematical features of phenomena in the world around us, making judgments about those phenomena informed by mathematical understanding, and generally using mathematics as a tool for dealing with the phenomena.

The above definitions, and other variations on them, all convey three important ideas. First, ML is much more than arithmetic or basic skills. Second, ML requires something quite different from traditional school mathematics. Third, ML is inseparable from its contexts. In this respect ML is more like writing than like algebra, more like speaking than like history. ML has no special math content of its own, but finds appropriate content for the context. Moreover, like writing and speaking, the standard of excellence increases with the sophistication and importance of the issue being analyzed. Mathematics plays a parallel role in mathematical literacy to that of language in literacy[2].

Unlike teachers of language, most mathematics teachers rarely try to link mathematics lessons to the everyday lives of their students who, consequently, don't expect it. Making ML a reality in most classrooms will need a revised '*classroom contract*'[3], supported by new and well-engineered classroom materials and professional development support – a major challenge for design and development.

In the rest of this section, we shall:

- outline the characteristics of a 'good ML problem';
- discuss what mathematical literacy looks like as it develops;
- touch on some areas of controversy.

Then, in the next section, we shall discuss how teachers and curricula may develop ML in the classroom as part, along with other modelling and applications, of the learning of other mathematical competencies.

2. WHAT KINDS OF PROBLEM?

Exemplification clarifies meaning so, after this general discussion of mathematical literacy and its curriculum implications, we offer a few examples of the kinds of task that one would expect students to tackle in curriculum and assessment that is focussed on ML. We begin with some assessment tasks – always good way to communicate learning goals, since they are brief and specific. For practical reasons, PISA uses a mixture of multiple choice and short answer items (OECD, 2003, pp. 57-92), illustrated by these first two tasks:

Example 3.4.2-1: Rock Concert

For a rock concert a rectangular field of size 100 m by 50 m was reserved for the audience. The concert was completely sold out and the field was full with all the fans standing. Which one of the following is likely to be the best estimate of the total number of people attending the concert?

2000, 5000, 20000*, 50000, 100000

Example 3.4.2-2: Robberies

A TV reporter showed the graph below and said: "The graph shows that there is a huge increase in the number of robberies from 1998 to 1999."

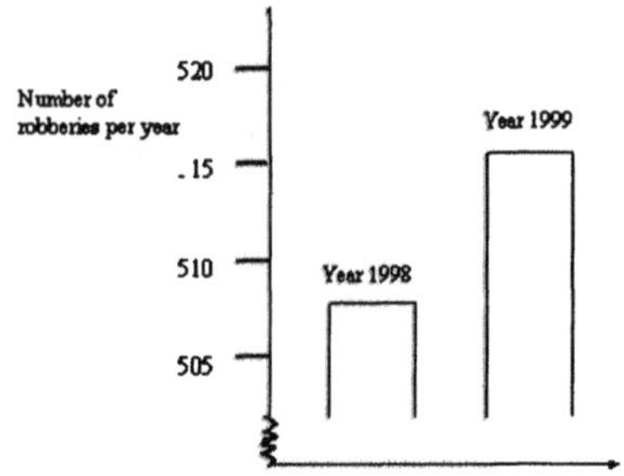

Figure 3.4.2-1. Graph of the number of robberies per year

Do you consider the reporter's statement to be a reasonable interpretation of the graph? Give an explanation to support your answer.

Of course, ML assessment or curriculum can and should have much more substantial and extended tasks than these. The *Numeracy through Problem Solving* (NTPS, 1987-89) modules provide examples of areas that motivate students to remarkably good, extended reasoning – for many students, work of much higher standard than their usual formal mathematics. The flavour is given by examples in the previous chapter and by *Design a Board Game.* Students begin a design process by critiquing and improving a number of badly-designed games. Both mathematical and non-mathematical faults are considered; the clarity of the rules, the fairness and interest of the game, the geometry of the board design and so on.

Example 3.4.2-3: Snakes and Ladders

Read the description of a game given in Fig. 3.4.2-2, then answer the questions below.

This is a game for two players. You will need a coin and two counters.

Rules

- Take it in turns to toss the coin.
  If it is heads, move your counter 2 places forward.
  If it is tails, move your counter 1 place forward.
- If you reach the foot of a ladder, you must go up it.
  If you reach the head of a snake, you must go down it.
- The winner is the first player to reach 'FINISH'.

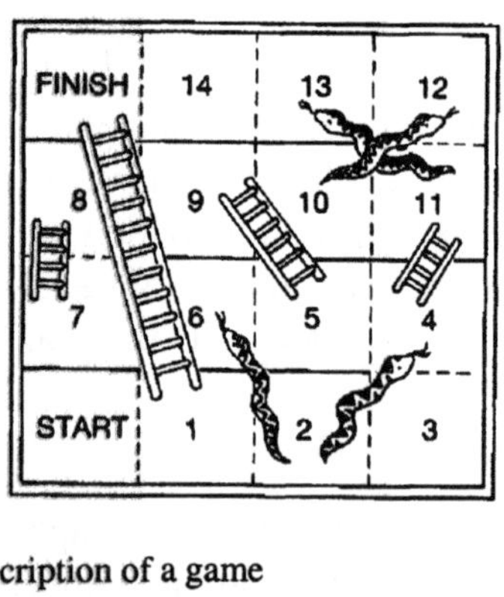

Figure 3.4.2-2. Description of a game

Questions:

1. Suppose you start by tossing a head, then a tail, then a head. Where is your counter now?

List and describe all the faults you notice with the board.

The goal here was to check that the student understands the basic principles of board game design, notably that the *board* and the *rules* must work together, and can apply a careful logical analysis. As in many well-engineered complex tasks, there is a 'ramp' of difficulty – some of the faults are harder to find than others.

At the end of each NTPS module there were written examinations at two levels, *Standard* and *Extension*, to assess how far the student can transfer the skills and insights they have developed in doing the three-week module to less- or more-distant contexts.

What are the general principles for task types for ML? Typical ML challenges involve real data, non-routine procedures, and complex reasoning, yet often require only relatively elementary mathematics. In contrast, school mathematics problems feature increasingly abstract concepts using simplified numbers, straightforward procedures and stylised applications.

> *Whereas school mathematics stresses elementary uses of sophisticated mathematics, mathematical literacy focuses on sophisticated uses of (often) elementary mathematics.*

Steen and Forman (2001) have summarised the characteristics of good ML problems as part of a list of "Principles of Best Practice" and we have adapted them for this book:

> *High quality mathematical tasks are authentic, intricate, interesting, and powerful*
>
> *Authentic*: they portray common contexts and honest problems; employ realistic data, often incomplete or inconsistent; meet expectations of users of mathematics; use realistic input and output; and, above all, reflect the integrity of both mathematics and the domain of application.
>
> *Intricate*: they expect students to identify the right questions to ask; require more than substitution into formulas; employ multi-step procedures and chains of reasoning; stimulate thinking that is cognitively complex; confront students with incomplete (or inconsistent) information; and demonstrate the value of teamwork.

Interesting: touch on areas of interest to students; appeal to a large number of students; they offer multiple means of approach; invite many variations and extensions; and provide horizontal linkages to diverse areas of life and work.

Powerful: they encourage and connect graphical, numerical, symbolic, verbal, and technological approaches; offer vertical integration from elementary ideas to advanced topics; propel students to more advanced mathematics; expand students' views of mathematics, its value and uses; demonstrate the importance of mathematics in the modern high performance work place, and in everyday life.

This book contains many examples of tasks that qualify as ML.

3. PATHWAYS TO MATHEMATICAL LITERACY

How do we recognise progress in ML? Essentially, students tackle more complex problems, in contexts less familiar to them, using more powerful mathematics – but only where it pays off in greater insight and/or more effective action. What is the micro-structure of this development?

PISA has investigated and described growth in mathematical literacy by focussing on a set of mathematical competencies that are based on the KOM framework. In conformity with the approach taken by PISA to report levels of proficiency in reading following the first round of assessment in 2000 (OECD 2001), the PISA project has developed and published six described levels of mathematical literacy (OECD 2004). A clear progression through these levels is apparent in the way in which the individual mathematical competencies specified in the PISA mathematics framework (OECD 2003) play out as mathematical literacy levels increase. They describe the *stages of development*, in increasing order, of the various competencies as:

- *Thinking and reasoning*: Follow direct instructions and take obvious actions; use direct reasoning and literal interpretations; make sequential decisions, interpret and reason from different information sources; employ flexible reasoning and some insight; use well developed thinking and reasoning skills; use advanced mathematical thinking and reasoning.
- *Communication*: follow explicit instructions; extract information and make literal interpretations; produce short communications supporting interpretations; construct and communicate explanations and argument; formulate and communicate interpretations and reasoning; formulate precise communications.

- *Modelling*: apply simple given models; recognise, apply and interpret basic given models; make use of different representational models; work with explicit models, and related constraints and assumptions; develop and work with complex models; reflect on modelling processes and outcomes; conceptualise and work with models of complex mathematical processes and relationships; reflect on, generalise and explain modelling outcomes.
- *Problem posing and solving*: handle direct and explicit problems; use direct inference; use simple problem solving strategies; work with constraints and assumptions; select, compare and evaluate appropriate problem solving strategies; investigate and model with complex problem situations.
- *Representation:* handle familiar and direct information; extract information from single representations; interpret and use different representations; select and integrate different representations and link them to real world situations; make strategic use of appropriately linked representations; link different information and representations and translate flexibly among them.
- *Using symbolic, formal and technical language and operations*: apply routine procedures; employ basic algorithms, formulae, procedures and conventions; work with symbolic representations; use symbolic and formal characterisations; mastery of symbolic and formal mathematical operations and relationships.

Like all models of problem solving, this model of stages is informative rather than definitive. For example, a given person will be at different stages, depending on the complexity, unfamiliarity, and technical demands of the problem they are tackling. Nevertheless such stages are characteristic of growth in these various competencies. They also provide a useful point from which further research may be directed to generating greater refinement in describing development in mathematical literacy.

4. CONTENTIOUS ISSUES

Chicken or egg – which comes first?

Many people believe that skills must precede applications and that once learned, mathematical skills can be applied whenever needed (in practice, for many students, in a future that never arrives). This is a false dichotomy. Considerable evidence about the associative nature of learning suggests that the skills-first approach works imperfectly, at best. For many students, skills learned free of context are skills devoid of meaning and utility. To be useful, skills must be taught and learned in settings that are both meaningful and memorable. One may observe that Pure Mathematics:

- grows vertically;
- climbs the ladder of abstraction to reveal, from sufficient height, common patterns in seemingly different things – abstraction is what gives mathematics its power, enabling methods derived in one context to be applied in others.

Mathematical literacy, on the other hand:

- grows horizontally;
- makes multiple connections, the core of understanding;
- clings to specifics in each context;
- marshals all relevant aspects of setting and context in order to reach conclusions that are reliable in practice.

Across contexts, *ML shows the pay-off of abstraction* – that the same mathematical tools can be powerful in a wide range of different areas.

Will ML undermine "real mathematics"?

Sceptics fear that modelling, if encouraged, will replace rigour and proof in mathematics classrooms. There are many legitimate reasons to ensure that reasoning and proof do not disappear from school mathematics. Students need to learn that justification is a distinctive part of mathematics; that proof is more than plausibility or confirmation; that among the levels of convincing argument, mathematical proof alone yields certainty; and that the rigor of mathematical proof makes lengthy chains of logical argument reliable. Although mathematical modelling rarely emphasizes formal proof, it does emphasise the value of:

- accuracy at the end of a long chains of inference and calculation;
- justifying findings, especially their applicability in relation to the problem context;
- explaining reasoning to team-mates and teachers;
- presenting conclusions coherently.

Through these means, modelling both demonstrates and rehearses the importance of rigorous logical argument.

Who should teach ML?

Many argue that mathematical literacy must be learned in context, while others believe that only mathematics teachers have the preparation and incentive to focus on it. The issue is complex. Since you can only model situations you already understand at least qualitatively, everyday life contexts are the obvious place for students to *learn active modelling with mathematics*. Teachers of the students' first language (English teachers in the Anglophone world) have long used everyday life problems as contexts for student work in their classrooms; ML asks mathematics teachers to do the same. For many this is a new challenge, needing good teaching materials and professional development support.

Ideally, students would also develop ML in other subjects – in history and geography, in economics and biology, in agriculture and culinary arts, and in social studies. Because contexts needing ML are ubiquitous, opportunities abound to teach it across the curriculum — in reading maps, designing art projects, understanding rules of grammar, analyzing scientific data, and interpreting legal evidence. Only by repeatedly using diverse aspects of ML in real contexts will students develop the habits of mind of a numerate citizen. Thus mathematics teachers should not, and can not, bear the entire burden of helping students become numerate. Like literacy, mathematical literacy is everyone's responsibility.

But do students understand other school subject areas well enough to model them autonomously? It is usually enough of a challenge to *learn models* in physics or economics – which is an important but very different process. Experience suggests that *mathematical literacy across the curriculum* will only happen if students first learn to model familiar practical problems in mathematics lessons. Cross-curricular teaching is an ideal, often tried but rarely sustained in schools; it only works when approached 'from both sides'.

If ML is to become a reality, it will probably depend on mathematics teachers carrying prime responsibility, with other subjects building on the foundations so laid. Perhaps, following a suggestion in the report *Making Mathematics Count* (Smith, 2004), mathematics should be seen as two subjects: *mathematical literacy* as the gatekeeper subject[4] for all students, and additional[5] *specialist mathematics* for those who want it for their future interests as scientists or traditional engineers – or simply find mathematics interesting enough to want further study. In some countries, *English Language* and *English Literature* are two subjects related in much the same way.

Is this mathematics?

ML is neither an expanded list of topics to be added to the mathematics curriculum nor is it just the basic skills part of a traditional mathematics program. Many basic mathematical skills (e.g., number sense and operations, proportional reasoning, estimation, logic, data analysis) are essential for ML – but so too are other concepts not much emphasized in school mathematics curricula (e.g., computer tools, statistical inference, mathematical communication). The open-ended thinking required to diagnose problems or to make decisions relies heavily on "newly useful" areas such as combinatorics, statistics, and geometry. In contrast, algebra and calculus, dominant features of today's curriculum, are used outside of school more as tools for calculation than as tools for reasoning, and only by specialists in certain fields.

Some worry that modelling problems are a time-consuming distraction that typically use relatively routine mathematical tools. While this may be an accurate description of some weaker programs, good modelling problems require a sophistication and precision that can push even the best students to attain mathematical results well beyond those achieved by most students in today's classrooms. As they secure a broad foundation of examples and concrete mathematics, students engaged in modelling build lasting connections between mathematics and the world in which they live. This grounding in specifics will lead naturally to subsequent generalizations and abstractions. Modelling parallels good pedagogy by moving from the specific to the general and from the concrete to the abstract.

---

[1] In some contexts and some nations, these terms are used narrowly to mean just "basic skills" – arithmetic plus a bit more. This is a corruption of the terms, just as *literacy* means much more than spelling, grammar and syntax. These skills are necessary but far from sufficient.

[2] The original definition, now often distorted, of the term *numeracy* – in the 1959 Crowther Report.

[3] The *classroom contract* is the agreement, usually unspoken and implicit, between teacher and students as to what each will do, what roles they will play, in the classroom. (The French, who first articulated the idea (Brousseau, 2003,p.24), call it the 'didactic contract' but in English 'didactic' is used to describe a specific, teaching style – lecturing that brooks no argument. That is the reverse of what we need here, or in any classroom focussed on learning.)

[4] Much more justifiable, as a universal requirement meeting a universal need, than current secondary mathematics which few adults can use in their lives beyond education.

[5] If it is offered as an *alternative*, it will surely remain the prestige track, with ML becoming a 'sink' subject, taken only by weak students, while the well-qualified adult population remains innumerate.

Chapter 3.4.3

CLASSROOM ACTIVITIES AND THE TEACHER

Soeren Antonius, Chris Haines, Tomas Højgaard Jensen and Mogens Niss with Hugh Burkhardt
University of Southern Denmark, Denmark, Email: soeren.antonius@dig.sdu.dk /
City University, London, UK, Email: c.r.haines@city.ac.uk /
Danish University of Education, Denmark, Email: thje@dpu.dk /
Roskilde University, Denmark, Email: mn@mmf.ruc.dk

Abstract: In this section we discuss a broad range of classroom activities, and of teaching style, that are required to produce the benefits that modeling can provide to the student learning mathematics. We discuss support for teachers and the shortage of good modeling tasks that need to be developed into effective curriculum materials.

1. INTRODUCTION

The development of performance in modelling processes[1] requires a much richer range of learning activities than the *explanation-example-exercises* 'ritual' that dominates traditional imitative curricula, and classrooms. These activities are similar to those needed for non-routine problem solving in pure mathematics and so, through them, students (and teachers) develop many other mathematical competencies. In previous chapters we have demonstrated that modelling provides concrete embodiments of mathematical concepts, develops reliable computation and checking, develops multiple connections inside and outside mathematics, and so on. In this section we discuss the broader range of classroom activities, and of teaching style, that this requires. What about support for teachers? There is a shortage of good modelling tasks that have been developed into effective curriculum materials. These need to offer sufficient support for typical teachers, who are mostly inexperienced in teaching modelling, without undermining student autonomy. Since most teachers are used to working with very supportive teaching materials, even in the familiar curriculum, few will do well in new and challenging teaching without similarly detailed support. Professional development activity will also be important, particularly in the early stages of teaching modelling.

2. THE PATTERN OF CLASSROOM ACTIVITIES

Among the characteristic differences from a traditional mathematics classroom, we have seen that students:

- spend much more time on each, more-substantial problem;
- discuss mathematics with each other;
- explore alternative solution pathways;
- choose appropriate mathematical 'tools' to employ;
- carry through extended chains of reasoning, reliably;
- use checking strategies to get their analyses technically correct;
- interpret and evaluate the reasonableness of arguments and solutions;
- explain both results and reasoning to others.

Overall, students take much more responsibility for their work and their solutions. (This is, of course, what they will need to do in life and work). This is a major shift in the beliefs of most students and teachers about the nature of "doing mathematics", and thus of the implicit 'classroom contract' of mutual expectations.

The diversity of applications and modelling activities in schools, colleges and universities is considerable. Their variety and complexity can be considered as follows:

- **Tasks in mathematics and applications** are shorter activities, often within a single lesson. Many are familiar in standard curricula: in the primary school tasks related to lengths, areas, volumes or to data collection, representation and analysis; in the secondary school, dealing with word problems and other illustrative applications; in further and higher education, short modelling exercises. Many of these will rightly be *illustrative applications* but *active modelling* is vital. At every level they should include smaller tasks on mathematising, handling the mathematics, and the other modelling processes, as well as more substantial problems like the following:
  Area of a Porsche asks the students to estimate the surface area of a car "for budgetting the paint shop". It is useful in showing alternative approaches and the model improvement process;
  Dead girl presents the stopping distance problem in a dramatic context –

at different speeds, how far will a car travel before being able to stop when someone steps out into the road;
Bus trip takes a different approach – it describes alternative answers under differing conditions for transporting a number of school children from one place to another. The social context also becomes important, introduced in contrast (Verschaffel et al., 2000) with the usual approach to word problems.

- **Investigations** are longer activities that may extend over periods from two or three classes to two or three weeks. The *Numeracy through Problem Solving* modules are well-supported examples of this kind of active modelling for secondary school.
- **Projects** are even more complex activities, extending over weeks or months, usually addressing a broader problem embedded in the real world. *Yatzy Oil Rig* (de Bock & Roelens, 1993) deals with the movement of an oil rig from its construction site down the River Scheldt to Rotterdam through hazards of overhead power cables, depth contours for the river bed, tidal variations and logistics.
- **Dissertations,** usually in the university sector, develop over periods up to one year. The subjects may be quite general, e.g. on cartography and conformal maps, with the direction of the work and the outcomes very much dependent on the student and on the supervisor.
- **Class/lecture demonstrations** are usually led by the teacher or lecturer, though the best examples involve the students in active roles. They are often physical demonstrations of mathematical models either in a laboratory situation or on a larger scale using everyday materials showing that models work. In a modelling cycle description they validate the model. In *Walking the plank* a scaffold plank is marked off in quarters along its length. It is then extended from the stage towards the audience. Participants from the audience come to the stage and counterbalance the demonstrator as he/she walks the plank. There has to be discussion and interaction during the activity on the number of children needed to balance the demonstrator. Strict safety measures need to be put in place. (Haines & Le Masurier, 1986)

Dimensional Analysis (for example, Giordano & Weir, 1991) is an extremely powerful, and neglected, technique in modelling and applications. It provides a way to introduce quite difficult physical models to those without a strong background in science and applied mathematics. For example, the motion of a simple pendulum can be discussed in terms of the physical quantities mass, length and time leading to the usual formula for its period. Similarly it is easy to establish that the wind force on a van travelling on a motorway or across a high bridge is proportional to the square of the speed at which the van is being driven. These two introductory models have been used to enthuse and engage pupils at upper secondary and undergraduate levels with some success.

For investigations, projects and dissertations it is usual for there to be a formal reporting stage; sometimes this is also done for shorter tasks in schools but more informally. Reporting methods include written reports (Berry & Davies, 1996), oral presentations (Crouch & Le Masurier, 1996) and posters (Berry & Houston, 1995).

Group work contributes significantly to the engagement of students, increasing motivation, and leading to better understanding of both the real world context and the mathematical concepts and techniques required for success. Recent work (for example, Ikeda & Stephens, 2001), has shown significant improvements in performance where discussion between members of the group takes place at the outset.

Assessment that recognises student achievement in applications and modelling has led to the development of innovative practices including: teacher assessment; self- and peer assessment; written reports; posters and oral presentations (Haines & Dunthorne, 1996). In addition, written examinations on specific project areas can assess students' ability to *transfer* their understanding and skills to less- or more-remote situations. They can also meet concerns about plagiarism. Preparing for such examinations brings challenging issues of transfer into the classroom.

3. TASKS AND TEACHING MATERIALS

The choice of mathematical tasks for students to tackle, in the classroom and in assessment, epitomises any curriculum. In the previous sections we have summarised and exemplified the essential characteristics of good modelling tasks, and illustrative applications. There are further examples of good tasks throughout this book.

The emphasis in teaching differs for the different types of task. *Modelling* can *only* be taught through a teaching approach that is investigative and student-centred, with the teacher playing largely consultative rather than directive roles. *Illustrative applications*, including stylised *word problems, can* be taught in traditional teacher-centred ways (as it largely has been) – though they too benefit from the same broadening of teaching style, which develops students' ability to work independently on less-routine tasks. While there is plenty of teaching material for teacher-centred teaching, there is still surprisingly little that offers detailed support for the greater challenges of teaching modelling. As in the early days of teaching problem solving within mathematics, much of the available material provides good problems with only general guidance on teaching strategies. While this may be sufficient for innovative teachers with some experience, it is not enough support for typical teachers who are new to teaching modelling.

The first fully-engineered materials for teaching modelling in schools were provided in the 1960s by *Unified Sciences and Mathematics for Elementary Schools* (USMES), the pioneering US project from the Education Development Center; even so, these provided rather general guidance and proved too challenging for most teachers. In the 1980's, the Shell Centre developed the much more supportive *Numeracy through Problem Solving* (NTPS) materials for lower secondary students, with an associated public examination.

The materials that have most directly and effectively addressed the development of other mathematical competencies through modelling have been developed over many years by the team at the Freudenthal Institute (FI). Their *Realistic Mathematics Education* program (RME) is based on *emergent modelling*, where students develop mathematical concepts through the modelling process, much as discussed in this chapter – but with well-engineered teaching material.

In recent years in both the UK and the US, the curriculum has again narrowed its focus, so investigative activities of any kind are rare. However, the current interest in mathematical literacy offers some hope of further progress, and the development of more well-engineered teaching materials to support typical teachers at all levels. They are sorely needed.

4. MODES OF WORKING IN THE CLASSROOM

These can be analysed in various ways. Basically, students can work individually or in groups; however, this dichotomy covers a multitude of important variations. There is a substantial literature on group working; here we have space for only a few comments.

Berry and Houston (2004, p. 36) use the concept *mathematical working styles* to capture how students work in modelling courses for undergraduates. It is a three-dimensional concept with the following components: the role of IT-tools; working with representations; and a sequence of actions that forms the approach to solving the problem. Ferri (2004, p. 47) uses the concept *mathematical thinking styles* and defines it as the way in which an individual prefers to present, to understand and to think through mathematical facts and connections by certain internal imaginations and/or external representations[2]. This has several elements organised in five stimuli strands: *environmental* (sound, light, temperature, school- and room-design), *emotional* (motivation, persistence, responsibility, need for structure), *sociological* (self-work, pair-work, team-work, adult support, varied), *physiological* (perceptual, intake, time, mobility), and *psychological* (global/analytical, hemisphericity, impulsive/reflective).

We shall focus on the sociological stimuli specified in the following question: What kind of interpersonal working modes can be observed in the classroom when students are working with modelling?

Individual work is (even when students sit in groups) the most common working style for traditional teacher-guided instruction. Students listen to the teacher, take notes, ask questions which the teacher answers, and do exercises. They also solve routine word problems at school and/or at home, which are later assessed by the teacher.

Group work seems to be an appropriate mode of working for modelling problems. Groups can be formed by the students themselves or by the teacher. They can be randomly or deliberately formed and may be homogeneous (group members have similar capabilities) or heterogeneous in composition. Groups may function in a variety of ways.

- Groups are formed and dissolved constantly. If a student has a problem, he/she tries to form a group (with another student) or tries to be admitted to an existing group in order solve this specific problem. When the job is done, the group is dissolved.
- The group functions as a pleasant sociological environment. Students small-talk – not about mathematical problems, but in parallel to their individual work. The group seems not to be a learning facilitator, but one should not neglect the fact that these environmental factors can play a significant role for some students' learning process (cf. the environmental stimuli).
- The group members work parallel to each other, on the same problems, but approaches, methods and results are constantly discussed, negotiated and checked in order to reach an agreement. Sometimes an agreement is reached, sometimes not (Matsuzaki, 2004) which could mean that different answers to the same problem are produced. This way of functioning involves the risk that a student, who is less able, un-motivated, frivolous, or lazy, can become a 'passenger' and not takes any responsibility for the group work.
- The problem is split into sub-problems and distributed to (or by) the group members in such a way that each group member is responsible for a specific part, and the complete answer to the problem is composed from the individual contributions. This is often the case with major problems and project work. It forces students to collaborate. The outcome is highly dependent on each student's work. If just one student fails, the work will not be completed in time.

Note that in all these cases individual work is also involved; as in everyday life, the balance between different modes of working is important.

Most students seem to prefer working collaboratively when modelling. As in everyday life, discussions are often helpful when tackling more complex problems. When one student is stuck, another may be able to supply a suggestion, which a third may be able to elaborate. In this, the teacher acts as a guide, asking questions to promote thinking and learning[3]. The group can do more than the sum of group members' contributions.

Group work can also be seen as a way of entering a *landscape of investigation* (Alrø & Skovsmose, 2003). This landscape is explored by a particular form of student-student and student-teacher relationship, and the theoretical background for this way of working and communicating is specified in the so-called *Inquiry-Co-operation Model.*

Group work is not unproblematic, however. If the group work relies on each group member, the work may fall apart if just one member is absent one day, or if he/she is not able to do his/her job. This may cause frustration and irritation, and these are just some of the problems that the group must be able to deal with. Group-work also raises the question of fairness; the outcome is highly dependent on the group members' willingness and ability to co-operate. The ability to co-operate in groups is not inherent in students; it has to be learned (Laborde, 1994).

5. THE ROLES OF THE TEACHER

When learning mathematical modelling, the traditional teacher's role as the prime source of explanation, demonstration and correct answers is no longer appropriate. The teacher can no longer micro-manage the students' thinking or they will not develop the strategic capabilities that are at the centre of modelling; guidance remains essential but it, too, must focus on strategic questions (Shell Centre, 1984), with:

- *More metacognitive prompts*: "What have you tried?", "What did you find?", "What are you going to try next?", "What will that tell you?".
- *Some prompts focused on specific strategies*: "Have you looked at some specific cases?", "Did you see any patterns that you recognize?", "It may help if you represent this in a different way", "Have you tried to check that using another method?"
- *Little detailed guidance*: "Isn't that the difference of two squares?", "Why don't you try a linear fit?", "That's wrong".

Developing the mathematics

While students work on a modelling problem, their main objective is to produce an interesting and useful analysis and report, not to develop particular mathematical techniques. The mathematics is used as a tool to facilitate this process and is not seen as an end in itself. These are appropriate priorities.

The teacher can, however, use the many opportunities provided by the work to motivate the learning of mathematical techniques in a more explicit way. This section offers suggestions, taken from the *Numeracy through Problem Solving* (NTPS) modules, as to how this can be achieved without destroying the essential flow of the modelling activity.

How and when may mathematical topics be introduced?

Mathematical activity may be initiated by either the student or by the teacher. For example:

A student may become aware of the need to acquire a particular skill in order to complete her consumer report. "What is the best way to present this data?" "Should I draw a pie chart, bar chart or what?" This kind of situation can lead to an invaluable learning experience because the student wants to know something. Such opportunities occur rather unpredictably. Also, if you have a large class, it is unwise to spend a great deal of time helping one person. One approach is to ask to the student to describe the problem to the whole class and invite help and advice from other students.

The teacher may wish to use some ideas from the problem to support a more intensive piece of work on a particular topic. 'Today, we are going to look at the topic of ratio, using the shopping surveys you carried out. Which type of chocolate gives you the most for your money?' Do not expect students to use, autonomously, mathematics that they have only recently been taught. There is a gap, typically of several years, between first 'learning' a skill and being able to use it with flexibility and fluency. Students will tend only to use skills that they have mastered. (There is evidence that this "few year gap" can be reduced – it requires a more 'rounded' approach to learning, with a variety of applications and non-routine problem solving to supplement and give meaning to technical exercises.)

Before, during or after – a timing dilemma

Teacher-initiated work on mathematical techniques related to the theme may occur before, during, or after the modelling itself. Each has advantages and drawbacks.

- *Before*: "I'll give them some practice at drawing pie charts now, so that they will be more inclined to use them later on, when they begin work on the module." This timing has the advantage that the student will, if all goes well, have the technique polished and ready to be used, but it seems artificial to learn a new technique just before you need for it. Students soon assume that the 'problem' is an illustrative application – merely a vehicle for practising the new technique, rather than developing their modelling competency.

- *During*: "They seem to be having difficulty in organising their data. We'll take a break from this module for a few lessons and do some work on this, together using data that I'll provide". This timing allows you to respond to needs as they arise. But, if students always expect you to produce the method or solution when the going gets difficult, you undermine autonomy. If such teaching is done too often then the work on the module will drag on too long and become boring.
- *After*: "When we have finished the problem, we will look at the techniques we have used in greater depth." This does not threaten student autonomy, and working on the module may motivate and enable them to perceive the value of techniques when they are taught. However, students may still not be able to use techniques autonomously unless they are given further opportunities to select and apply them in other modelling contexts.

Whatever you decide at each stage, it is important to be vigilant about sustaining the students' strategic control of their work; it is too easy to allow them to revert to the imitative role that the traditional curriculum encourages. If the balance of support is inappropriate for modelling, the students soon lose the feeling of 'ownership' of the problem to the teacher and revert to the traditional passive, imitative role that inhibits learning of all kinds.

Steen and Forman's "Principles of Best Practice" (2001) summarise all this, and is adapted here.

Effective teachers of modelling employ pedagogy that is:

Active:

- Encourage students to explore a variety of strategies.
- Stimulate discussion of available data in relation to what is being asked.
- Require students to seek out missing information needed to solve problems.
- Make hands-on use of concrete materials.

Student-centered:

- Focus on problems that students see as relevant and interesting.
- Help students learn to work with others.
- Developed strong technical communication skills among students.
- Provide opportunities for students to use their own knowledge and experience.

Contextual:

- Ask students to engage problems first in context, then with mathematical formalities.
- Suggest resources that might provide additional information.

- Require that students verify the reasonableness of solutions in the context of the original problem.
- Encourage students to see connections of mathematics to work and life.

This book illustrates these principles from work in various countries on the teaching of modelling. The published proceedings of the series of biennial ICTMA conferences (ICTMA) are also an important source of examples and analysis of the development of this field.

6. HOW MUCH GUIDANCE?

This question is at the heart of the teaching of modelling – indeed of all non-routine problem solving. However, as noted above, there are specific issues when other mathematical competencies are a learning goal. If the teacher allows students to select their own skills to deploy, then they are likely to choose only those with which they are most familiar and secure. They will tend to avoid the more challenging and difficult ideas, even when imitatively fluent[4]. On the other hand, if the teacher tells students which mathematical techniques to use, then the teacher removes the strategic demand and the problems become exercises in using the given techniques.

We illustrate this issue with two complementary case studies of classroom 'trajectories' starting from the same very broad task statement:

Which means of transport is the best?

They show that, while teachers face important strategic decisions, the situation does not need to be so polarised. The teacher doesn't have to choose between allowing complete strategic freedom and none. It is possible to manage the process so that students begin by tackling problems unaided, then compare advantages and disadvantages of the approaches and strategies used, and then refine these into more powerful methods or introduce new methods in a tentative 'maybe this will help' manner. In this way the teacher acts as a co-constructor of the mathematics. Some of the examples above show how this can be done – and the kind of support that helps teachers.

However, the constraints that the teachers applies through the initial problem is important. Here the first teacher takes a more open approach than the second. In the analysis we focus on two issues:

- How is the teacher dealing with the balance between mathematical modelling as a goal in itself, and as a means to develop some of the other mathematical competencies?
- What opportunities for the teacher does the approach offer, and what obstacles and difficulties does it carry with them?

First classroom trajectory

This comes from an experimental two-year teaching programme in a Danish upper secondary school (Jensen, to appear). The mathematics teaching was guided by the KOM set of mathematical competencies. The classroom activity structure enabled students to get experience with all aspects of the mathematical modelling process.

In the first phase of the work the teacher

- made the mathematical competencies explicit, but emphasised modelling;
- gave a list of interesting openly-formulated problem areas to choose from.

Within the task statement above, the students chose to focus on the following question:

What is the average use of energy in moving x people from one floor to another using respectively a staircase, an escalator and a lift? Which option uses the most?

This focus was much narrower than what the teacher had in mind – comparing cars, trains, planes etc from various perspectives: time; price; pollution.

The students continued the detailed specification of the problem by focusing on the relation between the height of the stairs of a staircase and the effectiveness of the use of energy in the leg muscles. However, after some discussion with the teacher, the physiology involved was deemed too demanding, so they simplified the problem to calculating the difference in potential energy between being on the two floors. The lift was dealt with satisfactorily, but the escalator turned out to be a real mathematisation challenge. How can one make a mathematical representation of the way a group of people enter and leave an escalator, the number of people on the escalator at a certain time and the use of energy this adds up to?

Modelling vs. other mathematics: In this case the teacher had a clearly specified focus: to develop the students modelling competency. It turned out that several other mathematical competencies were also used during the project, not least among these *communication* (through the negotiations in the problem specification and mathematisation processes) and *problem tackling* (through struggling with both formulating and solving the model problem). However, because of the openness of the task, one could not ensure the use of specific mathematical topics (Blomhøj & Jensen, 2003). For that, the teacher must take more control, reducing student autonomy.

Opportunities and obstacles: This teaching style helped the teacher to create an open, trustful and mutually supportive atmosphere in the classroom. This came from being honest about his role as mediator between the goals of the curriculum and the autonomous interests of the students. The second opportunity follows from this. The teacher helped develop modelling competency directly: *Here is a goal – go for it, and I will help you!* There seemed to be three difficult issues for the teacher to deal with:

- The demands, both personal and mathematical, of this style of teaching;
- the general difficulty of competency-guided teaching, not least due to the students' problems in grasping the essence of the different competencies.
- The distribution of time among the different projects and the various more teacher-controlled modes of teaching taking place.

Second classroom trajectory
In this example, the initial task was more closely specified by the teacher:

"Which is the best means of transport in the metropolitan area of Copenhagen amongst bicycle, bus, car, metro, or train for an individual living and working in this area?
It is up to you to specify the conditions for answering this question, including what "best" means, and what assumptions are being made. You have all four lesson-hours per week (plus homework time) for three weeks at your disposal."

The students tackled this task in five groups of 3 – 5, each of which made its own specification. Then each group consulted the teacher, who had to approve the specification and the approach chosen, bearing in mind the mathematical tractability at their level, the time frame at their disposal, and information and data to be identified and collected. As the teacher wanted to give the students as much independence as possible, she did not share her deliberations on these matters with the students; she made her own assessment of the task, the capabilities of the students in each group, and her opportunities to provide guidance at crucial points.

One group of students decided to keep a dual view of the term "best", because they found that both "time consumption" and "cost" were objective functions of relevance. They wanted to be able to look at the trade-off between the two, should they find (as they assumed) that different means of transport would be optimal.

At first this group wanted to tackle a well-defined simple problem: to select two points, A and B, in the Copenhagen area and then look at all realistic routes between the two for cars and bicycles, and at all the public transport options connecting small neighbourhoods of A and B, respectively. However, the teacher thought that this would not be challenging enough in mathematical terms – just comparison of a few costs and riding times. The work would be nearly all data collection – reading of time-tables, perhaps taking measurements etc. This educational challenge was not sufficient, so the teacher urged the group to specify its task in a different way.

Thus prompted, the group decided to *narrow* its task down to comparing transport by car with transport by public buses, *but* without specifying the points of departure and arrival. They did so by modelling the average cost and duration of car rides per km during rush hours and "quiet hours", respectively, taking into account the average number of stoplights per km and the expected waiting time at stoplights at different hours. Similarly, they used the official timetables of buses to compute the average fare and duration of a bus ride per km. In order to do all this they collected data samples of distance data and the places of stoplights, and they collected information from the traffic offices of the city of Copenhagen concerning the waiting times at different categories of stoplights in the city. They also attempted to model the time costs entailed by deceleration and acceleration at stoplights.

At the end of the second week, their model was almost complete, and conclusions were about to be drawn with regard to the main question. The teacher then asked students to *improve* the model by analysing the sensitivity of the model and its conclusions to inaccuracies and uncertainties in the data and assumptions made. This led students to pursue the effects of errors by means of interval arithmetic, a technique to which their teacher gave the group a brief introduction. Students finished their work by producing a poster, and defended this at an oral "poster session".

Modelling vs. other mathematics: In this trajectory, the teacher had a clear vision of the goals of the modelling activities:

- fostering students' competence in mathematical modelling in highly messy and blurred real world contexts;
- the making of decisions, idealisations, simplifications, and improvements;
- the collection of appropriate data (a primary goal); but also
- to activate, in the modelling process, mathematical knowledge and skills that they could deal with but at the edge of these students' ability.

Thus the project stimulated students' activation and consolidation, to different degrees, of the entire spectrum of mathematical competencies, as well as specific subject matter knowledge and skills. The emphasis was on the competencies of problem tackling, representation, communication and above all, of course, modelling. This teacher achieved this balance by guidance at a few crucial points so as to ensure that their work would be mathematically relevant and challenging, and relevant to the initial questions.

Opportunities and obstacles: As usual, the most significant challenge to the teacher is to strike a proper balance between student autonomy and teacher intervention. Too little feedback from the teacher can cause frustration and insecurity with the students – the teacher had to provide inputs to students that allowed them to take responsibility for their own work but also helped them over hurdles and decisions that were likely to carry them astray.

Achieving this balance presents every teacher of modelling with formidable demands. The Steen-Forman *Principles of Best Practice* (2001) has a useful list of pitfalls to avoid:

- Selecting tasks in order to cover mathematics rather than to explore and solve interesting problems.
- Overlooking interesting or challenging mathematics that lies embedded beneath the surface of many "real-world" examples.
- Imposing unwarranted structure on a contextually rich problem in the interest of ensuring appropriate mathematical coverage.
- Believing that complex problems require sophisticated mathematics and that there is something wrong with solutions that use elementary techniques.
- Choosing tasks that fail to help students prepare for higher achievement in mathematics.
- Presenting tasks in the form of mathematics worksheets, thereby sterilizing the context of everything that makes it problematic.
- Lacking conceptual continuity and intellectual growth in the sequencing of tasks in which mathematical activities are embedded.
- Failing to bring mathematical closure (including concepts, vocabulary, methods, generalizations) at the conclusion of an open-ended project.
- Not allowing sufficient reflection on the process of mathematical modeling.

In addition to reflective experience of supervising and guiding the processes involved, teachers also need a deep understanding of many different kinds of subject matter that allows them to predict the possible obstacles and outcomes of different paths students may follow. This requires new kinds of teacher decisions and interventions. The teacher has to be able to live with uncertainty, continually gathering and processing information about the state and development of each student, and making appropriate decisions on intervention. This is a *mathematics teacher competence* that cannot be acquired by means of pre-service preparation only. It has to be developed in service, but the seeds should be planted in pre-service education.

---

[1] *comprehension, formulation, transformation, interpretation, evaluation* and *communication,* through cycles of *improvement* from an initial *simple* model.

[2] c.f. learning styles of Dunn & Dunn (1998).

[3] In this way, group work is one way of giving substance to Vygotsky's *zone of proximal development.*

[4] Treilibs (1979) for example, invited 120 very able 17-18 year old students to solve a variety of mathematical problems set in realistic contexts. None used algebra.

Chapter 3.4.4

USES OF TECHNOLOGIES IN LEARNING MATHEMATICS THROUGH MODELLING

Daniel Pead and Bill Ralph with Eric Muller

University of Nottingham, UK, Email: Daniel.Pead@nottingham.ac.uk /
Brock University, St. Cathrines, Ontario, Canada, Email: bralph@brocku.ca

Abstract: This chapter focuses on the use of technologies in learning mathematics through modelling and applications. In the limited space available, we have chosen to concentrate, in the first part of the discussion, on some important uses of computer software to stimulate modelling development at the secondary level. In the second part of the discussion, at the tertiary level, we present an innovative mathematics program that makes extensive use of technologies and that has a core in applications and modelling.

1. INTRODUCTION

In mathematics there exists a substantial choice of technologies, and new alternatives are being developed every day. For some of these the *entry fee* – the time and effort spent learning the system – can be substantial. With modelling applications, this goes beyond simply learning how to operate the software but requires the user to learn a new set of concepts, terms and representations. The cognitive load should not be underestimated, however friendly the user interface. While this may contribute to competencies in the *representing* and *symbols* sectors of modelling, concepts that are too proprietary or esoteric are unlikely to be further built upon. Here we discuss the use of generic applications such as *spreadsheets, computer algebra systems, and programming* – where the entry fee is repaid in transferable, accredited information and communication technology skills. We also consider *applets, learning objects,* and *microworlds,* which achieve very low entry fees but often only have a narrow, fixed mathematical context. In the last part, we highlight a university program in which students develop mathematical concept knowledge using technologies and through applications and modelling.

2. MODELLING WITH TECHNOLOGIES IN SCHOOLS

We begin by considering the uses of spreadsheets, graphing tools, dynamic geometry software, applets, learning objects, and microworlds.

The dominant use of spreadsheets in schools is for data analysis and graphing. Their wide use arises because teachers of many different disciplines have become familiar with them. Thus students are able to carry their knowledge of the technology to different classes. Spreadsheets can be used for a class of tasks that combines *problem tackling* with modelling, for example, *optimising within constraints*. Fig. 3.4.4-1, is taken from Mathematics Assessment Resource Service project *Developing Problem Solving: Optimising* (MARS, 2002-04). It is developed to engage 11-14 year-old students into a deeper understanding of the concept of area and for them to explore alternative approximations. They are not expected to solve this analytically. The intention is that they calculate the area for a given number of floors and find the optimum by a systematic search. A spreadsheet is ideal for this as it reduces the amount of calculation, and encourages a systematic approach for calculating the total area – although having to work on paper might offer an incentive to adopt a more efficient search strategy.

Your task is to design a tower that meets the following constraints:

- The outside measurement will be 10 metres by 10 metres.
- The tower will have a number of floors.
- The more floors there are, the thicker the walls will have to be.
- The walls have to be 0.1 metres thick for every floor in the building (so if there are 5 floors, the walls must be 0.5 metres thick).

How many floors should you make the tower in order to have the greatest possible total floor area in the whole building?

Figure 3.4.4-1. Tower design

Many of the tasks used by the MARS project were prototyped using a spreadsheet. The following example Fig. 3.4.4-2 is actually an Excel spreadsheet used during the trialling of *Developing Problem Solving – Making Models*.

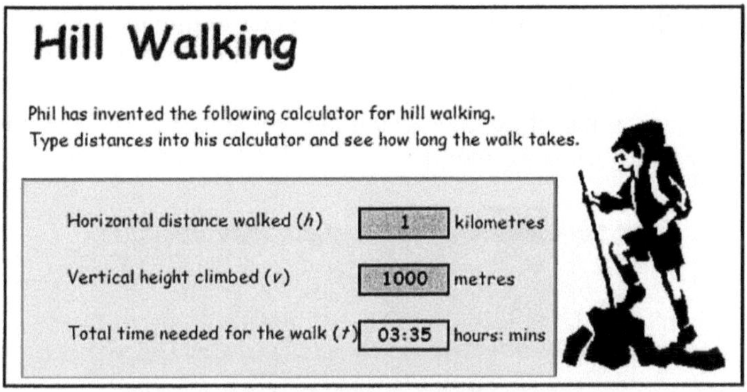

Figure 3.4.4-2. Hill Walking

The student can change the two shaded boxes and see the result – the rest of the spreadsheet is "protected" and the formulae are hidden. While this was replaced with a slicker, professionally programmed version for the finished product, producing similar materials is well within the grasp of a reasonably IT-savvy teacher (or student), given a few technical tips. The exercise here is for students to interact with the model and try and re-create it (either descriptively or as a formula, depending on their age/ability). In other words, one of the aims of the modelling situation is for the student to develop an understanding of formulating mathematically. The main challenge here is to tackle the problem systematically – many students will type in random values, always change *both* numbers, or fail to keep records.

There are also statistical softwares developed specifically for the classroom. Suurtamm and Roulet (this volume, Chapter 5.2), discuss an Ontario modelling course 'Mathematics of Data Management' that is built around a major modelling project. Examples of student projects (MDM4U, 2002) show use of spreadsheets and a dynamical statistical software (Fathom, n.d.).

Numerous educational graphing tools address the shortcomings of the business chart-oriented facilities of spreadsheets (for example, GT, n.d.). There is, however, a tendency towards "creeping featurism" and such packages offer more features than a typical teacher would ever use. On the other hand, some of these features can engage a creative student. Producing a graph tediously by hand – or using an inappropriate, presentation-focussed tool – could cause graphs to be seen as static illustrations, a poorly understood part of "writing up" the exercise. The applications discussed below use animation and interaction to reinforce understanding of the underlying representation and symbol/formalism competencies; forge links with other representations and to make plotting and dynamically changing graphs central to the modelling process.

Traffic (Swan et al., 1985) uses the computer to provide a visual link between abstract representations and 'real-world' situations. Here, students see an animation of cars driving along a road. Every second, a "photograph" of the road is taken and these are shown side-by-side. Finally, the photos, as shown in Fig. 3.4.4-3, are "morphed" into a distance-time graph.

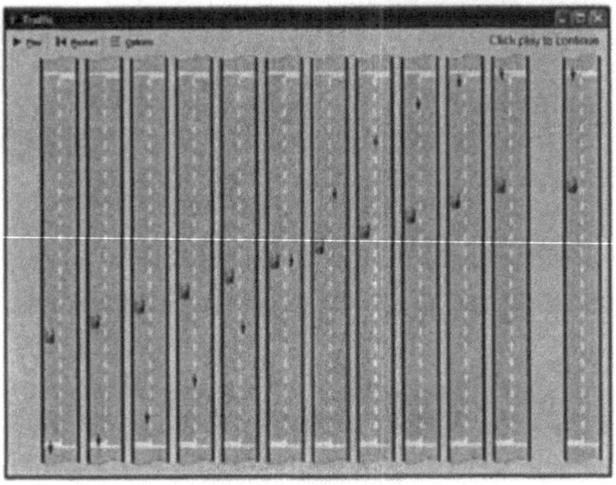

Figure 3.4.4-3. Traffic

Graphing software can be used interactively to improve students' understanding of a representation – for example, it is usually possible to animate the graph of a function by interactively varying one of the parameters. For example, Fig. 3.4.4-4 shows a screen from *Coypu* (Coypu, n.d.)

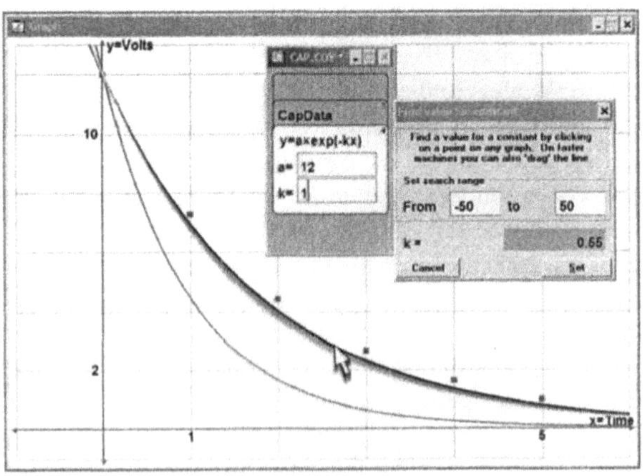

Figure 3.4.4-4. Coypu

The activity is to visually fit a line of the form $y=Ae^{-Kx}$ to a series of data points showing the voltage of a discharging capacitor. A has been chosen and fixed – now as the mouse is dragged K is calculated so that the line passes through the pointer, giving the user a strong "tactile" impression of how K changes the shape of the graph.

Students can now learn and apply geometry by using one of the engaging dynamic environments such as the *Geometric Supposer* (GSupposer, n.d.), *Cabri géomètre* (Cabri, n.d.) or *The Geometer's Sketchpad* (GSP, n.d.). Field-tested classroom activities for most school levels are available on the product websites, which also provide links to other sites dedicated to the use of the dynamic software. The range of topics that are listed in their publications underline the importance of visual representations for many learners across the mathematics curriculum.

Applets, learning objects and microworlds are all web-based activities. In general an applet is a small, limited-functionality piece of software often designed to be embedded in a web page in very much the same way as a picture or video. A learning object is a self contained instructional component that can be used in different learning contexts. In mathematics applets and learning objects usually focus on a specific mathematical concept. The best learning objects are designed to engage the learner in some activity or game and from there to progress through levels of conceptual development. Microworlds are intended to be little worlds in which a student explores alternatives, develops and tests conjectures while discovering properties and facts about this world. Some of these have been expanded to allow students to interact with classmates or with other students connected to the web. (The terms applet, learning object, and microworld are widely but not universally used with these meanings.)

A search of the web using 'mathematics applets' produces a list of several substantial collections. Fig. 3.4.4-5 shows an example from the Freudenthal Institute's site (WisWeb, n.d.). It illustrates Dijkstra's algorithm for finding the shortest distance between nodes on a connection graph.

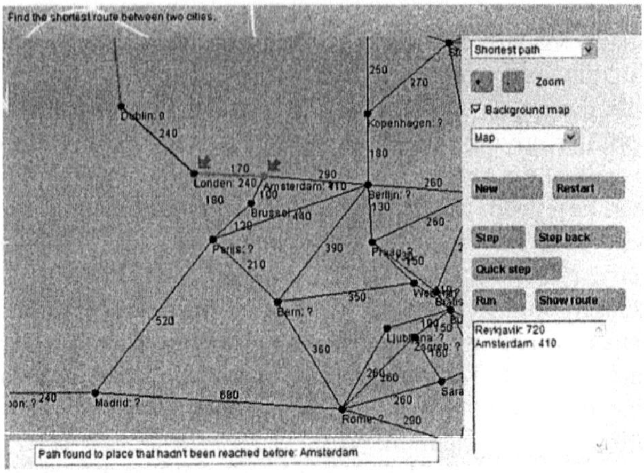

Figure 3.4.4-5. Dijkstra's algorithm

For other examples of sites that contain extensive references of mathematics applets see, at the school level (JavaMath, n.d.), and at the post secondary level (Mathematics Archives, n.d.).

The website (Brock-a, n.d.) provides examples of learning objects designed and implemented by university mathematics students. In *FireFire* the user develops the concept of trigonometric ratios, and in *Parabola Games* the learner explores the roles that parameters play in alternative formulations of a parabola. Examples of repositories of mathematics learning objects are Merlot and the National Science, Mathematics, Engineering and Technology Education Digital Library (NSDL, n.d.).

In the UK and in the mid/late 1980s, a large range of "microworld"-type software was produced by projects such as the ITMA group and SMILE (SMILE, n.d.). One challenge to the incorporation of such "small software" into everyday teaching has been the migration of computers from the regular classroom into the "IT Suite". The recent popularity of "electronic whiteboards" has seen a reversal of this and may allow a renaissance of the applet/microworld.

3. TECHNOLOGY AND MODELLING IN A UNIVERSITY MATHEMATICS PROGRAM

It is a little hard to imagine a group of first year university mathematics students choosing on their own initiative to spend hours of time on mathematics projects. But this kind of effort is what the faculty are regularly seeing in the new MICA program at Brock University. MICA stands for *Mathematics Integrated with Computers and Applications* (Brock-b, n.d.). It is a "hands on" approach to the teaching of mathematics that has engendered a remarkable change in the level of engagement of the students. In the context of this new program, we will look at some of the issues, rewards and problems that arise when introducing technology into undergraduate mathematics and when implementing applications and modelling into a set of core courses.

During the two years of intensive development of MICA, faculty sought to create a modern mathematics program that would foster creativity, and the mastery of mathematical concepts and their applications, while making the best possible use of modern technologies.

The seed of the program was planted twenty years ago when students in a half-course in applied calculus were placed in weekly Maple (Maple, n.d.) labs (Muller, 2001). This meant that not nearly as much time in lectures had to be spent on calculations, thereby leaving more time to discuss and develop the concepts of calculus and their applications. Furthermore laboratories provided opportunities for students to explore mathematical concepts using multiple representations, even before they were introduced in class.

Ironically, it became clear that students in this service course were having a richer mathematical experience than the mathematics majors because, for example, Maple allowed them to explore families of functions and more realistic applications, while majors were restricted to the small number of examples that could be calculated by hand. Consequently, the first year calculus course for majors was rewritten to incorporate Maple and the interactive CD-Rom called *Journey Through Calculus,* developed by Ralph (1999).

The central challenge of any mathematics program is to create an environment in which students become internally directed and personally invested in moving themselves along the long road to mastery. The problem with traditional undergraduate mathematics programs in this regard is that, if students try to take the initiative in creating and investigating problems and applications of their own devising, they quickly come up against difficulties that they cannot handle with purely analytical tools. For this reason, traditional programs must be very tightly choreographed around the applications and modelling that can be solved by hand.

Technology can offset the rigidity of a traditional mathematics program by providing students with access to an endless supply of problems and applications that can be investigated both computationally and analytically. It enables students to engage in a new level of creative discovery. It also plays a role in preparing students for new concepts by placing them *in a situation where they naturally raise the question before being shown the result*. This principle is fundamental to the MICA philosophy.

We start by looking at one component that will make or break a technologically oriented curriculum – the *computer laboratory activities and assignments*. The importance and difficulty of creating superb, friendly and yet challenging computer laboratory activities and assignments for mathematics courses cannot be overstated. The art of writing a good lab is to get the maximum mathematical punch with the minimum amount of programming while providing a clear framework within which the student can confidently develop the software to be employed in the mathematical investigation. Here are some of the faculty's experience in making effective lab activities:

- Build a parameter into any mathematics done in the lab. The best laboratory activities involve exploring a mathematical problem that shows interesting different behaviours across a range of parameter values. Encourage students to explore and give qualitative descriptions of the phenomena they are seeing.
- Make all computer activities syntax explicit. Mathematics students should not spend time thinking about computer syntax as nothing will frustrate students faster than not knowing which computer command to use.
- Encourage students to make conjectures and then write programs to test their conjectures.
- Reward students for making attractive self-contained, user-friendly interfaces for their programs – they then become personally committed to the creation of good programs.

When the activities are scheduled in a laboratory, with each student at their own computer, get sufficient expert staff. In courses where students are writing programs, experience has shown that there should be a minimum of one assistant for every 15 students.

The second issue of concern is the nature of the student's programming environment. This is particularly important because the MICA program requires no previous programming experience. Initially, first year students were given a half-course in the Java language (JAVA, n.d.). They found the syntax of the language so formidable that it overwhelmed the mathematical content – the *entry fee* was too great. To overcome this difficulty, the programming course was dropped, the programming language changed to Visual Basic.net (VB.net, n.d.) and the necessary programming was introduced into the first MICA course. The vast majority of the first year students pick up the language and graphics quickly (in three weeks).

The MICA mathematics program is built around a sequence of unique courses that emphasize the creative investigation, application, use, and presentation of mathematics using computers. In the limited space available we have chosen to provide some detail and reflection on the first MICA course only.

As students do not have experience in raising open-ended mathematical questions, the first part of the course describes two interesting areas of mathematics for student investigation. Examples are prime numbers and the Collatz conjecture. The purpose is to give the students a fertile arena for raising sensible conjectures that they can test by writing programs. The class is divided into small groups and asked to raise any interesting questions they can think of about in the areas investigated. For this session to work, the tone of the classroom has to be absolutely nonjudgmental so that all speculations are equally welcome. Conjectures are written on the board and are discussed. Special attention is paid to the feasibility of testing them, within the students' current level of programming skills. Each student is then asked to make a unique conjecture and then to write a program to test it. They hand in a functioning program and a full written report on their conjecture and what they found. The program's interface is expected to be self-explanatory, visually attractive and extremely user-friendly. It is truly unfortunate that these students have gone their whole intellectual life and have rarely been asked to raise a mathematical question. When this process is first initiated in class there is a certain "stiffness". But after a while the conjectures begin to come quickly and it turns out that students do indeed have a great capacity for raising interesting questions.

The goal of the second part of the course is very specific – to create a functioning encoding and decoding program based on RSA encryption that "spies" can use in the field. Students are taken through the theory of modular arithmetic, Euclid's algorithm for the greatest common divisor, the group of integers modulo n and Fermat's little theorem. They write programs that encapsulate each topic in preparation for the theory and coding of the full RSA algorithm.

The course then explores discrete and continuous dynamical systems. For example students write programs to exhibit the cobweb diagram and output the numerical data for the logistic equation. This done, the lectures are given right in the laboratory, with each student at their own computer, so that they can instantly use their own programs to verify the theory being developed in class. It is a very exciting way to teach this material.

The centerpiece of this course is the student's final project. Students work in pairs and select an area to work on. What they choose tends to indicate their major interest. Pure mathematics majors choose to explore and test conjectures. Students interested in applications and modelling choose areas that intersect other disciplines. Future teachers develop learning objects that create an environment for a learner to explore a mathematical concept. This final project has many of the characteristics of a modelling exercise. Students select an area that is of interest to them. They research it and determine those properties that are central and those that are secondary for its computer modelling. They develop a model for its programming. They have to work with great care – computer programming is unforgiving. They cycle through the process as they find that the results are not what they anticipated. And they communicate their results in an interactive environment. This last step is far more demanding than writing a report as the model must be robust. The quality and depth of the projects generally exceed the faculty's expectations. Students become very engaged in this work and the final projects are often elaborate, fascinating and a pleasure to mark. Examples of some of the more outstanding projects can be found on the Department's web site (Brock-a, n.d.).

The MICA program uses technology not only to reify and reinforce the learning of theoretical ideas but also to allow students to explore applications and modelling, that are well outside the usual boundaries of a traditional mathematics program. The act of writing a computer program promotes learning because it puts the student in a feedback loop where they are constantly checking to see if the output from their program agrees with the theoretically predicted values. Students quickly discover that it is virtually impossible to write such a program without first understanding its mathematical content. In this way, students are pushed passed a cursory understanding of the material to a new level of mastery. Finally, it should be noted that there is something very personal about writing programs. Students take considerable pride in writing robust programs with user friendly interfaces. Their level of engagement and enthusiasm for the MICA program has made all of the effort spent in its creation entirely worthwhile.

4. CONCLUDING REMARKS

Technologies impact many areas of mathematics education and their effective use requires new approaches by both teachers and students. Technologies challenge the hierarchical views of student access to mathematics as they allow students to work with mathematical concepts which are traditionally seen as too difficult for them. Technologies also challenge the view that applications and modelling can only be introduced after the student has developed all the required mathematical knowledge. Technologies in mathematics are evolving so that a single software application allows users access to multiple representations (for example numeric, graphic, and algebraic), and to several areas of mathematics (for example geometry, probability, and algebra) which are traditionally separated in the curriculum. In this way technologies can provide multiple connections in mathematics, supporting a student's holistic development of mathematical understanding.

Section 5.5

MODELLING PEDAGOGY

Edited by Hans-Wolfgang Henn

Chapter 3.5.0

MODELLING PEDAGOGY – OVERVIEW

Hans-Wolfgang Henn
University of Dortmund, Department of Mathematics, IEEM, D-44221 Dortmund, Germany, Email:wolfgang.henn@uni-dortmund.de

Abstract: Applications and modelling are an indispensable part of basic experiences students must have in mathematics lessons. Traditional teaching forms, contents of curricula, and methods of assessment often stand in the way of this goal.

1. BASIC EXPERIENCES

Following Heinrich Winter, the well-known German mathematics educator, three basic experiences are necessary in order that mathematics lessons will convey general education principles on every school level (Winter, 2004).

(BE 1) To realise in a specific way, and to understand phenomena in the world around us, which we are and should be concerned with. (Fundamental contributions of mathematics towards acquiring important knowledge about our world).

(BE 2) To learn about and to understand mathematical issues represented in language, symbols, pictures, and formulas as intellectual creations, as a deductively ordered world of its own kind. ("Inner world of mathematics", mathematics shows that a rigorous science is possible).

(BE 3) To acquire problem-solving (heuristic) skills by analysis of tasks which go beyond mathematics. (Mathematics as a school of thought).

Mathematics proves to be an inexhaustible pool of mathematical models, which allow us to understand better the world around us. However, for concrete examples, both of the two other basic experiences play central roles, especially when the important demand for interconnectedness according to the spiral principle is taken seriously. For example, "problem solving" (Pollak: "Here is the problem, solve it") is inherent to BE 3. On the other hand, "modelling"(Pollak: "Here is a situation, think about it, find out what the problem should be") relates to BE 1. The deductive aspect of mathematics (Pollak: "Here is a theorem, prove it") belongs again to BE 2. In any case, applications and modelling are an indispensable part of the Basic Experiences of Winter and thus contribute to conveying a balanced image of mathematics in school at every level. These basic experiences also set goals for teaching applications and modelling. To reach these goals an adequate Modelling Pedagogy is necessary, which includes adequate problems, adequate teaching methods and instructional modes, adequate tools and adequate modes of assessment.

Teaching affects the image that students will take with them into their future life as responsible citizens and future decision makers. This image should contain both the beauty *and* the functionality of mathematics. But applications of mathematics in other fields should not be studied for its own purpose alone. Reflecting on what relates mathematics with the rest of the world is indispensable, ethical issues of mathematical actions have to be highlighted, and students have to be sensitised for it (Skovsmose, 1989).

2. THE REALITY IN SCHOOLS

Unfortunately, as a rule, reality-oriented teaching on applications outside mathematics is covered only to a limited extent in teaching although there is a long-standing agreement on the importance of creating relations between realistic situations and mathematics teaching.

Many 'real' problems in mathematics teaching are only mathematical problems 'in disguise' and not genuine real (life) problems. For the students 'uncovering' these problems 'in disguise' is reduced to finding out the algorithms that have been hidden by the teacher, and immediately 'real' mathematics takes over. True modelling, involving the transition from reality to mathematics, mathematical analysis, and the transfer back of the results into the real situation is rarely discussed seriously - students are not sensitised for the adequacy of the argumentation.

Curricula packed with teaching content, and traditional teaching methods (especially prevailing teacher-centred approaches), are further conditions that hinder applications and modelling. Finally, assessment methods affect possibilities for applications and modelling: Centralized assessment for final examinations or even tests at the end of each school year often reduce teaching to mindless drill and practice of procedures and calculation techniques. Often, teaching can only be described as "teaching for the test".

3. APPLICATIONS AND MODELLING AT ALL LEVELS

Mathematical modelling is the mutual fertilization of mathematics and the rest of the world (Pollak, 1979). Through modelling, students are enabled to build a bridge between mathematics as a tool to understand better the world around them, and mathematics as abstract structure. For this, suitable teaching situations are indispensable. Lyn English (2003) demands "rich learning experiences", i.e. authentic situations, chances for own exploration, multiple possibilities for interpretations, and social competence to take up the responsibility for one's own model up to communicating it to other students. Teachers are often reluctant to include mathematical modelling in their teaching. Katja Maaß (2006) points out that the complete modelling process is time-consuming and difficult. She also shows, however, that modelling activities can be started successfully in a normal teaching situation. Further positive examples for integrating applications and modelling into teaching are presented in this section. Chapman (Chapter 3.5.1) investigates the strategies of a sample of teachers teaching mathematical modelling; Lingefjärd (Chapter 3.5.2) discusses the Swedish reality courses in mathematical modelling in teacher education; Matsuzaki (Chapter 3.5.5) presents a video-taped case study of cooperative mathematical modelling.

Students should not be spared the difficulties and effort related to applications and modelling, including important activities such as data collection (sampling), writing reports on the work done, and justifying and defending the results. Often group work is an appropriate working style.

The vertical interconnectedness from primary to tertiary level is indispensable: One cannot start early enough with simple modelling examples. An excellent example how such a teaching can be integrated already at primary level is the *mathe2000 project* of the Dortmund group of researchers around G. N. Müller and E. Ch. Wittmann (Wittmann, 2001). This project shows the importance of interactions, between development and design of curricula, teacher education and assessment, where the last especially depends on the educational policy of the "decision makers" of a country.

For assessment methods, (which include applications and modelling) there are proposals and concrete experiences (see Section 3.7 of this volume). Excellent approaches have been developed by Peter Galbraith (see for example Galbraith, this Volume, Chapter 2.5).

4. THE ROLE OF COMPUTERS

Today's available computer technology can contribute in a special way to aid in the learning process, and is equally helpful and important for all three BEs. Here, I am thinking especially about dynamic geometry software (DGS) and computer algebra systems (CAS). Firstly, the computer is a powerful tool to aid in modelling and simulation, therefore related to BE 1. Secondly, the computer can positively influence the generation of adequate basic concepts ("Grundvorstellungen") of mathematical ideas – especially through dynamical visualisations – and therefore contributes to BE 2. Lastly, the computer furthers heuristic-experimental work in problem solving, thus contributing to BE 3.

Recent experiences with using computers in teaching indicate that the currently widely accepted aims of mathematics education (such as learning to solve problems, getting to know heuristic strategies, concept formation, proving, mathematizing) are still fully valid when computers are used. More than that, using computers can help to reach these aims better (Henn, 1998). Two chapters of this section deal with questions around the role of computers in mathematical modelling. Stefan Hußmann (Chapter 3.5.3) develops a concept of technology-based open learning arrangements, and Djordje Kadijevich (Chapter 3.5.4) discusses standards of computer-based modelling.

REFERENCES

Englisch, L. (2003). Mathematical Modelling with Young Learners. In Lamon, S. J. et al. (Eds), *Mathematical Modelling: A Way of Life*, (pp. 3-17). Chichester: Ellis Horwood.

Henn, H.-W. (1998). The Impact Of Computer Algebra Systems On Modelling Activities. In Galbraith, P, W. Blum, G. Booker, & I. Huntley (Eds.), *Teaching and Assessing in a Technology rich World*, (pp. 115-123). Chichester: Ellis Horwood.

Maaß, K. (2006). Modelling in classrooms: What do we want the students to learn? In Haines, Ch. et. al. (Eds.), *Mathematical Modelling (ICTMA 12): Engineering and Economics*. Chichester: Ellis Horwood.

Skovsmose, O. (1989). Models and reflective knowledge. *International Reviews on Mathematical Education, 21*(1), 3-8.

Winter, H. (2004). Mathematikunterricht und Allgemeinbildung. In H.-W. Henn, & K. Maaß (Eds.), *ISTRON Materialien für einen realitätsbezogenen Mathematikunterricht, Band 8.* (pp. 6-15). Hildesheim: Franzbecker.

Wittmann, E. Ch. (2001). *Report of the PME 25 Research Forum "Designing, Researching and Implementing Mathematical Learning Enviroments – the research group Mathe 2000"*. (Coordinator Erich Ch. Wittmann, IEEM, University of Dortmund, Germany).

Chapter 3.5.1

MATHEMATICAL MODELLING IN HIGH SCHOOL MATHEMATICS: TEACHERS' THINKING AND PRACTICE

Olive Chapman
University of Calgary, Canada, Email: chapman@ucalgary.ca

Abstract: Modelling in the mathematics classroom is discussed based on the thinking and practice of a sample of exemplary high school mathematics teachers. These teachers held conceptions of mathematics, word problems and problem solving that placed importance on real world connections and influenced the creation of a classroom culture to support modelling. Their teaching strategies that allowed students to develop flexibility to engage in modelling are highlighted.

1. INTRODUCTION

Word problems, in a broad sense, are an integral aspect of school mathematics. They can be used as a basis for application and a basis of integrating the real world in the learning of mathematics. They can motivate students to understand the importance of mathematics concepts, and help students to develop their creative, critical and problem solving abilities. Verschaffel (2002) described their goals as "to bring reality into the mathematics classroom, to create occasions for learning and practising the different aspects of applied problem solving, without the practical ... inconveniences of direct contact with the real world situation [p. 65]." Traditional word problems and traditional instruction, however, have been critiqued for not accomplishing these goals in a meaningful or effective way. For example, there is concern expressed about suspension of sense making (Schoenfeld, 1991; Verschaffel, Greer, & DeCorte, 2000) by students when working with contextual problems, particularly in connecting school mathematics and reality. One way that has been promoted to improve the situation is incorporating a modelling approach in designing and dealing with word problems. For example, Greer (1997) proposed a conception of word problems that calls for mathematical modelling that takes the real world knowledge into account. Boaler (2001) showed that a modelling approach encouraged the development of a range of important practices in addition to knowledge situations. Verschaffel (2002) provided evidence to support the importance and feasibility of applying the modelling perspective successfully in mathematics education for all students. However, whether or how modelling gets implemented in the classroom will depend on the teacher.

Research on the mathematics teacher suggests that an understanding of teacher thinking/beliefs/conceptions and practice (e.g., Thompson, 1992) is important to improve the teaching of mathematics. This paper is based on a study of teachers' thinking and practice that relate to their teaching of word problems, in a broad sense. The focus here addresses the question: In the context of courses not based on a modelling and application curriculum, what is the thinking of teachers who include modelling in their teaching and what pedagogical strategies do they use that facilitate modelling?

2. RESEARCH PROCESS

This paper is based on data from a larger study on teacher thinking in teaching word problems. The focus here is on six experienced junior high [JH] and senior high [SH] school teachers who participated in the study and integrated modelling in their teaching. They were from local schools and considered to be exemplary mathematics teachers in their school systems.

Data sources consisted of open-ended interviews of, for example, the teachers' thinking about mathematics, problem solving and word problems and their experiences with teaching word problems; of classroom observations and discussions of their actual instructional behaviors during lessons involving word problems; and of role-play to capture the nature of their thinking in a different mode. The overall goal for the data was to capture what the teachers did, how they did it, and their thinking behind it. Role-play scenarios allowed the teachers to act out a situation, for example, presenting a word problem to a class of students. Interview questions were framed in different ways including a phenomenological context to allow the teachers to share *their* way of thinking and to describe their behaviors as lived experiences (i.e., stories of actual events).

The researcher and two research assistants, working independently, reviewed the data to identify attributes of the teachers' thinking and actions that were characteristic of their perspectives of teaching word problems. These attributes were grouped into themes and validated by comparison of findings by the three reviewers and triangulation among the different data sources. Modelling became a focus as one of the emerging themes and was then elaborated on by scrutinizing the data for situations that conveyed it.

3. MODELLING IN THE TEACHERS' CLASSROOMS

The teachers' courses that formed the basis of this study were framed by the official mathematics curriculum, implemented in 1997-2000, that does not explicitly emphasize modelling as a topic. However, "the program rationale and philosophy" section of the curriculum document explains that students will be expected to apply mathematical knowledge to non-routine, real-life problems. It also mentions the critical skill of using mathematics to find solutions to real-life situations and the need for connections, which illustrate the subject's usefulness in solving problems, describing and modelling real-world phenomena, and applying mathematical thinking and modelling to solve problems that arise in other disciplines. Unlike the teachers who interpreted and implemented this curriculum in a traditional way that routinized applications and problem solving, the teachers in this study incorporated practices in their teaching that mirrored the philosophy of the curriculum. However, most of these practices predated this curriculum and included situations with aspects peculiar to modelling.

Modelling was reflected in the teachers' practices for problem solving and the use of problems with real or realistic situations for students to develop and apply mathematical models. Problem solving activities included understanding an existing model, constructing a new model, and applying a model to solve a problem in a non-mathematical field. Problems used in these activities were not typically open-ended from a real-world perspective but were often approached as if they were. Following are three examples of modelling-oriented tasks the teachers assign to students:

i. As part of the topic of similar triangles, students work in groups to find the height of tall objects in the schoolyard that they cannot measure directly, e.g., the flagpole, a tree, and the school building where it is the tallest.

ii. Students are asked to bring a spoon to class. The task is for the student to draw a top, front and side view of the spoon, locate these views in a coordinate plane, and determine the equations of the curves in each view. This task is assigned as part of a grade 12 curve-sketching unit.

iii. Students work in pairs, choose a doughnut, a Vernier calliper, and grid paper and are asked to find the volume of the doughnut as part of the grade 12 topic of volume of revolution about the x-axis of various curves.

4. INFLUENCES ON MODELLING PEDAGOGY

The teachers held conceptions about mathematics, word problems, and problem solving that placed importance on the real world in a way that implied the need for modelling in learning mathematics. For example, their conceptions of mathematics included: "Mathematics is a discipline that seeks to understand our world and as a study, it focuses on analyzing things and quantifying things and looking for logical appearances on things and is a subject that continues to grow, as society demands that it grows. ... I just see it as a way to understand our world" [SH3]. "Mathematics is a language of patterns, but it isn't just discovering the patterns on their own. It's actually being able to articulate them, to mathematize them, and then to use these articulated patterns as tools. Well, then, to see these patterns repeated in the world and having the mathematization handy so that you can impose it on that next pattern that you see" [SH1]. "Mathematics is ... a logical, rational way for us to deal with the things that we do in our world" [JH3].

The teachers conceptualized word problems as being most valuable and meaningful when they are located in the world in terms of actual or realistic situations. For example, "In the real world, people ... build bridges and buildings and fly planes and they do all kinds of things that use mathematical knowledge and those are problems that need to be solved ... problems that relate to the world" [SH3]. "I want the problem to be as much as possible, real-world problems and not contrived things that nobody cares about" [SH4]. "Real-world context is an optimal condition for situations to be a problem ... [The problem] starts with the real world, but then forces you to do or encourages you to go to some kind of mathematical world where mathematical rules apply and then go back and correct your world" [SH6]. The teachers also viewed word problems as being most valuable and meaningful when: "The solver has to impose a structure on the problem to create the solution" [SH1]. "The problem makes you see things differently" [SH3]. "The problem should be one, which captures their attention, invites them, intrigues them and prods them to want to solve it" [SH2].

Finally, these teachers conceptualized problem solving as a thinking process and a life skill. "[Students must] have some kind of mathematical images, models, or lenses, which will help them to interpret the data in some kind of form that will help them organize all this information in the problem and ultimately start come up with some kind of solution. ... They have to learn how to unfold their models on the situations, ... to draw them out in the appropriate situations. ... They have to understand when they read a problem, they look for clues in the problem, which will link up with mathematical models that they already know, or mathematical tools that they can use, to translate that problem into the model" [SH1]. "Problem solving is a skill you have to use throughout your life, so we have to ... teach problem solving in our mathematics classrooms so that we can have students truly believe and truly feel that they have a skill that is something they are going to use for the rest of their lives, a life long skill" [JH3].

These conceptions were instrumental in influencing the teachers' choices of tasks and pedagogical approaches that favored modelling in their teaching.

5. STRATEGIES THAT SUPPORT MODELLING

The following is representative of the teachers' pedagogical perspective: "Focusing on problem solving strategies ... encouraging the kids without doing the thinking for them ... questioning, supporting children as they are doing the work ... There are times when we need to do work as a large class and ... times when you will break them off or vice-versa or start in small groups and bring to large group, but that the focus is for kids to become problem solvers, not solvers of this type of problem" [SH2]. Only selected aspects of the teachers' teaching are discussed here. The focus is on two strategies that seem to be important in creating a classroom climate for students to engage in modelling. These strategies were common to all of the teachers but each teacher differed in how he or she executed them. The following are abbreviated examples of what these strategies look like.

Strategy 1: Learning about problem solving. There was a meta-cognitive focus in the teachers' teaching in terms of getting students to learn about problem solving as in the following examples:

Example 1: Variations of this approach were used by the SH teachers. Students are presented with a word problem framed in a real-world context to which they can relate. They work in small groups to try to solve the problem on their own. More importantly, they are required to analyze their process by considering, e.g., what did they do, why did they use that method, why/how did it work, how did they make sense of the problem, how did they make sense of the solution. The teacher's role while students work in groups includes circulating and listening to "how are they developing the strategies that they are going to use and try and get some clues on what they understand" [SH4]. After a specified time, students share their solutions or how far they got and how they perceived the process. The teacher helps them to unpack the process they went through to identify their heuristics and, if necessary, discusses what could be a more "sophisticated solution"[SH1] in terms of a mathematical model, i.e., students are helped to see how the context suggests particular mathematical models or procedures. This process is repeated for a few problems. The result of this activity is a set of questions that reflect students' understanding of approaching contextual problems, for example: What does the problem mean to you? What mathematical tool can you use? Does the tool work? Does the result make sense? How does your solution compare to others? The meaning of the problem and whether the results make sense, both require examining them in relation to the real-world context. This allows for multiple interpretations and solutions for a problem.

Example 2: This teacher [JH3] describes her approach that was validated during the classroom observations: "I present situations where the students can learn a structure for doing the problem solving. ... [Students work in groups to solve the problem, however, one student in each group takes on the role of observer.] ... They look at the process that is going on to solve the problems.... They [also] are looking at how the actual group interaction occurs so they do become a good cooperative learning group and they support each other in that group. ... The students who are the observers, I take them aside and I talk to them separate from the rest of the class. So they have a little bit more information about what they need to observe than the group they are observing. I ask them to look for what kinds of things are happening in the groups ... that promotes the process of reaching the answer to the problem ... [and] that are hindering them from reaching a solution. ... I ask them to listen to the kinds of questions that are being asked in the group. ... What kind of an approach they use for the problem solving. Is there some kind of an organized process? ... Do they read all of the problem at once? Do they go back and reread the problem? ... What is their initial plan? ... Do they go back and check if the answer was reasonable? ... [Each student gets a turn at being observer. There is a whole class discussion after each round of observations.] ... They [students] keep a record of it [the list of things that come up] initially, and then we go back and we go through and review and refine and then eventually we come up with a list that is a list of what are the steps that we need to use for problem solving." The teacher usually organizes the list into categories, such as, "understand the problem, decide on a strategy, try the strategy and check for reasonableness."

Example 3: This teacher [JH2] gets students to reflect on the assumptions they bring to the problem, then, to examine the assumptions that are in the problem and to look ahead to what the possible solutions could be in interpreting the problem. His goal is to get students to visualize or think about what the possible answers could look like. His initial example to discuss assumptions is: "A tree at the side of the road broke off and fell across the road. The road was 20 meters in width. How long do you think the tree is?" He asks students what they consider to be an answer. The initial response is usually 20 meters. He then leads a discussion on the assumptions about the different possibilities of how a tree falls and the implication for the solution. This process is repeated for other word problems and forms part of the approach for dealing with future problems.

Strategy 2: Integrating word problems in course. The teachers generally integrated word problems throughout the course and not treat them as an isolated topic in the course. For example, they would use word problems or real-world situations to start, develop and end each mathematics topic. The following is an abbreviated version of one teacher's [SH3] approach for teaching systems of linear equations that reflects this integration.

SH3 asks students to collect pictures of graphs that intersect and represent real-life situations from any source other than a mathematics textbook. In a whole-class setting, each student shows his or her picture and talks about what the graph represents, what it means when graphs intersect, what the intersection shows, and why the intersection is important. SH3 then assigns the problem: "You have a part-time job in sales. Is it better to have a straight commission of 7% on the sales you make or a fixed weekly salary of $250 plus 2% commission on the sales you make?" Students are required to consider the problem based on their real-life constraints and to solve it in any way they can. The next class starts with sharing and discussion of the students' solutions to the problem. SH3 poses questions about how the problem could be solved graphically, how the point of intersection is useful to the analysis and solution of the problem, and the relevance of solving systems of two linear equations to the students' real-life experiences. This leads to an exploration of intersecting lines using the graphing calculator. It begins with students returning to the graphs they had collected and, for graphs involving two intersecting straight lines, to work in groups to set up the equations for each set of lines using data they read off the graphs. Following this, SH3 facilitates a whole-class discussion of the meaning of the algebraic representations in the context of the applications and why they are useful. After working in groups to investigate different approaches to solve systems of equation, each student is required to make up a word problem that can be solved with the method he or she explored in his or her group. After classes in which each group teaches the method it explored to the other groups, students are assigned "to bring in real-world word problems that they find else where, other than the textbook, to talk about how they use or can use systems" [SH3].

The two preceding strategies represent behaviors by the teachers that allowed students to develop flexibility to engage in modelling and application in a meaningful way in a course that is not explicitly about modelling and application. These teachers have learned through experience how to help students to overcome their initial resistance to think for themselves. Thus, their students learnt to embrace their pedagogical approaches and became highly motivated to work with non-routine problems in non-routine ways.

In conclusion, then, the paper suggests that the nature of teachers' thinking is important in creating a classroom culture to support modelling. It illustrates how teachers are able to implement strategies that allow students to experience modelling in a meaningful way in regular mathematics courses, thus supporting the view that modelling can and should be part of them.

Note: This paper is based on a research project funded by the Social Sciences and Humanities Research Council of Canada.

REFERENCES

Boaler. J. (2001). Mathematical modelling and new theories of learning. *Teaching Mathematics and its applications, 20*(3), 121-127.

Greer, B. (1997). Modelling reality in mathematics classrooms: The case for word problems. *Learning and instruction*, 7(4), 293-307.

Schoenfeld, A. H. (1991). On mathematics sense making: An informal attack on the unfortunately divorce of formal and informal mathematics. In J.F. Voss, D.N. Perkins, & J.W. Segal (Eds.), *Informal reasoning and education* (pp. 311-343). Hillsdale, NJ: Lawrence Erlbaum Associates.

Thompson, A. (1992). Teachers' beliefs and conceptions: A synthesis of the research. In D.A. Grouws (Ed.) *Handbook of research on mathematics teaching and learning*. NY: Macmillan.

Verschaffel, L. (2002). Taking the modelling perspective seriously at the elementary school level: promises and pitfalls. In A.D.Cockburn and E.Nardi (Eds.), *PME 26 Proceedings*. Vol.1, 64-80. Norwich, University of East Anglia.

Verschaffel, L., Greer, B., & DeCorte, E. (2000). *Making sense of word problems*. Lisse, The Netherlands: Swets and Zietlinger Publishers.

Chapter 3.5.2

MATHEMATICAL MODELLING IN TEACHER EDUCATION – NECESSITY OR UNNECESSARILY

Thomas Lingefjärd
University of Gothenburg, Sweden, Email: thomas.lingefjard@ped.gu.se

Abstract: Although mathematical modelling provides excellent opportunities to teach and learn mathematics, many departments of mathematics hesitate to develop and teach courses in mathematical modelling. I build this statement on a survey I did, in which I was communicating the obstacles and possibilities in mathematical modelling, with all different departments of mathematics and/or mathematics education in Sweden.

1. INTRODUCTION

The mathematics curriculum changes slowly, regardless if we speak about the intended curriculum, the implemented curriculum, or the attained or realized curriculum. The curriculum in any mathematics classroom is powerfully constrained by the school culture. Sometimes it takes a driving force from outside school to change the mathematics curriculum, like for instance technology. The vivacious advances in technology that has taken place during the last decades has called both for appropriate changes in the mathematics curriculum as well as in the teaching of mathematics.

Mathematical modelling is an example of how a subject can become possible – or at least easier – to teach early in school because of the technical evolution around us. Mathematical modelling is furthermore an interdisciplinary subject bringing together mathematics and many other fields, where it is possible to illustrate how mathematics is used in products and processes all around us.

Nevertheless, it was not until the middle of 1990 when the term mathematical modelling started to appear explicitly in the Swedish curriculum. As a matter of fact, it became visible both in the curriculum for compulsory school and for the gymnasium about the same time.

The importance of mathematical models has increased in the age of information society. Everything that happens inside a computer is the result of a mathematical model, as one example. It is important that this area is acknowledged in mathematics education. (Skolverket, 1997, p. 19)

The school in its teaching of mathematics should aim to ensure that pupils:

- develop their ability to design, fine-tune and use mathematical models, as well as critically assess the conditions, opportunities and limitations of different models, (English version of the Swedish curriculum for the gymnasium, 2000, p. 61.)
- develop their knowledge of how mathematics is used in information technology, as well as how information technology can be used for solving problems in order to observe mathematical relationships, and to investigate mathematical models. (English version of the Swedish curriculum for the gymnasium, 2000, p. 61.)

It is somewhat ironical that the topic of mathematical modelling has been in the focus of research in mathematics education for much longer than it has been mentioned explicitly in the curriculum. Many researchers and mathematics educators have addressed the benefits and obstacles of teaching, learning, and assessing mathematical modelling (Engel, 1968; Pollak, 1970; Mason, 1988; Blum & Niss, 1989; De Lange, 1996; Noss & Hoyles, 1996; Lingefjärd, 2000; Ottesen, 2001; Lingefjärd, 2002a; Lingefjärd, 2002b, Doerr & English, 2003; Holmquist & Lingefjärd, 2003).

2. TEACHER EDUCATION

During the nineties all education systems in Sweden have changed to a decentralized system with goal- and achievement-oriented structures, and with national authorities as evaluating systems on different levels. The parliament decides which universities and university colleges that may exist. There are 34 at the moment. The government decides which degrees that are going to be established. The National Agency for Higher Education decides which universities that have the right to issue a certain degree. The universities have freedom to arrange e.g. the teacher education within the frame given through the regulation for the degree. At present time, teacher education programs in mathematics are given at 26 geographically different places. Some of these are within large universities, others within small community college universities or situated at local branches of large universities.

If exposure and training is a significant factor in teachers strategic decisions about what topic to emphasize and teach in school, then it might be expected that, compared to experience teachers for whom technology and mathematical modelling may represent an intrusion into established practice, newly graduating teachers would be better prepared and more likely to teach mathematical modelling and use modern technology when doing so. In order to find out if courses in mathematical modelling actually were given and managed at the 26 different departments, I decided to undertake a small survey in the spring of 2003. Faculty members at departments of mathematics, departments of mathematics education, and at departments of education, as well as general administrators, were given the possibility to answer to the following questions:

Do you, and your department, organize and arrange a course or courses in mathematical modelling for prospective teachers?

- <u>If Yes.</u> What training in mathematics and mathematics educations is needed for a future mathematics teacher, so that she or he will be able to teach mathematical modelling with the help of modern technology?
- <u>If No.</u> Since you, and your department, at present time do not teach the prospective teachers mathematical modelling, what are the major reasons not to do so?

The survey was sent by e-mail to all the faculty members at the 26 different campuses who, according to their web pages, were involved in mathematics education. In addition I also mailed the same questions to administrators that I considered would have some insights over the content in their programs. Within two weeks I received about 200 answers, some short and concise like "No, we don't", others more elaborating on the subject, and some even sending me course materials, syllabus, and web page addresses where I could see how the course was organized and examined, and so forth. All universities, or their equivalents, responded and many of the responding faculty members were interested in participating in a discussion about the survey, its purpose and result.

Even if the response rate was good, the results were disappointing. Four universities could answer yes: we give a course in mathematical modelling, although two of them offered the course in mathematical modelling as an eligible or voluntary course that only some few students chose to follow. Two more universities were planning courses in mathematical modelling that should start in the fall of 2003. The remaining 20 departments did not offer any course in mathematical modelling. The main argument from the faculty at these sites was that the curriculum was too crowded, the students should first study algebra; calculus; discrete mathematics; geometry; linear algebra; statistics, and so forth. The underlying argument often showed to be the lack of insight in mathematical modelling among the faculty staff, the feeling that mathematical modelling by nature is an interdisciplinary subject, and therefore not "real mathematics". Another argument was that mathematical modelling often involves technology, which is considered to be "unfair" and "fuzzy" mathematics by many "hard mathematicians".

It is important to acknowledge that the opinion about mathematical modelling not being pure mathematical is true. Mathematical modelling is not a precise body of mathematical knowledge in the same way that calculus or linear algebra is. Mathematical modelling is a process, and as most processes it has a variety of definitions. Mathematical modelling can be seen as using a complex web of knowledge related to different branches of mathematics, in order to solve an applied problem by mathematical methods. Mathematical modelling also needs translating abstract solutions to concrete reality; it then travels outside the domain of mathematics. This is probably one major reason why many mathematics departments believe that mathematical modelling is less useful than other branches of mathematics in the preparation of teachers of mathematics.

Ottesen (2001, pp. 337 – 338) argues why mathematical modelling could be seen as a way to learn more mathematics. Blum and Niss (1989, p. 5) define five arguments termed: formative, critical, practical, cultural, and instrumental. The instrumental argument is similar to what Ottesen advocates for:

> Assist students' *acquisition* and *understanding of mathematical concepts*, notions, methods, results and topics, either to give a fuller body to them, or to provide motivation for the study of certain mathematical disciplines. (p. 5)

In my communication with mathematicians and mathematics educators throughout the survey, I used another argument that might often be left out. Mathematical modelling can be used as a way to summarize and assess the mathematical competencies the students possess. Let me give a short example.

3. THE CATWALK PROBLEM

Bob Speiser presented the catwalk problem in a research seminar (spring 2002) at Gothenburg University, and a colleague and I decided to use it in a modelling course in the fall of 2002. One aspect of this problem is that it tests how much calculus students actually know or understands. Speiser and his colleagues have tried the problem with college students as well as high school students. We decided that it could also be used with students who have taken several courses in calculus, and designed a study inspired by, and partly similar to the one Speiser and his colleagues reported on.

Basic calculus is a way to study change and motion. In the catwalk problem, one challenge is to build connections between local rates of change and total changes, based on real-world data. The problem was originally designed to expose some of the complexity inherent in the use of mathematics to examine motion. Work on this problem by college calculus students has been reported in three papers by Speiser and Walter (1994a, 1994b, 1996).

The problem is illustrated by a series of photographs that there is not space enough to reproduce here in full. The photographs consist of 24 frames of a single cat, entitled *Cat in Walk Changing to a Gallop*. Eadweard Muybridge made the photos in 1880, by using 24 cameras that were activated successively at intervals of 0.031 second. They show the cat against a background grid, composed of lines spaced 5 centimeters apart. Every tenth line is darker. The 24 photographs show the cat over a total time of action of 0.71 second. We gave our students copies of the photos, the information described above, and asked them to construct one or two mathematical models describing how the cat moved over that time period. They were specifically asked to answer the following two questions: How fast is the cat moving in Frame 10? How fast is the cat moving in Frame 20? Fig. 3.5.2-1 illustrates two consecutive frames out of the 24.

Figure 3.5.2-1. catwalk

In the modelling class at Gothenburg University, only 2 students out of fifteen in the class managed to reason analogical to the calculus they had studied. Most of the others created mathematical models (with the help of technology) that were unable to give a good description of the transforming from walk to gallop in frame 10. The two successful students plotted the movement versus time in a xy-diagram first and measured the change in slope in frame 10 and frame 24 by a ruler. By that basic approach, they knew a good approximation to the answer before they started to construct a mathematical model for the catwalk. The calculus needed for this procedure is taught already at the Swedish gymnasium level. The catwalk problem proved to be an excellent tool to illuminate how different concepts of calculus connect and how important it is not to forget the basics. After the course, the students who were technology oriented in their problem solving approach and who got lost in their mathematical models, had to admit that they had forgotten the most basic way of measuring the speed of change.

With excerpts from this study which I do not have enough space to fully report on in this paper, and by using these arguments – that student's conceptions and misconceptions in calculus, as well as their beliefs about calculus, can be lifted up to the surface by mathematical modelling exercises – I managed to convince a handful of the mathematicians I communicated with in the survey that it could be both useful and interesting to give a course on mathematical modelling in their teacher training program. The problem was of course that they in general do not have faculty members who master the necessary technology as or is prepared to create or find enough challenging and complex problems at the right level.

4. DISCUSSION AND CONCLUSION

To create, maintain, and sustain a course in mathematical modelling is indeed a difficult task. Even if university teachers are interested in the subject, there are many different hurdles to pass. There is competition from other branches of mathematics, branches that are considered to be more "natural" in a teaching training program. There is a need of university teachers who master appropriate use of modern technology. There is a substantial demand of skills needed for both mastering a variety of problems and subjects, as well as procedures for handling the teaching and assessment of the mathematical modelling process.

Nevertheless, if one is able to overcome all these hurdles and difficulties along the way, there are rewards waiting along the way. One major benefit will come from a better understanding of what the students actually understands of the mathematics they have studied. Many students in the mathematical modelling classes I have studied seem to have forgotten many topics, even those studied at the gymnasium level, many of the students introduced and defended contradictory ideas despite their records of satisfactory mathematical achievement. These results challenge many of the foundations of how prospective mathematics teachers are taught and assessed. The findings I have done can be seen as supporting the position that prospective mathematics teachers for the secondary grades may not need more mathematics courses as much as they need different learning experiences. Such experiences should engage them in reasoning and in constructing mathematical models, in assessing the extent to which a mathematical argument is valid, and in developing, comparing, and evaluating alternative solution processes.

It is clear that the progress of computing technology is far from ended. We can expect the calculator of tomorrow to know as much as and maybe more than what the computer software of today does. And courses in mathematical modelling are important for prospective mathematics teachers as well as for other students who study mathematics. It is obvious that teachers of courses on mathematical modelling must pay great attention to they way they set up, conduct, and grade their assessments. With technology, it is sometimes very easy, much too easy, for students to provide answers, sometimes even the correct answer, without really understanding what the problem is about. Without assessment situations that make use of the technology and involve the students in critical thinking about what the technology offers in terms of possibilities and solutions, we may very well create students who are dependent on technology and not critical and insightful users of it.

REFERENCES

Blum, W., & Niss, M. (1989). Mathematical problem solving, modelling, applications, and links to other subjects – state, trends and issues in mathematics instruction. In W. Blum, M. Niss, & I. Huntley (Eds.), *Modelling, applications and applied problem solving: Teaching mathematics in a real context* (pp. 1-21). London: Ellis Horwood.

De Lange, J. (1996). Using and applying mathematics in education. In A. J. Bishop, K. Clements, C. Keitel, J. Kilpatrick, & C. Laborde (Eds.), *International handbook of mathematics education* (pp. 49-97). Dordrecht: Kluwer.

Engel, A. (1968). Systematic use of applications in mathematics training. *Educational Studies in Mathematics*, 1, 202-221.

Holmquist, M., & Lingefjärd, T. (2003). Mathematical modelling in teacher education. In Q. Ye, W. Blum, S. K. Houston, & Q. Jiang (Eds.), *Mathematical Modelling in Education and Culture ICTMA 10: Applications in Science and Technology* (pp. 197-208). Horwood: Chichester.

Lingefjärd, T. (2000). *Mathematical modelling by prospective teachers.* Electronically published doctoral dissertation, University of Georgia. *Can be downloaded from http://maserv.did.gu.se/matematik/thomas.htm*

Lingefjärd, T., & Holmquist, M. (2001). Mathematical modelling and technology in teacher education – Visions and reality. In J. Matos, W. Blum, K. Houston, & S. Carreira (Eds.), *Modelling and Mathematics Education ICTMA 9: Applications in Science and Technology* (pp. 205-215). Horwood: Chichester.

Lingefjärd, T. (2002). Mathematical modelling for preservice teachers: A problem from anesthesiology. *The International Journal of Computers for Mathematical Learning* 7(2), 117-143.

Lingefjärd, T. (2002). Teaching and assessing mathematical modelling. *Teaching mathematics and its Applications 21*(2), 75-83.

Mason, J., (1988). Modelling: What do we really want pupils to learn? In D. Pimm (Ed.), *Mathematics, teachers and children* (pp. 201-215). London: Hodder & Stoughton.

Noss, R., & Hoyles, C. (1996). The visibility of meaning: Modelling the mathematics of banking. *International Journal of Computers for Mathematical Learning, 1*(1), 3-31.

Ottesen, J. (2001). Do not ask what mathematics can do for modelling. In D. Holton (Ed.), *The teaching and learning of mathematics at the university level. An ICMI study* (pp. 335 – 346). Dordrecht, The Netherlands: Kluwer.

Pollak, H. O. (1970). Applications of mathematics. In E. Begle (Ed.) *The sixty-ninth yearbook of the National Society for the Study of Education* (pp. 311-334). Chicago: University of Chicago Press.

Skolverket. (1997). *Kommentar till grundskolans kursplan och betygskriterier i matematik. [Commentary to the mathematics curriculum for the compulsory school*] Stockholm: Liber.

Speiser, R. & Walter C. (1994a). Constructing the Derivative in First-Semester Calculus, *Proceedings, International Group for the Psychology of Mathematics Education, North American Chapter (PME-NA XVI)* (pp. 116-122). Baton Rouge, LA.

Speiser, R. & Walter C. (1994b). Catwalk: First-Semester Calculus. *Journal of Mathematical Behavior, 13*, 135-152.

Speiser, R. & Walter C. (1996). Second Catwalk: Narrative, Context, and Embodiment. *Journal of Mathematical Behavior, 15*, 351-371.

Chapter 3.5.3

BUILDING CONCEPTS AND CONCEPTIONS IN TECHNOLOGY-BASED OPEN LEARNING ENVIRONMENTS

Stephan Hußmann
University of Dortmund, D-44221 Dortmund, Germany,
Email: stephan.hussmann@math.uni-dortmund.de

Abstract: Self regulated learning and technology-based open learning environments show much interdependence. Illustrated by an example of a learning environment dealing with integral calculus, we look at the question to what extent the use of computers support independent concept formation and actually increases the educational value of mathematics teaching.

1. INTRODUCTION

Two of the areas within mathematics teaching that have changed a lot over the past years are self-regulated learning and the use of the computer. These developments have been pushed forward and promoted by the demand for an education that enables human beings to understand the role of mathematics in the world, and to use mathematics in order to respond to the demands of the present and future life. Special emphasis is no longer placed on the knowledge of rules and calculation routines, but on building up mathematical competences such as problem-solving, modelling, reasoning and concept formation. Therefore, there is a demand that pupils receive the opportunity to deal with relevant and reality-oriented problems during mathematic lessons.

This shift of emphasis could be significantly promoted by purposeful computer use and by strong emphasis on self-regulated forms of learning. In comparison with traditional education, it is possible for example, to solve tedious arithmetical routines virtually by the push of a computer button, and to visualize effortlessly complex data and facts. With this capacity it is possible to experience the role of mathematics in the world with reality-orientated problems in an exemplary way. Free space is created that can enable the shaping of real situations, the solving of complex problems, and the presentation of suitable arguments. As a result of learning processes, these process-based skills are now equated with the previously preferred content-based skills. This belief that mathematics is a process implies the fortification of mathematics teaching so that lessons are increasingly orientated towards the learner, and therefore self-regulated forms of learning are necessary. Heinrich Winter formulated the growing task of mathematics in the form of three basic experiences ("Grunderfahrungen") that the learners should be able to experience (see this volume, Chapter 3.5.0).

2. A LEARNING ARRANGEMENT FOR INTEGRAL CALCULUS

In this section a learning arrangement for integral calculus is introduced, based on the basic experiences of Winter, and additionally process related competencies are built to complement content related ones.

2.1 Intentional problems

Cornerstones of this concept are open, complex and reality-oriented problems, so-called *intentional problems*, which are dealt with by self-organized groups. Special designations mean that these problems are problem-based and structure-oriented. *Problem-based* means in this context that the problems build the orientation for the learning process, and through them the content and process oriented competencies are developed. When something is problem-oriented it means that the usual lesson sequence, with the phases introduction, formulation, retention, and practice in the common teaching style is irrelevant - the organization of these phases is the learners' business. They work the problem out together in different groups, secure their results through documentation in so-called *research journals*, and create examples and the necessary practice material either alone or with the help of a teacher.

The creation of examples and exercises are oriented in the first step through the different context of each problem, and the individual's prior knowledge, but this kind of productive practice also aims towards structural aspects of mathematics. The structure of the examples and exercises build the size and the content of the net of concepts that must be built in order to solve the problem successfully.

This independent discovery and exploration of specific mathematical structures, typical for each subject area, by each learner characterizes the intentional problems that are structure-oriented (Hußmann, 2004).

With such independence comes discussion that in common lessons is more or less ignored. Furthermore, it goes beyond the independence that is induced through the use of open tasks, being influenced by the organization that is open as well. Intentional problems allow exactly this independent concept formation. The ideas cannot develop, however, if they are not included in the problem as a possibility, and this property is described by the attribute "intentional". The intention with respect to the integral-concept is the introduction of the central concepts and ideas of integral calculus. This includes the building of a wide field of conceptions of the integral, the knowledge of different integral ideas, the development of some first calculation rules, and a basic idea of the fundamental theorem of calculus. The central conceptions are those of cumulation and balancing, and different integral ideas are, for instance, not restricted to the Riemann-Integral. Integral aspects include, for example, the area aspect or the mean value aspect.

Altogether, the students received three problems of which two will be introduced here (Hußmann, 2002). One of the situations deals with a female lorry driver who was checked by the police and during this normal check, the shown speedometer chart was observed (Fig. 3.5.3-1).

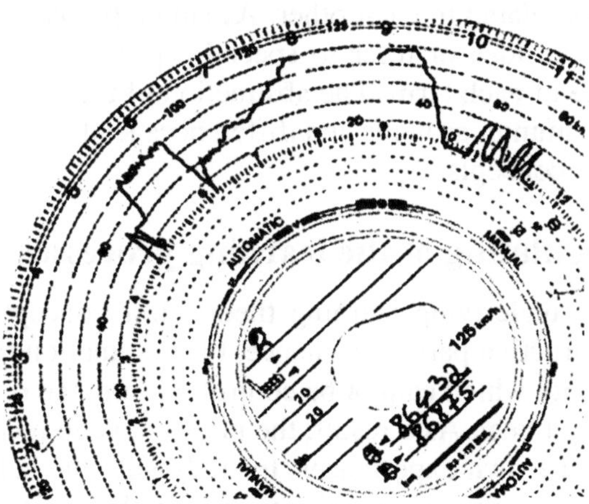

Figure 3.5.3-1. Speedometer chart

The period between 8 am and 9 am obviously raises questions. The driver claims that she took a break during that time, but the police remain sceptical.

The second situation illustrates the growth of girls and boys until the age of 18 (Fig. 3.5.3-2). In which time segments are the girls taller (or smaller) than the boys? For this task the students are given the diagram and – on request – the terms.

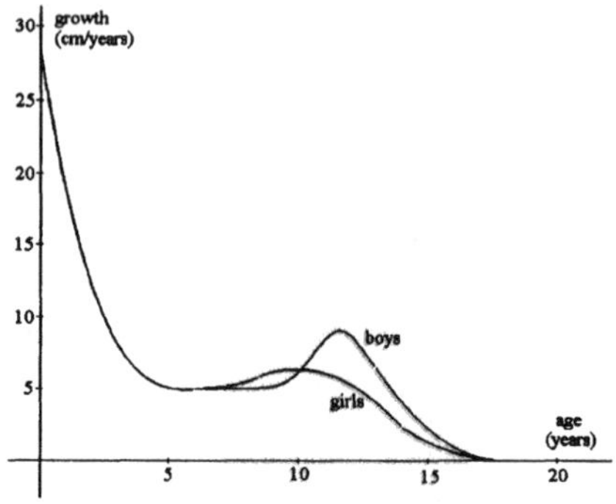

Figure 3.5.3-2. The growth of girls and boys

The following term K_B describes the body heights of the boys:

$$K_B(x) := \begin{cases} -0.0973 \cdot x^3 + 1.8312 \cdot x^2 - 11.407 \cdot x + 28.4909 & 0 < x \leq 6 \\ 0.00017733 \cdot e^x + 4.92845 & 6 < x \leq 8.5 \\ -0.243877 \cdot x^2 + 4.84906 \cdot x - 7.7548 & 8.5 < x \leq 14 \\ \frac{23}{160} \cdot (x-18)^2 & 14 < x \leq 18 \end{cases}$$

Intentional problems are not simply a line up of different interesting problems for the opening up of a mathematical context, but are with regard to the central terms related to each other. As single problems, they show the different facets of an idea or refer to special aspects. As a whole, they all contain the essence of each term. The abstraction of the concrete by recognition of structural common grounds and differences leads the learners towards general sustainable concepts.

2.2 Self regulating of the learning processes

Concerning the underlying learning theory, this conceptual approach is based upon a constructive paradigm; hence, the assumption that learning is a self-regulated activity which cannot be controlled from the outside but which can be encouraged at best. As a result, the pupils have to receive diverse opportunities to activate own experiences, to study with the help of their interests, to use their own speech, in order to decide the topic and much more (Steffe & Gale, 1995). Of course, each institutional general condition is an obstacle, but it is still possible to give responsibility to the learners. Goals that are prescribed from the outside are "only" documentation, reflection and presentation of the problem, and the mathematical theory.

Since the degree of self-adjustment is different from person to person, individual support for the learners is required, which is organized through the build-up of a dialogical communication structure between teachers and pupils, and which is, at the same time, characteristic of an open lesson organization (Ruf & Gallin, 1999; Hefendehl-Hebeker, 2004). Central aspects of this dialogical principle are the confidence in the learner's efficiency, and the concentration on the ideas and products made by the students. With this in place the right balance between the learner's construction and the teacher's instruction is found, and becomes the key for successful learning and teaching.

2.3 Use of computers

The conception of the learning arrangement calls for the computer as an integrative component; i.e. that it can always be used as a tool. The use of the computer promises long-term relief of calculating-skills, promotes the construction of procedural competencies, and makes Winter's basic experiences possible. The computer – as it is used here – intensifies cognitive abilities and skills like saving, organizing, comparing and distinguishing, so that abilities like mathematising, interpreting and concluding can be used, deepened and can be qualitatively trained (BE 1). Furthermore, it displays thinking processes through iconic and symbolical visualizations and supports with it the conception (BE 2). It also allows, in connection with the problems, heuristic approaches (BE 3), all referring to a lesson arrangement that supports self-controlled learning.

3. SELF-REGULATED CONCEPT FORMATION THROUGH COMPUTERS

The central aim of this learning arrangement is the construction of a sustainable understanding of the integral concept. For this, the problems require the construction as necessary, and are responsible for an adequate input at the beginning of the learning process. By using some parts of a typical learning process within this arrangement as an example, the computer's role will be reflected within concept formation (Hußmann, 2002).

Most of the pupils tried to start the topic by using the problem about the growth of the respective sexes. The given graphs and terms continue with the existing knowledge of differential calculus. They use this knowledge in order to reverse the differentiation as an action in order to transfer size in growth, with which they develop a first idea of anti-derivatives. For that, they do not immediately study the given complicated terms as a whole, but they first subdivide them into simple types of functions. They differentiate linear, square, and cubic functions, and then they put them together with the help of addition and subtraction to obtain the given terms. The computer as a calculator-menial enables the learners to explore and experiment many different possibilities. As the topic is "Introduction to Integral Calculus", the learners also experiment with the command "integral" and they detect that the algorithm of this command does what the students have learned about reversing the differentiation. They vary their terms, and the symbolic arithmetic ability of the computer allows them to formulate first rules that they assume lie behind the "black box" of the integral. They link the monomial to polynomial functions and develop first ideas about the sum-rule and the factor-rule of integral calculus. The provision of manifold possibilities for a symbolic manipulation of terms provides as a result, not only a contribution to the third, but especially to the second basic experience.

This symbolic manipulation with the computer, however, also shows the limits of this first idea. The second sub-function showing the male graph cannot be described with the help of the anti-derivative as a closed term. The computer outputs a term, which still consists of an integral. In fact, this term can be plotted but it is not clear what the computer actually calculates. This is a special feature of intentional problems. They contain barriers and irritations that make the subject reflect the developed concepts, and that make the subject change it if necessary. This would not be possible here without the use of the technology. As far as that goes, the CAS' contribution to Winter's second basic experience goes beyond the many possibilities of visualization.

The change of conceptions is first carried out by changing to another problem – and for this the speedometer offers the greatest motivation. The pupils try to find out whether the driver lied, for which purpose, it is necessary to determine the distance covered during the missing period of time. The speed-time-diagram and the over-all distance that is recorded in the speedometer are useful clues for that. The activation of the coherence of speed, time and distance is not difficult for the learners, who form sums of products and generate a timeline, which becomes more and more detailed. The more detailed the axis is, the more exact are the results, so you can estimate the upper limit if you argue in favour of the woman, and the lower limit if you think the woman made an error. Here, this intentional problem offers possibilities to develop the integral as generalized sums of products in terms of a Riemann-Integral, particularly with the construction of the conceptions of the area aspect. The input of a spreadsheet can probably simplify the arithmetic here, however this does not justify the use of a computer. In regard to the transfer of the process towards general operations, the use of a spreadsheet is clearly more effective, because the spreadsheet supports the development of the central idea of cumulation.

The values that were determined in this way, differ, however, so much that consideration of accuracy is called for. Many pupils enlarge the display window of their CAS with the objective of increasing the amount of the small boxes under the graph, in order to obtain a more precise value for the distance. This procedure does not reduce the inaccuracy however, for while the number of small boxes increases, the respective amount in the examined area remains constant. This example hints at risks that are connected with the use of a computer as a builder of concepts. The emerging display of the visualisations has to be reflected upon very critically, and this usually requires the assistance of the teacher. Accordingly, the teacher is also essential during the self-regulated learning in such a learning arrangement.

The use of a spreadsheet, however, allows for an increase in exactness. Through the integration of new staging points as rows in the spreadsheet, and through viewing the effect on the sum that is built from the individual products of time and speed, the experience of cumulation can be demonstrated. These values are finite sums.

The next step, which is to understand the integral as a limiting value of the sums of products, is only partly practicable with the computer, and dissociation from the technology is necessary here. A non-critical view and too close proximity to the technology can be counterproductive. The boys especially, often show behaviour while working with a computer that can be described as trial-and-error-conception - one experiments as long as something suitable appears on the screen. Even though this strategy is often successful, it also often prevents the pupils from thinking about the subject, and it has just a little in common with notions of digital competence. Lack of competence distinguishes itself, among other things, by *not* using the computer in a suitable situation.

After the pupils have experienced the ideas of cumulation and of limiting value of the sums of products by using the idea of the area aspect, they usually try to implement the developed algorithm into the computer, in order to determine rapidly the limiting value of the sums of products of any functions. This takes place with self-selected examples, mostly by using the mentioned trial-and-error conception. A successful use of this method is very promising, but problems are also noticeable. Any real numbers can be inserted into the lower boundary of the sum, and this creates a productive basis for discussion, in case different learners get varying results. Nevertheless, the threat, in regard to the build up of singular concepts that are not sustainable, is also obvious. Here, it is the teacher's duty to identify possible misconceptions in a timely way, and to integrate them into the discussion processes of the group.

With this result and the conceptional change, many learners return again to the problem about the "gender growth", determine respective antiderivatives, comprehensible approximations, and display them diagrammatically. The resulting picture again has the function of a productive irritation, even though it first causes disappointment when the students look at it for the first time (Fig. 3.5.3-3).

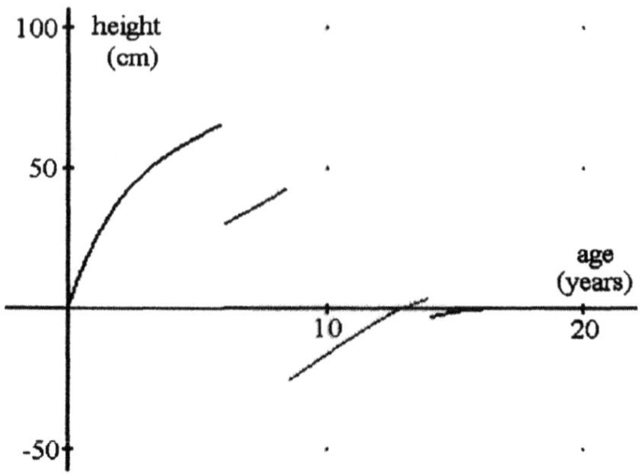

*Figure 3.5.3-3.*The growth of the boys – What happens?

Different from expectations, the sub-functions do not build a steady graph. What has happened? The reason is that there exists no one-to-one correspondence, and for this the students invent the term of a "*zero-anti-derivative*". This means the anti-derivative, whose constant is zero while integrating. This function – a pupil mentioned – "has to be shifted high and low until it fits."

Here, visualisation also initiates ideas and conceptions, from which sustainable notions arise. With the help of these two extracts, it is possible to identify two – in the sense of the constructivist paradigm – central tasks of the technology that supports independent concept formation. On the one hand, it is the *function of construction* by contributing to building ideas, and on the other hand, there is the *function of irritation* by initiating a change of concept. These are two aspects that help to emphasize the special status of the use of technology for the development of independent conception during mathematic lessons.

REFERENCES

Gallin, P., & Ruf, U. (1998). *Sprache und Mathematik in der Schule.* Seelze: Kallmeyer.

Hefendehl-Hebeker, L. (2004). Lernen im Dialog. *Der Mathematikunterricht, 50*(3), 45-51

Hußmann, S. (2002). *Konstruktivistisches Lernen an Intentionalen Problemen.* Hildesheim: Franzbecker.

Hußmann, S. (2004). Off the beaten track – Mathematics as concept-mongering. *Nieuw Archief voor Wiskunde 5*(5), Nr. 1, 64-68.

Steffe, L.P., & Gale, J. (Eds.) (1995). *Constructivism in education.* Hillsdale, NJ: LEA.

Winter, H. (2004). Mathematikunterricht und Allgemeinbildung. In H.-W. Henn, & K. Maaß (Eds.), *ISTRON Materialien für einen realitätsbezogenen Mathematikunterricht, Band 8,* (pp. 6-15). Hildesheim: Franzbecker.

Chapter 3.5.4

TOWARDS A WIDER IMPLEMENTATION OF MATHEMATICAL MODELLING AT UPPER SECONDARY AND TERTIARY LEVELS

Djordje Kadijevich
Megatrend University of Applied Sciences and Mathematical Institute of the Serbian Academy of Sciences and Arts, Belgrade, Serbia, Email: djkadijevic@megatrend-edu.net

Abstract: Despite its unquestionable educational value, mathematical modelling has had so far mostly a marginal role in everyday mathematics education. To overcome such an inappropriate state, we should help mathematics educators realize the full power of computer-based mathematical modelling, develop suitable standards of such a modelling, and ensure their proper utilization.

1. INTRODUCTION

Even when computers are available, mathematics teachers rarely use them in their educational practice, probably because they do not have (enough) knowledge and skill related to what can be achieved by using these tools (see Manoucherhri, 1999). It seems that most mathematics educators do not realize the full power of computer-based mathematical modelling and, because of that, the wider inclusion of modelling in everyday mathematics education is simply not attainable (Kadijevich, 2004). How may this state be improved? First, there is a tendency for the standardization of technology-based mathematics education.[1] Second, contrary to most mathematical courses at upper secondary and tertiary levels with more or less known content and teaching method, such courses on modelling may (and probably do) considerably differ from institution to institution. We may thus develop suitable standards of computer-based modelling[2] and ensure their proper utilization. By doing this, we should achieve a better coordination among different perspectives on the reality of modelling (see Fig. 3.5.4-1).

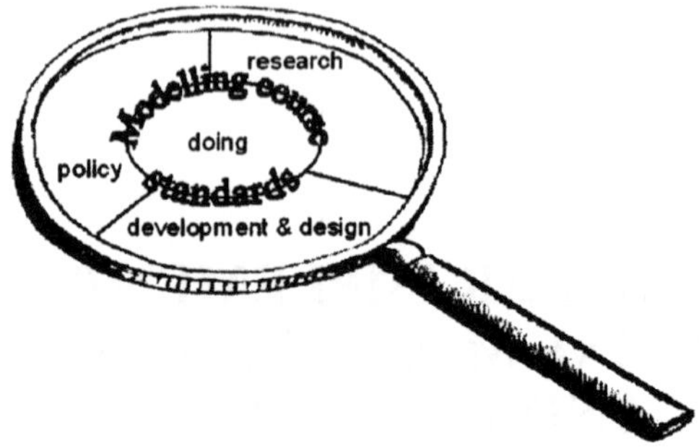

Figure 3.5.4-1. Coordinating perspectives on the reality of modelling (adapted from Blum et al., 2002)

2. REALIZING THE FULL POWER OF COMPUTER-BASED MODELLING

While modelling in general empowers a modelar's thinking and learning, computer-based modelling amplifies this empowerment through utilizing computers as versatile mindtools.

Instead of developing adequate mental models mirroring the presented conceptual models, students often memorize these conceptual models to use them in school or academic settings, while they exploit their mental models in informal (everyday) settings (e.g. Vinner, 1983; Greca & Moreira, 2000). According to *Ibid*, as presented in Fig. 3.5.4-2, mental models can incrementally be developed in the direction of the desired conceptual models through explicitly-taught modelling (developing formal mathematical models and testing, refining or upgrading already built models).

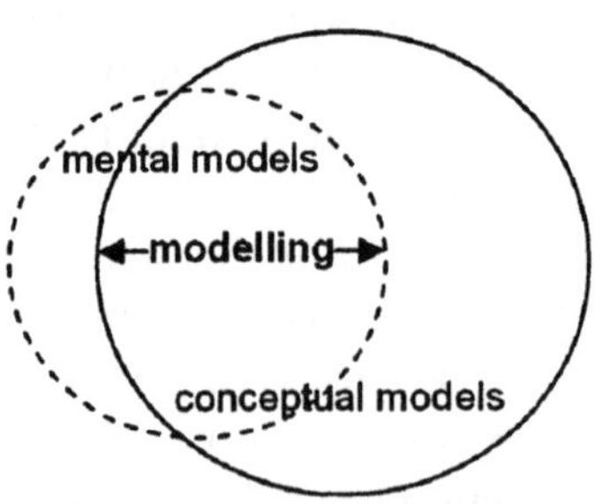

Figure 3.5.4-2. From mental models to conceptual models through modelling

Although the use of a sophisticated device and the transition from tool (impersonal device) to instrument (personal device) is achieved through a long process of instrumental genesis (Trouche, 2003; Guin et al., 2005), the use of computers as mindtools (Jonassen, 2000), expanding our mental function[3], can indeed be achieved in computer-supported modelling. Consider, for example, the utilization of Microsoft Excel and its various add-ins like SimTools (for simulations and iterative processes[4]) and RISKOptimizer (for simulation with optimization[5]) that enables very sophisticated modelling.[6]

3. DEVELOPING SUITABLE STANDARDS OF COMPUTER-BASED MODELLING

Bearing in mind the outcome of educational research relevant to modelling[7], one may propose the following five standards of computer-based modelling in the upper secondary and tertiary levels of mathematics education (derived from Kadijevich, 2003).

- *Recognize a humanistically-oriented context of modelling*. Realize that, no matter how mathematically good a model may be, the applied data quantifications may be arbitrary, the selected optimization criteria subjective, and the chosen applications questionable. Be aware that the developed models just give decision makers additional information to help them become better informed, and that it is always a human who decides on the course of action and takes full responsibility for its consequences.
- *Present modelling as a complex process.* Present several incrementally-developed models concerning the same real life situation. Realize the complexity of modelling arising from interplay among modelling steps, and from interactions among modelling actors whose ways of thinking, values, attitudes, preferences etc. may be quite diverse. Be aware that the development of an institutionalized model may require very much more time and effort than the development of a prototype model.
- *Use modelling to empower thinking and learning*. Be aware of existing mental models and desired conceptual models. Help students realize the validity and possible limitations of their mental models. Help students incrementally develop their mental models in the direction of the desired conceptual models by means of modelling. Use modelling to promote better understanding.
- *Recognize and empower cognitive, metacognitive and affective issues of modelling.* Be aware that modelling is based on a demanding interplay of modelar's cognitive, metacognitive and affective domains. Help students carry out matematizations (clarify a real problem, generate variables, select variables, and set up conditions) appropriately and confidently. Help them set up those conditions that enable an easy (easier) solution to the mathematical problem. Help them evaluate models critically.[8] Promote positive affective contexts about mathematics and the problem domain.

- *Use computers as mindtools for modeling.* Apply able tools such as Casio ClassPad, Microsoft Excel or Texas Instruments Derive.[9] Avoid, whenever possible, the "black box" view of the applied tool, by giving or requiring conceptual and procedural explanations of the performed actions and calculations. Require students to solve tasks involving routine calculations, conceptual conclusions and links between procedural and conceptual knowledge, not limiting them to a particular technological tool. View computers as tools that expand our mental function.[10]

Like the ISTE standards of educational technology, each of these or other agreed standards may be described by a list of suitable indicators mirroring the sentences used in its initial description.

4. ENSURING A PROPER UTILIZATION OF STANDARDS OF COMPUTER-BASED MODELLING

Teachers may be aware of current teaching ideas, may believe that they have implemented them in day-to-day teaching, but the real practice of such reformers may not substantially differ from everyday practice of those viewed as non-reformers (see NCES, 1999; p. 124).[11] To avoid a discrepancy between intended and implemented standards of computer-based modelling, pre-service and in-service professional development of mathematics teachers should successfully deal with various critical issues. One of them is related to realizing the full power of computer-based modelling. The other three are summarized below.

- *Selecting basic indicators for the official standards.* Let us suppose that we utilize 8 modelling standards comprising 30 indicators. Many teachers, especially those less-experienced and not so technology-minded, may find these indicators quite demanding. Thus they may base their teaching practice just upon several basic indicators, still bearing in mind the broader context.[12] Utilizing an opportunity to select one's own indicators is particularly advantageous to those involved in teacher professional development as they can focus on issues that are subject to change (see Kadijevich, 2002).
- *Making the selected indicators alive.* Just as learning through multimedia design can be beneficial to students in many ways, resulting in better understanding of and more interest in mathematical, didactic and technological issues (see Kadijevich & Haapasalo, 2004), so should (future) mathematics teachers become designers of multimedia lessons.[13] Because the suggested or targeted set of professional, psychological or didactic guidelines may be too demanding to be implemented successfully[14], a multimedia project instructor should encourage project participants to

choose their own subsets of these guidelines and help them implement these subsets successfully.

- *Reinforcing the context of the official standards.* Having in mind the advantages of Web-based professional development for mathematics teachers (Shotsberger, 1999), a critical, balanced and well-designed implementation of modelling standards should be achieved through such a support.Educational institutions and professional organizations ought thus to maintain Web sites where continuing computer-based modelling experiences are provided. Such a requirement is in accord with Kilpatrick (2003) who underlines that to improve the practice of mathematics teaching, we need "the creation of new forms of continuous professional development" (p. 326).

5. CLOSING REMARKS

As regards modelling pedagogy, Blum et al. (2002, p. 164) call for "appropriate pedagogical principles and strategies for the development of applications and modelling courses and their teaching". There is no doubt that the presented standards considerably help us define and successfully utilize such principles and strategies. Although good modelling standards will not guarantee good modelling practice and expected educational outcomes, they would – primarily applied as a useful framework not as a dogmatic recipe – confidently help us spread the agreed modelling philosophy, recruit its followers among skilled and open-minded educators, manage their professional development, and assess the effects of educational outcomes enabling adequate further steps. Researchers in the modelling community may thus focus on an elaboration of these standards involving issues of assessment and teacher's professional development as well as on their adjustments to different educational levels. Research may also empirically focus on critical variables that influence (future) utilization of such standards. To achieve this end, it may refine the approach of Kadijevich et al. (2005) who examined mathematics teachers' interest to achieve educational technology standards in terms of their computer attitudes and of the professional support, concerning these standards, that they received during their pre-service professional development.

ACKNOWLEDGEMENT

This contribution resulted from the author's work on the project "Geometry, Education and Visualization with Applications" funded by the Ministry of Science and Environment Protection, Republic of Serbia.

REFERENCES

Blum, W. et al. (2002). ICMI Study 14: Applications and modelling in mathematics education – Discussion document. *Educational Studies in Mathematics, 51*(1-2), 149-171.

Burghes, D. N., & Wood, A. D. (1984). *Mathematical Models in Social, Management and Life Sciences.* Chichester: Ellis Horwood.

Crouch, R., & Haines, C. (2004). Mathematical modelling: transitions between the real world and the mathematical model. *International Journal of Mathematical Education in Science and Technology, 35*(2), 197-206.

Galbraith, P. (2002). Life wasn't meant to be easy': Separating wheat from chaff in technology aided learning. *Proceedings of the 2nd International Conference on the Teaching of Mathematics at the Undergraduate Level* (Hersonissos-Greece, 1–6 July 2002), University of Crete, Crete. From www.math.uoc.gr/~ictm2/Proceedings/invGal.pdf

Galbraith, P., & Haines, C. (2001). Conceptual and procedural demands embedded in modelling tasks. In Matos, J. F. et al. (Eds.), *Modelling and Mathematics Education: ICTMA9: Applications in Science and Technology* (pp. 342-353). Chichester: Horwood Publishing.

Greca, I. M., & Moreira, M. A. (2000). Mental models, conceptual models, and modeling. *International Journal of Science Education, 22*(1), 1-11.

Guin, D., Ruthven, K., & Trouche, L. (Eds.). (2005). *The Didactical Challenge of Symbolic Calculators: Turning a Computational Device into a Mathematical Instrument.* New York: Springer.

Heugl, H. (1997). Experimental and active learning with Derive. *Zentralblatt für Didaktik der Mathematik, 29*(4), 142-148.

Ikeda, T. (1997). A case study of instruction and assessment in mathematical modelling – 'the delivering problem'. In Houston, K. et al. (Eds.), *Teaching & Learning Mathematical Modelling* (pp. 51-61). Chichester: Albion Publishing.

Ivic, I. (1989). Profiles of educators: Lev Vygotsky. *Prospects(UNESCO), 19*(3), 427-436.

Jonassen, D. H. (2000). *Computers as Mindtools for Schools.* Upper Saddle River: Prentice Hall.

Kadijevich, Dj. (1999). What may be neglected by an application-centred approach to mathematics education? *Nordic Studies in Mathematics Education, 7*(1), 29-39.

Kadijevich, Dj. (2002). Four critical issues of applying educational technology standards to professional development of mathematics teachers. *Proceedings of the 2nd International Conference on the Teaching of Mathematics at the Undergraduate Level* (Hersonissos-Greece, 1–6 July 2002), University of Crete, Crete. Frome www.math.uoc.gr/ ~ictm2/Proceedings/pap196.pdf

Kadijevich, Dj. (2003). Basic requirements for the design and assessment of a computer-based course on mathematical modelling. In Mladenović, N., & Dugošija, Đ. (Eds.), *SYM-OP-IS 2003 – Zbornik radova* (pp. 485-488).Beograd: Matematički institut SANU. From www.mi.sanu.ac.yu/~djkadij/djksymopis03.pdf

Kadijevich, Dj. (2004). How to attain a wider implementation of mathematical modelling in everyday mathematics education? In Henn H.-W., & Blum W. (Eds.), *ICMI Study 14: Applications and Modelling in Mathematics Education, Pre-Conference Volume* (pp. 133-138). Dortmund: University of Dortmund.

Kadijevich, Dj. (2004a). Improving mathematics education: Neglected topics and further research directions (doctoral dissertation). *Publications in Education* (No. 101). Joensuu: University of Joensuu. From www.joensuu.fi/research/index.html.

Kadijevich, Dj., & Haapasalo L. (2001). Linking procedural and conceptual mathematical knowledge through CAL. *Journal of Computer Assisted Learning, 17*(2), 156-165.

Kadijevich, Dj. & Haapasalo, L. (2004). Mathematics teachers as multimedia lessons designers. In Lagrange J.-B. et al. (Eds.), *Actes du Colloque Européen ITEM Reims 20-22 juin 2003* (on CD). Reims : IUFM.

Kadijevich, Dj., Haapasalo, L., & Hvorecky, J. (2005). Educational technology standards in professional development of mathematics teachers: An international study (contribution to ICMI Study 15: *The Professional Education and Development of Teachers of Mathematics*, São Paulo-Brazil, 15-21 May 2005). From http://stwww.weizmann.ac.il/G-math/ICMI/strand1.html.

Kadijevich, Dj., Haapasalo, L., & Hvorecky, J. (2005a). Using technology in applications and modeling. *Teaching Mathematics and its Applications, 24*(2-3), 114-122.

Keune, M., & Henning, H. (2003). Modelling and spreadsheet calculation. In Ye Q-X. et al. (Eds), *Mathematical Modelling in Education and Culture: ICTMA10* (pp. 101-110). Chichester: Ellis Horwood.

Kilpatrick, J. (2003). Scientific solidarity today and tomorrow. In Coray, D. et al. (Eds.), *One Hundred Years of L'Enseignement Mathématique: Moments of Mathematics Education in the Twentieth Century, L'Enseignement Mathématique: Monographie N° 39* (pp. 318-330).

Lambert, P., Steward, A. P., Manklelow, K. I., & Robson, E. H. (1989). A cognitive psychology approach to model formulation in mathematical modeling. In Blum, W. et al. (Eds.), *Applications and Modelling in Learning and Teaching Mathematics* (pp. 92-97). Chichester: Ellis Horwood.

Manoucherhri, A. (1999). Computers and school mathematics reform: Implications for mathematics teacher education. *Journal of Computers in Mathematics and Science Teaching, 18*(1), 31-48.

Moore, J. H., & Weatherford, L. R. (2001). *Decision Modelling with Microsoft Excel.* Upper Saddle River: Prentice Hall.

National Center for Education Statistics (1999). *The TIMSS Videotape Classroom Study: Methods and Findings from an Exploratory Research Project on Eighth-Grade Mathematics Instruction in Germany, Japan, and the United States.* Washington: U.S. Department of Education. From http://nces.ed.gov/pubs99/1999074.pdf.

Niss, M. (2003). Application of mathematics '2000'. In Coray, D. et al. (Eds.), *One Hundred Years of L'Enseignement Mathématique: Moments of Mathematics Education in the Twentieth Century, L'Enseignement Mathématique: Monographie N° 39* (pp. 273-284).

Ossimitz, G. (1989). Some theoretical aspects of descriptive mathemathical models in economic and management sciences. In Blum, W. et al. (Eds.), *Modelling, Applications and Applied Problem Solving: Teaching Mathematics in a Real Context* (pp. 43-48). Chichester: Ellis Horwood.

Schoenfeld, A. H. (1992). Learning to think mathematically: Problem solving, metacognition, and sense making in mathematics. In Grows, D. A. (Ed.), *Handbook of Research on Mathematics Teaching and Learning* (pp. 334-370). New York: Macmillan.

Shotsberger, P.G. (1999). The INSTRUCT project: Web professional development for mathematics teachers. *Journal of Computers in Mathematics and Science Teaching, 18*(1), 49-60.

Stigler, J.W., & Hiebert, J. (2004). Improving mathematics teaching. *Educational Leadership, 61*(5), 12-17.

Trouche, L. (2003). Managing the complexity of human/machine interaction in a computer based learning environment (CBLE): Guiding student's process command through instrumental orchestrations. Paper presented at the third CAME symposium *Learning in a CAS Environment: Mind-Machine Interaction, Curriculum & Assessment* (Reims-France, 23-24 June 2003). From www.lonklab.ac.uk/came/events/reims/2-Presentation-Trouche.doc.

Vinner, S. (1983). Concept definition, concept image and the notion of function. *International Journal of Mathematics Education in Science and Technology, 14*(3), 293-305.

---

[1] See the NCTM Principles and Standards of School Mathematics at http://standards.nctm.org/ and the ISTE Educational Technology Standards at www.cnets.iste.org.

[2] Throughout this article word *modelling* refers to *mathematical modelling*.

[3] Vygotsky's view; as F. Bacon (1561 – 1626) said *"Nec manus, nisi intellectus, sibi permissus, multum valent: instrumentis et auxiliis res perficitur."* ("Left to themselves, neitheir hand nor intelligence is of much worth; the work is elaborated by using tools and aids." – Ivic, 1989; p. 430)

[4] See http://home.uchicago.edu/~rmyerson/addins.htm.

[5] See www.palisade-europe.com/riskoptimizer/.

[6] Technology can enrich learning possibilities for modelling, especially when combined with learning through multimedia design and on-line collaboration (Kadijevich, 2004a).

[7] See Burghes & Wood (1984), Ossimitz (1989), Lambert et al. (1989), Schoenfeld (1992), Heugl (1997), Ikeda (1997), Kadijevich (1999), Greca & Moreira (2000), Galbraith & Haines (2001), Kadijevich & Haapasalo (2001), Moore & Weatherford (2001), Blum et al. (2002), Galbraith (2002), Niss (2003), and Kadijevich et al. (2005a). These standards should be viewed as a summary of relevant findings that primarily emerged within the ICTMA research circle. This circle (International Community of Teachers of Mathematical Modelling and Applications) is presented at www.infj.ulst.ac.uk/ictma/.

[8] As a next step, each of these *help* indicators should be clarified in the following way: "By applying ... help students ..." For example, "By assisting students realize basic features of appropriate models, help modelers evaluate the developed models critically."

[9] ClassPad is a calculator but it can be emulated at a computer screen by using the ClassPad Manager software. For these tools, see, for example, www.classpad.org/overview.html, www.microsoft.com/office/excel/ and www.ti.com/derive/.

[10] No matter which learning environment is utilized, many students experience difficulties in moving between the real and the mathematical world (see Crouch & Haines, 2004). However, despite different students' views of the utilized technology (potentially) promoting different kinds of learning (Galbraith, 2002), technology can help reduce such difficulties (see Keune & Henning, 2003), enabling us to concentrate on subtasks that cause the most difficulties in moving between the two worlds.

[11] According to Stigler & Hiebert (2004), teachers may use good problems (e.g. those that focus on concepts and connections among mathematical ideas), but implement them in a wrong or inadequate way, e.g. as problems that call for basic computational skills and procedures.

[12] Possible extended by their own sound indicator(s).

[13] Through, for example, the development of HTML files (with Microsoft Word and its "Save As Web Page" command) comprising Java applets downloaded from the Internet.

[14] "You get frightened when thinking of all these requirements!" – a student's reaction.

Each of the hyperlinks given in this contribution was tested and found active as of April, 23, 2006.

Chapter 3.5.5

HOW MIGHT WE SHARE MODELS THROUGH COOPERATIVE MATHEMATICAL MODELLING? FOCUS ON SITUATIONS BASED ON INDIVIDUAL EXPERIENCES

Akio Matsuzaki
Junior High & Senior High School at Komaba, University of Tsukuba, Japan,
Email: makio@sakura.cc.tsukuba.ac.jp

Abstract: The focus of this chapter is on models which are made and shared by modellers through cooperative mathematical modelling, and to consider their potential for advancing process. I see changes to models as one of the transitional factors in mathematical modelling. For able-to-be-shared models, it is necessary to have close-to-common *situations* for students as a means for advancing process. In this case, a teacher's indication would be to emphasize the sharing of *situations* for each of the students. For not-able-to-be-shared models, it is considered that students should try to share their respective *situations,* even if their associated models are unable to be shared. Through sharing their respective situations students may be able to agree to adopt some common representations of those situations.

1. INTRODUCTION

Schemas of mathematical modelling involve advanced processes based on ideal mathematical modelling (c.f. Blum, 1985), in which several models are generated and revised through the processes. I describe 'facts' of mathematical modelling as variables in transition (Stillman & Galbraith, 1998) because variables, which structure problems, change through problem solving. When several modellers carry out mathematical modelling together, modelling interaction, seen as generating and revising models, is important in modelling pedagogy. In this study I adopted cooperative paired learning as a research setting, and drew up the following two research questions: *'How might mod-*

els be made and shared in the case of cooperative paired learning?' and 'What could be considered influences on the process of making and sharing models?'

The aims of this study are to follow the development of models which are made and shared by students through cooperative mathematical modelling, and to inquire concerning influences that help in advancing this process. I would see changes in models as one of the transitional aspects of interest in mathematical modelling, and additionally focus on *situations* based on each modeller's learning or living experiences. Feedback is required on the original situations involving real models in terms of a schema for mathematical modelling. The definition of the *situations* is in terms of pictures that are imaged or recollected based on individual learning or living experiences. One of the findings in my prior research was that the *situations* of individual modellers had strong effects on mathematical modelling (Matsuzaki, 2004).

2. CASE STUDY

In this research setting students are required to undertake cooperative paired learning and communicate using a think-aloud method. A video-taped case study session took place on two occasions – 52:37 minutes and 49:10 minutes respectively. The subjects were 10th grade female students (Meg & Yoshi), and data referred to in this paper are protocols of the students' conversations and descriptions recorded on their worksheets.

2.1 Deciding Some Variables

At first the teacher presented an open problem: *'How much brightness is needed to read a book?'* One of the students suggested some 'variables' from reading sentences of a problem (Fig. 3.5.5-1).

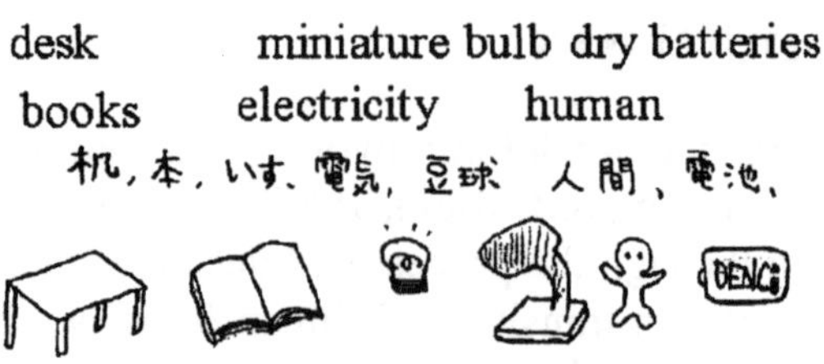

Figure 3.5.5-1. Variables suggested by Yoshi

At that time the teacher asked them to discuss the 'variables' together, because they remembered the lighting in their own rooms, which differed from each other.

[Protocols (1)] First Occasion:

11:24	Teacher	What situations did you image?
11:30	Yoshi	My own...I don't read books often. I imaged when I read books.
11:43	Teacher	Where? Put it more clearly?
11:46	Yoshi	My own desk in my room.

For the students a common situation was needed to solve the problem co-operatively, and they decided on the following five artefacts: chair, desk, a fluorescent tube, worksheet, and themselves. (These are common 'variables' from each student's selection). They were now visualizing in a laboratory situation, and discussed methods of measurement to solve their problem.

2.2 Measurement of the Real Data

The students used a tape measure and an illuminometer to measure brightness (Fig. 3.5.5-2). A fluorescent tube is different from a miniature bulb and has greater length. They looked into this point and measured several types of brightness (○,1-○,3 in Fig. 3.5.5-3). Here the figure of the laboratory room in which the research has been implemented is common to the real model generated by the students. (The laboratory room is a common situation for the students because they are now in this room.)

[Protocols (2)] The First Occasion:

38:33	Meg	We'll investigate brightness at the centre & the end of a fluorescent tube.
38:38	Yoshi	The centre & the end....
38:46	Meg	The centre of a fluorescent tube is brighter than at the end

Figure 3.5.5-2. Illuminometer

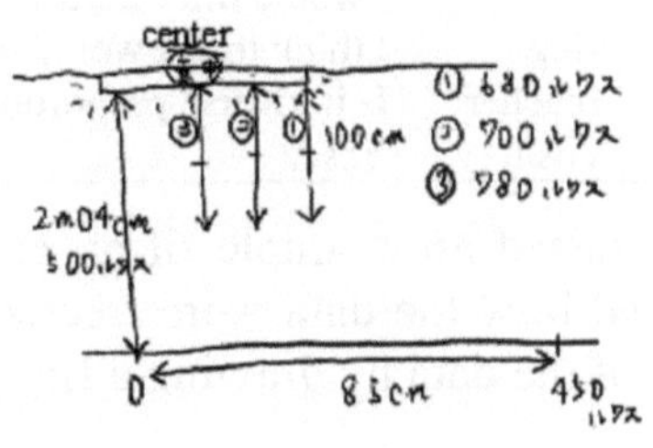

Figure 3.5.5-3. Model of laboratory room

The teacher asked the students about the relationship between brightness and distance from ceiling. Here variables are brightness and distance measured at the centre of a fluorescent tube (Fig. 3.5.5-4). The students said that these data could be used to find a formula for brightness.

[Protocols (3)] The First Occasion:

52:16	Teacher	What will you do with this data?
52:22	Meg	I want to develop a formula, develop a formula!
52:25	Teacher	Really?

52:27	Meg	Yes....Can you find a formula?
52:30	Yoshi	A formula for brightness!

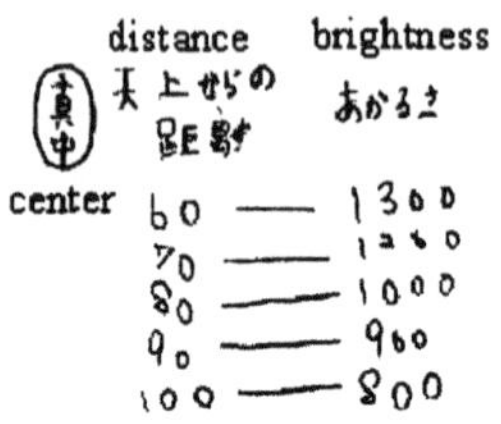

Figure 3.5.5-4. Real data

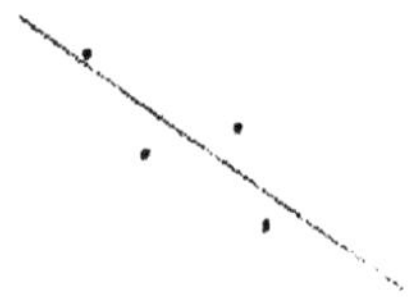

Figure 3.5.5-5. Idea of approximation

2.3 Finding Functions that Fit the Real Data

The students plotted the distance from the centre of the tube on the x-axis, and the brightness on the y-axis, and plotted five points in order to explore the relationship between brightness and distance. Then Meg said she expected that x is inversely proportional to y. The teacher asked why the relationship was inversely proportional, but neither student could provide a reason. So the teacher asked them to confirm the relationships between x and y.

[Protocols (4)] The Second Occasion:

08:51	Teacher	Draw a line now, Meg? What did you draw, Yoshi?
08:55	Meg	I think this is inverse proportion.
08:57	Yoshi	I would draw a line, but....
08:58	Meg	I think that it would be also inverse proportion.
08:59	Teacher	Is it inverse proportion?
09:00	Yoshi	Yes.

Meg raised an example of an experiment in her science classroom, and explained how the data were treated. She explained the method of approximation of the data by drawing a line of the best fit (Fig. 3.5.5-5).

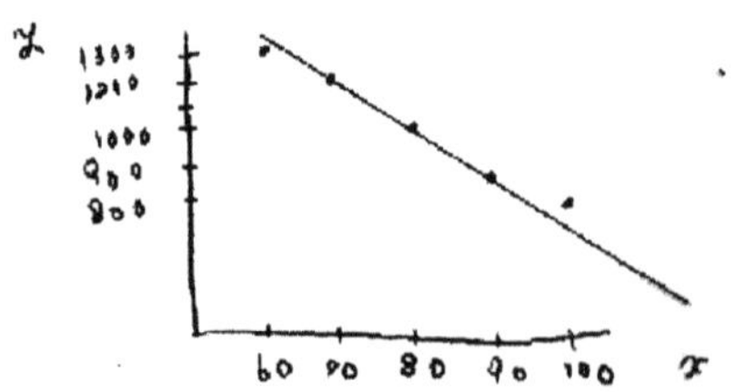

Figure 3.5.5-6. Approximation by proportion

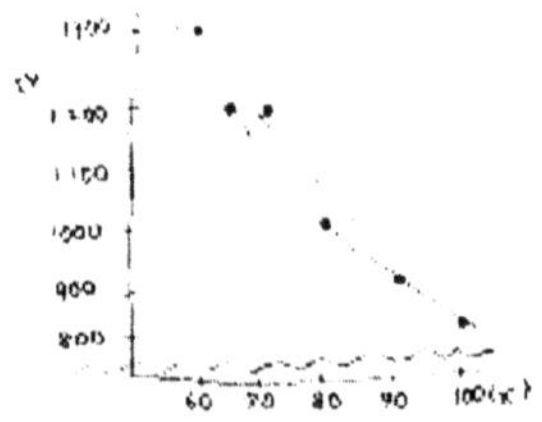

Figure 3.5.5-7. Approximation by inverse proportion

Here is an experiment with a common situation, because they attended the same science classroom. However the ways they approximated the present data were different: Meg approximated the data by a line (Fig. 3.5.5-6) and Yoshi approximated them by a curve (Fig. 3.5.5-7). These graphs are both mathematical models generated from a measured data set arising from variables on real models. So the teacher asked the students to decide whether the relationship was proportion or inverse proportion. Trying to answer the question, the students devised the idea of 'combining' proportion and inverse proportion instead of deciding between proportion and inverse proportion.

[Protocols (5)] The Second Occasion:

10:02	Meg & Yoshi	I would combine proportion with inverse proportion.
10:03	Teacher	Oh, combine them? What do you mean?
10:05	Meg	I would combine proportion whose gradient is negative with inverse proportion.
		(Some lines omitted)
11:58	Yoshi	If we match out of data to other data, the relationship would be inverse proportion....?
12:01	Meg	Let's match all the data, shall we?

The students obtained the following two equations from their data:

$$\text{(I)}\ \ y = -200x + 20{,}800, \qquad \text{(II)}\ \ y = \frac{80{,}000}{x}$$

They equated the one with the other (Fig. 3.5.5-8) and got the answer x = 4, 100.

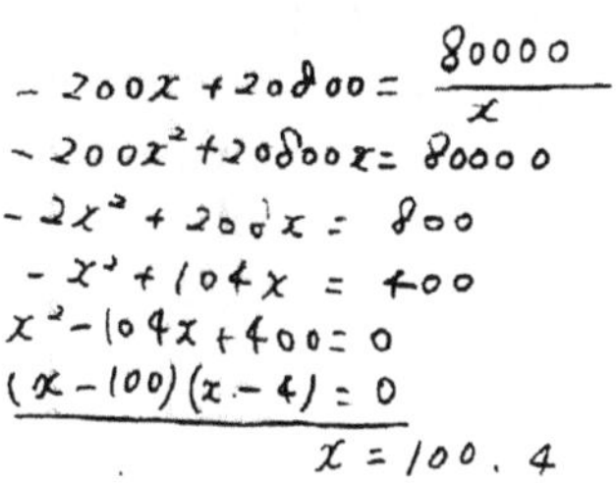

Figure 3.5.5-8. Mathematical Workprocess

Solving a quadratic equation is a common situation when 'combining' ideas generated by two equations. Here the 'variables' are the two equations situated in a mathematical model, and the answers situate a mathematical result. Although the students were able to solve the above simultaneous equations by their 'combine' idea, they were unable to interpret their answer, as they didn't understand what the answer meant in the real world context.

[Protocols (6)] The Second Occasion:

22:38	Yoshi	*x* indicates the distance from ceiling.
22:39	Meg	We combined two equations....What did we answer to?
22:44	Yoshi	Proportion and inverse proportion....
22:47	Meg	What's this answer!?
23:01	Meg	It's so wrong....

23:05	Yoshi	x indicates the distance from ceiling, isn't it?
23:13	Yoshi	x indicates the distance from floor, isn't it? (Laughing)
23:17	Meg	Why did we get the answer $x = 4$?

2.4 Revising Approximate Functions

The teacher asked the students again whether the relationship was one of simple proportion or inverse proportion, and instructed them to express their images of 'combine' on the worksheets.

[Protocols (7)] The Second Occasion:

25:02	Meg	I can't combine the graphs....
25:06	Meg	Inverse proportion until one position, and proportion from one position.
25:10	Teacher	Did you do it?
25:11	Meg	No!
25:14	Yoshi	There are several patterns....
25:16	Meg	One half of them is a curve, and another half is a line.
25:19	Meg	I don't know....
25:21	Yoshi	This part is inverse proportion, but in this part I can draw a straight line.

Meg drew another 'in-between' graph that took a middle position between a proportion graph and an inverse proportion graph (Fig. 3.5.5-9). On the other hand, Yoshi drew a graph that simply connected the proportion part to an inverse proportion part (Fig. 3.5.5-10). For these different images of 'combine', the teacher asked the students to develop an agreed common opinion. They had previously drawn several functions in divided domains in their mathematics classroom, but in this case they found it too difficult to represent the combined graph by a formula, because they couldn't recollect such a representation. So they focused on treatment of the data again.

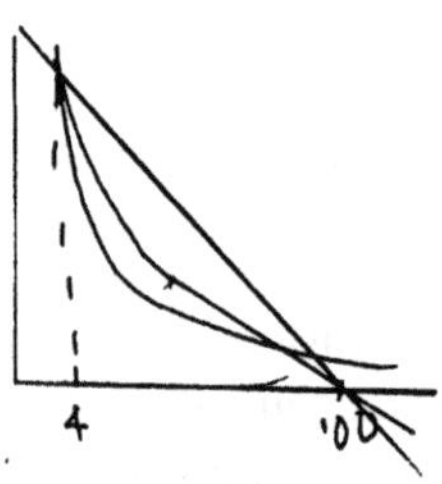

Figure 3.5.5-9. Graphs drawn by Meg

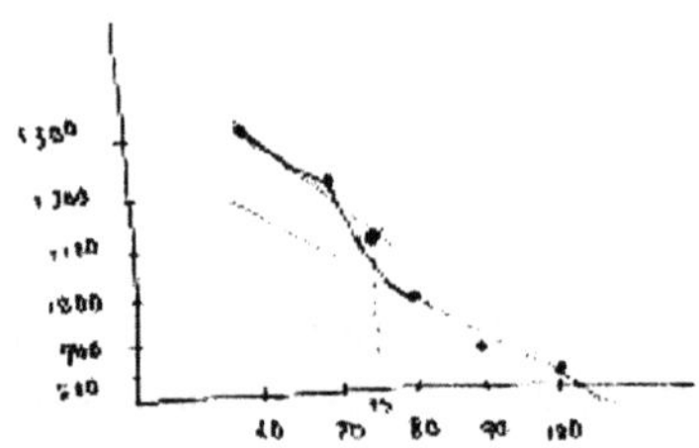

Figure 3.5.5-10. Graphs drawn by Yoshi

[Protocols (8)] The Second Occasion:

35:44	Teacher	Meg, did you say something about measurement error about these values?
35:49	Meg	Yes....
35:50	Teacher	What do you think about it?
35:54	Yoshi	It must be error, but....
35:55	Meg	If this value is an error, the relationship is one of inverse proportion. Do you think so?

35:59	Yoshi	Yes! (Some lines omitted)
36:23	Yoshi	Unn...It's funny. I've never seen such graphs ('Combined' Graphs).
36:25	Meg	Such graphs are funny.
36:27	Teacher	It's OK? Is this a result?
36:29	Meg	Yes....But I think that it is difficult for us to make a formula.
36:31	Yoshi	To make a formula is difficult.

Finally the students agreed the relationship was one of inverse proportion because they confirmed the equations by using a first set of data that they had measured from the end of a fluorescent tube again (Fig. 3.5.5-10). Both of them said the relationships must also be inverse proportion in the case of the centre of a fluorescent tube, provided they took out the data that they had recorded in error.

3. DISCUSSION

In this study's research setting, the students are required to tackle a problem *together*. In this part, I discuss identifying the able-to-be-shared models and not-able-to-be-shared models from the perspective of changing variables: able-to-be-shared models can be used in cases of changing or modifying models based on common variables. Not-able-to-be-shared models are evident when students use different variables or cannot see how to bridge their differences through adopting some common representation of the situation being modelled.

3.1 Able–to-be-Shared Models for Each of the Students

Able-to-be-shared models are seen in the *Idealization & Mathematical Work* processes. At first the model shared in the *Idealization* process is used in deciding some 'variables' (see 2.1). Each student spoke the same keyword 'reading' recollected from the open problem (see also [protocols (1)]). The 'reading' situations recollected by each individual student were not the same 'reading' situation, because they each recollected reading in their own home, and so would consider different 'variables'. So the students have set common situations to solve the problem cooperatively, and this common *situation* is based on a laboratory room (see Fig. 3.5.5-3). Secondly the model shared in the *Mathematical Work* process is seen to be finding functions that fit the real data (see 2.3). Simultaneous equations are one of the available mathematical models, and the students got mathematical results by solving a quadratic equation derived from simultaneous equations (see Fig. 3.5.5-8). Here both students used the 'combine' idea (see also [protocols (5)]) of solving two equations simultaneously. It is necessary for able-to-be-shared models to be close to common situations for both students, and in this case a teacher's directions would put emphasis on promoting a sharing of the *situations* for the students.

3.2 Not-able-to-be-Shared Models for Each of the Students

Not-able-to-be-shared models were seen in the *Mathematization & refining mathematical models* processes. At first, these models are seen in a *mathematization* process in finding functions that fit the real data (see 2.3). The students recollected 'science classroom' situations when fitting the real data. Although these situations involved common variables, the graphs used by each student were different (see Fig. 3.5.5-6 & Fig. 3.5.5-7). Secondly, not-able-to-be-shared models are seen in *refining mathematical models*, in revising approximate functions (see 2.4). In this process, involving recollected common situations, the use of 'combine' ideas is indicated by solving two equations simultaneously. So the teacher confirmed this 'combine' idea to the students by drawing graphs (see Fig. 3.5.5-9 & Fig. 3.5.5-10). It was found at first that neither student could agree on a common representation that would allow them to move forward based on the different graphs that each had drawn (see also [protocols (8)]). These graphs are themselves refined models, and these refining processes are promoted by the teacher's indications. Here 'combine' ideas are shared for each of the students, since these ideas are now included earlier. This shows that it is not only common models or situations that are sometimes unable to be shared. So the teacher needs to help students to identify those aspects which are capable of being shared, and those that are not. In this case study, the teacher's feedback indication was effective in helping the students to decide what could be shared and so helped students to agree on a model that worked for them.

REFERENCES

Blum, W. (1985). Anwendungsorientierter Mathematikunterricht in der didaktischen Disukussion. *Mathematische Semesterberichte, 32*(2), 195-232.

Stillman, G., & Galbraith, P. (1998). Applying Mathematics with Real World Connections: Metacognitive Characteristics of Secondary Students. *Educational Studies in Mathematics, 36*, 157–195.

Matsuzaki, A. & Matsuzaki, M. (2001). Analysis of Projected Figures in the Process of Recreating a Work of Art in *KADO:* Through the Assessment of Mathematical Modelling. In Burry, M., Datta, S., Dawson, A., & Rollo, J. (Eds.), *Proceedings of Mathematics & Design 2001* (pp. 285–293). Geelong, Australia.

Matsuzaki, A. (2004). 'Rub' and 'Stray' of Mathematical Modelling. In Henn, H.-W. & Blum, W. (Eds.), *ICMI Study 14: Applications and Modelling in Mathematics Education: Pre-Conference Volume* (pp.181–186). Dortmund, Germany: University of Dortmund.

Section 3.6

IMPLEMENTATION AND PRACTICE

Edited by Thomas Lingefjärd

Chapter 3.6.0

IMPLEMENTATION AND PRACTICE – OVERVIEW

Thomas Lingefjärd[1]
University of Gothenburg, Sweden, Email: thomas.lingefjard@ped.gu.se

Abstract: Implementation and practice of mathematical modelling and applications has obviously many definitions, depending on whose definition we use. No book is large enough to give a fair view of what it might mean around the world. This chapter will nevertheless try to illuminate different definitions and views on both theory and practise.

1. INTRODUCTION

To assemble a chapter which can adequately summarize and interpret the complexities connected with the vast field of implementation and practise of mathematical modelling and applications and at the same time try to incorporate different views expressed by people from around the world is, evidently, a vastly difficult and overwhelming task. I would like to ensure all readers of this chapter, that any mistake or extraordinary decision I have done in my interpretation and realization of these views is exclusively mine.

The word implementation stands for realization, to accomplish, to introduce. Our working group, which met three times during the Dortmund conference, made a huge effort to accomplish an overall agreement about what implementation and practise of mathematical modelling and applications might mean in different countries at different educational levels and especially for the participators in the implementation and practice group. It would be an overstatement to claim that we succeeded. Nevertheless, we did agree about the fact that our part of the final Study Volume should address good and convincing examples, possible to accept and interpret by "normal" teachers in "normal" classrooms to correspond satisfactory with prescribed curriculum at any level. We also found a mutual understanding about the need to give insight in views from different countries and from different contextual approaches.

Our discussions tried very hard to penetrate and interpret especially the following open questions from the Discussion Document.

- To what extent do teaching practices within applications and modelling courses draw on general theories of human development and/or learning?
- What criteria are most helpful in selecting methods and approaches suggested by such theories?
- What criteria can be used to choose (e.g. between individual and group activity) the most desirable option at a particular point within an applications and modelling teaching segment?

We added another question, imminent from the growing notion that the process of implementation and practise must exist elsewhere than just in the teaching.

- Where does modelling take place, apart from the various class rooms at various levels?

The four papers, which were selected to represent and mirror this variety of opinions, are a small but well-balanced selection of views on what implementation and practice might represent in different contexts. They also reflect the importance to focus at both theoretical and practical levels to inform future research (by researchers) and implementation (for teachers and administrators). Only one of the papers addresses the fact that many modelling problems on an advanced level quite often require advanced technology for handling the modelling process, a fact that for a time might exclude this kind of experience in some countries.

2. PAPER SUMMARIES

There is no doubt that we humans use a lot of mathematical reasoning when we perform all kinds of everyday life routines and actions. But do we observe and acknowledge our daily life mathematics? In Chapter 3.6.1, Michèle Artaud (France) illustrates the way mathematics often is hidden in everyday life practises and how difficult it might be to see it. Her theoretical approach is essentially based on the "anthropological approach" in the didactics of mathematics. The basic concept of the anthropological approach is that of *praxeology* (Chevallard, 1999), a meta-theoretical construct used to model what is involved in mathematical as well as other human activities. A praxeology consists, abstractly, of a class of similar *tasks* which can be executed by applying a set of *techniques*, situated in and enabled by a larger system called *technology*, and with an overarching discourse of justification and regulation of practise (*theory*). The "anthropological" part exists in the primacy of *tasks* understood as "requirements for human action to achieve rather precise goals".

Through an illustrative and useful example with a wooden floor whose floor-boards are laid on backing strips with an angle of 45 or 60 degrees, Artaud illustrates the interplay between the different stages in the praxeology, and concludes that mathematical modelling is an exceptional way to illuminate and highlight the mathematics which so often is implicit in the real world, something we all must strive for making through for our students.

Many mathematical techniques and conceptual considerations we use in daily life are imprecise and rough approximations. In Chapter 3.6.2, Wilfried Herget and Marlene Torre-Skoumal (Germany) illustrate how a real world modelling task change the view of mathematics as a precise and accurate subject into the reality of imprecise estimations and inaccurate measuring methods. The underlying theme of their chapter is based upon the philosophy of letting students experience mathematical situations, so well-defined by Henry Pollack:

> The heart of applied mathematics is the injunction "Here is a situation; think about it. "The heart of our usual mathematics teaching, on the other hand, is: "Here is a problem; solve it" or "Here is a theorem; prove it." We have very rarely, in mathematics, allowed the student to explore a situation for himself and find out what the right theorem to prove or the right problem to solve might be. (Pollak, 1970)

Herget and Torre-Skoumal try their modelling situations and exercises with primarily secondary students, although other targets for open-ended discussions are invited to give opinions. The open-ended problems or situations are introduced with an image of a physical object and the modelling exercise is how to measure for instance the size of a giant shoe or the volume of air in a hot-air balloon. How would you consider the challenge to make a qualified guess on the latter question? How precise would you consider it important to be? A throughout discussion about how different students at different mathematical levels might solve the problems is presented.

Are there mathematical branches that are more suitable and even more incentive for implementation of modelling and applications than others? Is possibly probability and statistics such a natural domain for fostering mathematical modelling competences? Eva Lakoma's (Poland) Chapter 3.6.3 about teaching probability and statistics, illustrates the steps students has to go through when facing a real world phenomena and simplifying this phenomena into a mathematical model. The process of creating mathematical models, posing hypothesis and verifying conjectures is her personal list of aims for mathematics education. Lakoma guides us through an experiment: A phenomena from the world of sports, namely competing with shooting a basketball through a basketball basket. When analyzing students' explorations, she highlights four main steps of the process of students' thinking and acting: *discovering and formulating of a problem; constructing a model of 'real' phenomenon; analysing the model; interpreting results obtained from the model with the 'real' situation.* These steps are further discussed and illuminated through other examples from *stochastics* (arbitrary for probability and statistics). Lakoma claims that probability and statistics give students natural motivation to develop their competency of mathematical modelling.

What do we know about mathematical modelling activities at different work places? In the fourth and final Chapter 3.6.4, Geoff Wake (United Kingdom) asks himself why it is that although mathematical models loom large in accounts of workers' activities there has been so little emphasis on what we can learn about mathematical modelling from workplace based research and the implications we can draw for the teaching and learning of mathematics in schools is yet not well developed.

In this chapter, we are given the possibility to follow the modelling work by two different workers, namely Alice (office worker) and Alan (railway signal engineer/designer). Both these two professionals use sophisticated mathematical models to carry out their duties according to "industry standard". Alice build her spreadsheet models to calculate the agreed measure "debtor days" in order to be able to communicate the performance level of the company within the company. Alan is in the security business, he overlooks calculations by other workers of where speed restrictions indicators and other signals should be placed along railway tracks so that train drivers might respond safely.

Wake demonstrates that neither Alice nor Alan have freedom in the way they set up their mathematical models and that their day-to-day activity in their companies often is focused on interpretation of the results of their application of the model. So Alice and Alan do really not control their own models, instead it is as if the models often control them. This is possibly a normal situation in industries, offices or private enterprises.

REFERENCES

Chevallard, Y. (1999). L'analyse des pratiques enseignantes en théorie anthropologique du didactique. *Recherches en Didactique des Mathématiques 19*(2), 221-266.

Pollak, H. O. (1970). Applications of mathematics. In E. Begle (Ed.), *The sixty-ninth yearbook of the National Society for the Study of Education,* (pp. 311-334). Chicago: University of Chicago Press.

---

[1] Valued contributions to a discussion that formed a basis for the introduction to the chapter were provided by Olive Chapman, Rod Nason, Hironori Osawa, Lothar Profke, Akihiko Saeki, Christine Suurtamm, Ross Turner, Pauline Vos, and Michael Wheal.

Chapter 3.6.1

SOME CONDITIONS FOR MODELLING TO EXIST IN MATHEMATICS CLASSROOMS

Michèle Artaud
IUFM d'Aix-Marseille, France, Email: m.artaud@aix-mrs.iufm.fr

Abstract: The first section of the discussion document, "Rationale for the Study" (Blum et al., 2002), mentions on the one hand that there exists a lot of works in mathematics education centered on applications and modelling and, on the other hand, that these works have had little effect on what happens in classrooms. The purpose of the present paper is to propose some explanations of this phenomenon by considering conditions for modelling to exist in mathematics classrooms.

1. AN EXAMPLE: LAYING A WOODEN FLOOR

Let us consider Mrs. Smith: she has just bought a new house and she wants to lay a wooden floor in the living room, a rectangular room whose dimensions are 6 m × 5 m. She has chosen a "Point de Hongrie" wooden floor, that is, a wooden floor whose floor-boards are laid on backing strips with an angle of 45 or 60 degrees (see Fig. 3.6.1-1).

Figure 3.6.1-1. "Point de Hongrie" wooden floor

After a quite intense reflection, she decides to take an angle of 60 degrees. Unfortunately, in the technical book she has, only the case of the 45 degrees angle is treated, the author saying that in this case, if L is the length of the boards and E the distance between two backing strips, then $E = L/\sqrt{2} = L/1{,}414$ or $L = 1{,}414 \times E$. Given that the door is on the 6 m dimension side and that she wants a whole number of rows there, she decides to set E at 30 cm. To know the measure of L, she takes a sheet of her daughter's paper and, after some trials with pieces of wood, she draws the following picture (Fig. 3.6.1-2) in real size and then measures the length of the floor-board, finding L = 59,8 cm.

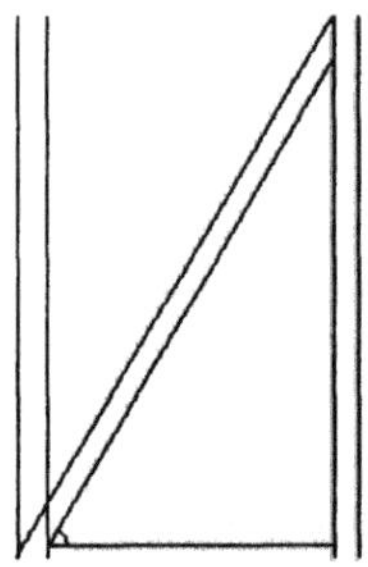

Figure 3.6.1-2.

When her daughter Leila arrives, she recognizes a problem she has just studied at school. While drawing a little picture (see beside), she explains that there is a right-angled triangle with *L* as the hypotenuse and *E* as the adjacent side of the angle of 60 degrees (Fig. 3.6.1-3). She calculates then:

$$\cos 60° = E/L$$
$$0{,}5 = E/L$$
$$L = 2E = 60 \text{ cm.}$$

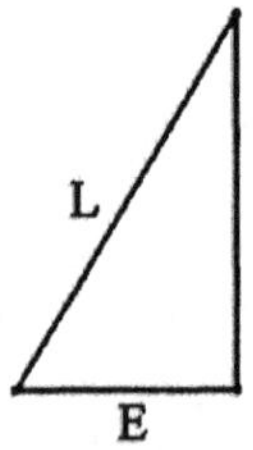

Figure 3.6.1-3.

As in the previous example, mathematics is often *implicitly involved* in described practices (such as the one found by Mrs. Smith in the book) and therefore it is not seen that practices are produced with mathematics (Chevallard, 1988). This can explain why making links between these apparently "non mathematical" practices and mathematics is often not likely to happen. This is a very important point to which we are coming later. Let us first introduce some theoretical frameworks that will be useful for identifying conditions for modelling to exist in classrooms.

2. MODELLING IN MATHEMATICAL AND DIDACTIC PRAXEOLOGIES

For many years now, mathematics teaching and learning activities have been studied in France within the theoretical framework of the Anthropological Theory of Didactics (ATD) as created by Yves Chevallard (Chevallard, 1991, 1999, and 2002). In this approach, two main aspects are considered. The first one regards *what is learnt and taught* and is modelled in terms of *mathematical praxeologies*. The second one concerns *learning and teaching activities as such* and is modelled in terms of *didactic praxeologies*. The concept of praxeology allows one to describe and analyse the gist of human activity, be it mathematical or not. More precisely, a praxeology consists of four main components: a number of *types of tasks*, that is what one has to do (e.g. solve quadratic equations, teach Pythagoras' theorem, etc.); with each type of tasks, a *technique* which provides a way to achieve tasks of the given type (e.g. a way to solve effectively quadratic equations); for every technique, a *technology*, that is a "discourse"[1] able to justify, explain, and "generate" the technique; and finally, a *theory*, that is a discourse that plays the same role towards technology that technology does towards technique, and makes it possible to interpret techniques and set up technological descriptions and proofs.

In the previous example, we could identify a mathematical *type of tasks*, T, "in a right-angled triangle calculate the hypotenuse L when an angle α and the adjacent side to the angle E are known". The *technique* that Leila brings into play is to write that $\cos\alpha = E/L$; to calculate $\cos\alpha$; to write $L = E/\cos\alpha$ and then to calculate $E/\cos\alpha$. The *technology* is based on technological elements such as the definition of $\cos\alpha$ in a right-angled triangle.

Praxeology always arises as an answer A to a question Q: when, like in the example of laying a floor, question Q is a non mathematical one, the answer A, if including mathematics, is a *mixed mathematical praxeology*, that is a praxeology in which mathematics is mingled with the "real world area" that can be other academic science[2]. In our example for instance, one must know what backing string, floor-board, a "Point de Hongrie" wooden floor are as well as knowing about the cosine of an angle in a right-angled triangle.

Modelling is then a process which allows coming from the initial question Q to a mathematical question Q_M to which an answer A_M will be produced – this answer could be well-known and mathematical praxeology is then brought into play or might be produced by using didactical praxeology (Artaud, 1993, 1994). In our case, Q is: "For a "point de Hongrie" wooden floor, what is the length of a floor-board when the distance between two backing strips is 30 cm and the angle that the floor-boards make with the backing strips is 60 degrees?". The process of modelling brought into play by Mrs. Smith and her daughter introduces the mathematical question Q_M "How do you calculate the hypotenuse of a right-angled triangle when an angle α is of 60 degrees and the side adjacent to the angle measures 30 cm?" of which the answer A_M is given by the above-described mathematical praxeology.

This above-mentioned process of modelling is also valid in mathematics: The geometric situation could be for instance modelled in an analytical or in an algebraic way. It is then of great interest to enlarge the sense of modelling to mathematical situations (Chevallard, 1985, 1989a&b; Bolea, Bosch, & Gascon, 2003).

In the ATD, the teacher is considered as the *director of the study process* the students carry out, a process which is structured along six dimensions or *didactic moments:* the moment of the *first encounter*, the *exploratory* moment, the *technical* moment, the *technological-theoretical* moment, the *institutionalisation* moment, and the *evaluation* moment. These six moments are six types of tasks that allows one to describe the teacher's practice in terms of *didactic praxeologies*. A didactic praxeology is constructed by a person when he or she studies a mathematical organisation or helps others to study it. According to Chevallard, each one of the six *moments* of the study process has a specific function essential to the successful completion of this process (Chevallard, 1999, p. 250-255) quoted by Bolea, Bosch, and Gascon (2003):

> The *first moment* of study is that of the *first encounter* with the organisation O at stake. Such an encounter can take place in several ways, although one kind of encounter or "re-encounter", that is inevitable unless one remains on the surface of O, consists of meeting O through at least one of the types of tasks T_i that constitutes it. [...] The *second moment* concerns the *exploration* of the type of tasks T_i and *elaboration of a technique* τ_i relative to this type of tasks. [...] The *third moment* of the study consists of the *constitution of the technological-theoretical environment* $[\theta/\Theta]$ relative to τ_i. In a general way, this moment is closely interrelated

to *each* of the other moments. [...] The *fourth* moment concerns the *technical work*, which has at the same time to improve the technique making it more powerful and reliable (a process which generally involves a refinement of the previously elaborated technique), and develop the mastery of its use. [...] The *fifth moment* involves the *institutionalisation*, the aim of which is to identify what the elaborate mathematical organisation "exactly" is. [...] The *sixth moment* entails the *evaluation*, which is linked to the institutionalisation moment [...]. In practice, there is always a moment when a balance has to be struck, since this moment of reflection when one examines the *value* of what is done, is by no means an invention of the School, but is in fact on a par with the "breathing space" intrinsic to every human activity.

It is clear that a "complete" realisation of the six moments of the study process must give rise to the creation of a mathematical praxeology that goes beyond the simple resolution of a single mathematical task. It leads to the creation (or re-creation) of at least the first main elements of a *local* mathematical praxeology, that is a praxeology which included several types of tasks and techniques structured around a technological discourse.

3. CONDITIONS FOR MODELLING TO EXIST IN MATHEMATICS CLASSROOMS

From the teaching point of view, on the one hand, modelling seems a well-adapted means for achieving emergence of mathematical praxeologies that is for realizing at least part of the moment of the *first encounter*, the *exploratory* moment, the *technical* moment, the *technological-theoretical* moment. On the other hand, in which way modelling will be treated in didactic praxeologies would be an important condition for modelling being a part of pupils' *topos*, that is types of tasks that pupils have to do on their own, without a teacher's help. An important point then is that for existing in students' *topos*, modelling kinds of tasks must be explicitly outlined in types of tasks around which mixed mathematical praxeologies have to be made. "Calculate magnitude g related to system S" is for instance such a modelling type of tasks: an undergraduate level technique consists for instance in determining a functional relation between magnitude g and other known magnitudes related to system S and then calculate the value of the function. When classrooms are observed, specimens of this type of task are solved, but the type of tasks itself is not in the mathematical praxeology studied.

If real world situations or other science situations are used for mathematical praxeology to arise these must not be seen like a dressing up which must be get rid of quickly as a real part of the process of study and of what to study, then in the *institutionalisation moment*, the teacher has to institutionalise not only the mathematical praxeology itself but the whole mixed praxeology. In this way, some modelling practices will live explicitly in mathematics classrooms and will be in pupils' *topos*. This point is a very important one because in most situations mathematics is only implicit. As seen in the above example of laying a floor, real world situations to be studied are considered in the real world as non mathematical: if somebody has to lay a "Point de Hongrie" wooden floor, he or she will decide that, in fact, an angle of 45 degrees is of better effect; or that Mrs. Smith's graphical technique is sufficient (and in practice, few small corrections will be made when actually laying the floor-boards). It is sometimes difficult to recognize that mathematics is relevant when analyzing such real world situations. In this respect, speaking of applications of mathematics contributes to obscuring the modelling that has to be done: this leads for instance to thinking that teaching a mathematical answer A_M and solving some problems in which A_M is brought into play will be sufficient because there will then exist for students an ability to transfer.

For being effectively realized in classrooms, previous conditions need to be supported by academic mathematicians, and on this point mathematicians are somehow ambivalent. In effect, the situation of mathematicians' community is well described by Verdania Masanja in her invited conference at ICTM2 in Greece:

> As a consequence of the new approach to mathematics, pure mathematicians drifted away from applications and saw no need to collaborate with other scientists, even their traditional neighbours, and the physicists. On the other hand, application of the highly abstract modern mathematics could not be easily visualised by the traditional users of mathematics. The period 1930's to 1970's saw a divergence within mathematics itself and between mathematics and other applied sciences. Mathematics became more inward looking, and the distinction between pure and applied mathematics became much more pronounced. (Masanja, 2002, p. 3)

At least in some European countries, pure mathematicians exert strong influence on mathematics teachers and on applied mathematicians. On the one hand then, applied mathematicians' work includes modelling, but this part of their work is not emphasized; on the other hand, the *majority* of what mathematics teachers have studied in their own studies is pure mathematics. This situation is very far from that which was prevailing in France as well as in Europe when mathematics teaching was being developed: mathematics teachers had to know how to teach pure mathematics (arithmetic, geometry) and mixed mathematics (mechanics, art of fortification, topography, pyrotechnics, etc.) and mathematicians such as Descartes, Pascal, Huygens, or Newton took care of mixed mathematics (Artaud, 1989, or Artaud, 1999).

Social and political conditions and didactic conditions are both important. To develop robust didactic praxeologies using modelling that allow constructions of some mathematical praxeologies, and for these to exist in classrooms depends on conditions that are outside of schools: the mathematicians' discourse on what mathematics is, is one of these conditions.

4. THE DIDACTIC TRANSPOSITION OF MODELLING TASKS

The above development is predicated on the point of view that modelling tasks must be transposed into the "usual" mathematical teaching, the ecological conditions of which will then be modified as mentioned above. There is another way to transpose modelling tasks into classrooms: this consists of creating a new didactic system, mathematical modelling teaching.

It is perhaps easier for this other way of transposition to occur as it allows us to avoid consideration of the prevailing relationship between mathematics and modelling.

On the other hand, if mathematical modelling teaching is added to the ordinary didactical system, then the teaching process must be accorded extra time. This would be difficult to obtain in the general teaching system; modelling could perhaps be provided for students who are supposed to need it, such as engineering students for instance.

Modelling is nevertheless an excellent method by which to make obvious the mathematics that is implicit in the real world; and it is therefore very important for all students.

REFERENCES

Artaud M. (1999). Conditions et contraintes de l'existence des mathématiques dans l'enseignement général: permanences et évolutions. *Petit x 50*, 23-38.

Artaud M. (1995). La mathématisation en économie comme problème didactique – La communauté des producteurs de sciences économiques: une communauté d'étude. *Les débats de didactique des mathématiques,* ouvrage coordonné par Claire Margolinas. Grenoble : La Pensée Sauvage Éditions. pp.113-129.

Artaud M. (1993). *La mathématisation en économie comme problème didactique*. Thèse pour l'obtention du grade de docteur en mathématiques de l'Université d'Aix-Marseille II. Marseille: IREM d'Aix-Marseille.

Artaud, M. (1989). *Conditions, contraintes et discours apologétique dans l'émergence de l'enseignement des mathématiques à l'âge classique – Etude de didactique historique.* Mémoire pour le DEA de didactique des disciplines scientifiques, Lyon.

Bacon, F. (1) The advancement of Learning. Adelaide : eBooks@Adelaide, 2004. Available on the Internet at http://etext.library.adelaide.edu.au/b/bacon/francis/b12a/index.html [ref. February 2006].

Bolea, P., Bosch, M., & Gascon, J. (2003). Why is modelling not included in the teaching of algebra at secondary school? *Proceedings of CERME 3.*

Blum W. et al. (2002). ICMI Study 14: Applications and modelling in mathematics education – Discussion document. *Educational Studies in Mathematics 22*, 37-68.

Chevallard, Y. (2002). Organiser l'étude – 1. & 3. Écologie et régulation. *Actes de la XIe école d'été de didactique des mathématiques.* Grenoble: La Pensée Sauvage éditions.

Chevallard, Y. (1999). L'analyse des pratiques enseignantes en théorie anthropologique du didactique, *Recherches en Didactique des Mathématiques 19*(2), 221-266.

Chevallard, Y. (1991). *La transposition didactique – Du savoir savant au savoir enseigné*, deuxième édition. Grenoble: La Pensée sauvage éditions.

Chevallard, Y. (1989). Implicit Mathematics: Its Impact on Societal Needs and Demands. Communication au groupe thématique 7, The Mathematics Curriculum:Towards the Year 2000, du sixième congrès international sur l'enseignement des mathématiques (ICME 6, Budapest). In J. Malone, H. Burkhard, & Ch. Keitel (Eds.), *The Mathematics Curriculum: Towards the Year 2000.* Perth, Australia: Curtin University of Technology.

Chevallard, Y. (1985, 1989a, 1989b). Le passage de l'arithmétique à l'algébrique dans l'enseignement des mathématiques au collège. *Petit x 5*, 51-94 ; *19*, 45-75; *23*, 5-38.

Descartes, R. (1637). Le Discours de la Méthode. Available on the Internet at http://classiques.uqac.ca/classiques/Descartes/discours_methode/discours_methode.html [ref. February 2006].

Majansa, V. (2002). Mathematics and other Disciplines: The Impact of Modern Mathematics in other Disciplines. *In Proceedings of the 2nd International Conference on the Teaching of Mathematics (at the undergraduate level)*, University of Crete 1 – 6 July 2002. Available on the Internet at http://www.math.uoc.gr/~ictm2/Proceedings/invMas.pdf. [ref. February 2006].

---

[1] The meaning of "discourse" here is akin to Descartes' use of the word in his *Discourse on method* (1637).

[2] Mixed is used here in its classical sense, as we find it for instance in this quotation of Francis Bacon: "The mathematics are either pure or mixed. To the pure mathematics are those sciences belonging which handle quantity determinate, merely severed from any axioms of natural philosophy; and these are two, geometry and arithmetic; the one handling quantity continued, and the other dissevered. Mixed hath for subject some axioms or parts of natural philosophy, and considereth quantity determined, as it is auxiliary and incident unto them. For many parts of nature can neither by invented with sufficient subtilty, nor demonstrated with sufficient perspicuity, nor accommodated unto use with sufficient dexterity, without the aid and intervening of the mathematics; of which sort are perspective, music, astronomy, cosmography, architecture, enginery, and divers others".

Chapter 3.6.2

PICTURE (IM)PERFECT MATHEMATICS!

Wilfried Herget and Marlene Torres-Skoumal
Martin-Luther-University of Halle-Wittenberg, Germany, Email: herget@mathematik.uni-halle.de / Vienna International School, Austria, Email: mskoumal@vis.ac.at

Abstract: In this paper some unusual open-ended problems are presented, which have been "tried and tested" in secondary school. The main focus is not on calculation but rather on all the steps necessary before the calculations can begin. "Here is a situation. Think about it!" (Henry Pollak) Such exercises are indispensable toward the introduction of skills inherent in mathematical modelling where the emphasis is not on algorithmic procedures but rather on the higher order skills of translation, interpretation, and evaluation of the real life problem in terms of the mathematical model and its solution(s).

1. HERE IS A SITUATION – THINK ABOUT IT!

In the minds of the masses, "Doing Math means calculating". True, but this is certainly not the whole story. There is far more to mathematics than "mere" calculation! In this paper, some unusual open-ended problems will be presented which we successfully used in secondary school. In these tasks, calculating is not at the forefront, but rather all the thinking and planning skills necessary before the calculations can begin. "Here is a situation. Think about it!" (Henry Pollak) In the following exercise a newspaper article depicting a giant shoe is used as a starting point. "What size is this giant shoe?" Everyone seems to find a task like this rather unusual, and it is always intriguing to hear the different ways of solving the problem.

The person who fits this giant shoe must have enormous feet!

Antal Annus, a 73-year-old shoemaker from the Hungarian village of Csanádapáca, is depicted here, proudly presenting his hitherto most impressive "creation". To this very day, we still do not know whether he really made the shoe for one of his customers.

Goslarsche Zeitung, 7.1.1995

Figure 3.6.2-1. What size is this giant shoe?

2. MANY DIFFERENT WAYS OF SOLVING THE PROBLEM

The standard approach is to use an object in the picture as an estimator or yardstick, e.g., the man's glasses, his head, the width of the apron he is wearing, etc. It is quite easy to measure these things, both in the picture and in reality. A few simple calculations suffice to give us the length of the shoe. Once we have obtained the length of the shoe, however, we still do not know its real size! Have you ever thought about the relationship between the length of a shoe and the various parameters indicating a shoe's size? This could well turn out to be an interesting research project!

A colleague came up with another idea about how to solve the problem at hand. "A shoe is about the same length as a human face!" Assuming 42 to be the standard size shoe in Europe, we simply have to do our sums, providing the relationship *shoe length to shoe size* is linear. A school girl came up with another fascinating solution. Imagine that we turn the shoe at 90 degrees around the man's naval, we will then discover that the shoe is a little smaller than the man. If the man is about 1.70 m tall then the shoe must be approximately 1.5 m in length. Two students who conveyed their idea very clearly using body language put forth another possible solution: If we imagine the man in real life with his arms stretched out, he spans at least the length of the shoe. In the case of an average-sized human, this would be about 1.60 m. In reality, therefore, the shoe is approximately 1.5 m in length.

How reliable, however, are the different approaches to the problem? The results lie somewhere between 1 m and 2 m, so how accurate are in fact the various measurements and estimates? In the end, a critical comparison of each method might well reveal a slight difference but we still haven't come up with "the right solution"!

Finally, our task is to look at the relationship between normal shoe sizes and the length of the foot, in centimetres. Where do we start? One way would be to collect data by measuring various shoes. (Measuring big and small shoes lends itself well to homework since there are bound to be some "giants" and "dwarves" in the family and neighbourhood!) Another possibility, one which is rather unusual in math lessons, would be to make some enquiries in local shoe shops. If we are lucky, we might find some data on the shoe boxes themselves. Finally, we will get the type of relationship *normal shoe sizes ↔ length of the foot in centimetres* and some respective formula.

3. VERY PRECISE ... AND VERY ROUGH!

Math lessons are typically characterised by precision. For example, if three sides of a rectangular box are 3 cm, 5 cm, and 7 cm respectively (and precisely, of course!), then find the volume of the box. But this obsession becomes an exercise in futility the moment mathematics becomes involved with "the rest of the world". There, most of the numbers which crop up are only approximately correct. This is inevitable and unavoidable! Likewise, the results are only rough estimates.

In our lessons, therefore, one of our tasks, indeed obligations, should be to bridge the gap between these two different worlds: the world of accuracy so typical of mathematics, and that of lack of precision in the rest of the world. This is imperative because both worlds are important and both are indispensable. How can we possibly learn the true value of the precision and certainty of mathematics if we have not yet learnt that, in the "rest of the world", this precision and reliability is something which is very difficult to achieve? On the other hand, one can only learn to cope well with this inaccuracy and blatant lack of precision if one has learned to exploit the many possibilities offered by the very precise field of mathematics.

4. A PICTURE TELLS A STORY OF (WELL OVER) 1,000 WORDS!

How is it then possible to bridge the gap between mathematics and the "rest of the world"? How do we carefully and sensitively introduce the young student to the uncertain world of mathematical modelling? At this point we propose a very special method: setting tasks mostly based on rather unusual newspaper cuttings that we are apt to call "Pictorial Problems" or "Picture Mathematics". Many tasks based on real-life situations are often far too cluttered with text to be truly effective for the young mathematics student. This is where a picture, supplemented by the students' general knowledge and imagination, comes in handy: "A picture can indeed say far more than a thousand words!"

► **How many litres of air does this hot-air balloon hold?**

Stern, 44/2001

Figure 3.6.2-2. The hot-air balloon

How many litres of air does this hot-air balloon hold?

At first, of course, to solve this task a model of the hot-air balloon should be made as accurately as possible, with the help of an object that is easy to describe. The wider the range of mathematical instruments available, the more instruments can be used to solve this task.

In a Calculus course, the interpretation of the set task would be based on rotated solids. Modern pocket calculators, with built-in programmes for regression analysis, have no limits whatsoever with regard to the type of function used. Indeed students left to their own resources showed untiring efforts in their quest for the "best" approximation. At the tender age of 15, 9th graders were researching how to calculate volumes of revolution using a computer algebra system, although not fully understanding, of course, why such a procedure of integrating the squares of functions between two parameters and multiplying by π would be an effective method of approximating a volume since this topic formally comes later in their mathematical experience.

Other students were satisfied with much simpler solutions. They simply looked at familiar geometrical shapes in their model building kit and selected something suitable.

For example, one model of the balloon is made of the upper section using the shape of a hemisphere and the lower part is then made using a cylindrical cone. Another model consists of "chiselling" the balloon into a hemisphere, a frustrum of a cone for the middle section, and then a cone. In both cases the person in the photo is used as a yardstick. Other solutions may use even simpler models of the balloon: Let us take a big sphere as a suitable substitute, or even a cube – and it works, indeed!

When taking measurements we are obviously very much aware of how inaccurate these values are. There is little point in making note of the figures which appear after the decimal point as shown on the pocket calculator. Some calculations using upper and lower values should be made arriving at an "interval of tolerance" as an answer. Finally we obtain for the total volume of the balloon roughly 7,000 cubic meters.

5. DIFFERENT WAYS BUT COMMON IDEAS

Let us now itemize the steps inherent in the process of seeking solutions to these examples:

- "Real world" mathematics remains the focal point for the duration of the activity until a solution is reached – the problems do not exist merely as a desperate attempt to superimpose a real world problem on analytical techniques previously learned.
- The facts are analyzed and the mathematically relevant details are filtered out while the perhaps interesting, but irrelevant information (for the solution's sake) is laid aside.
- An appropriate object is chosen to serve as a yardstick for the necessary measurements which have to be made in the solution process.
- Necessary simplifications are performed.
- The interesting measurements are taken from the picture; through the measurement process one is constantly conscious of the unavoidable element of uncertainty yielded through the approximations.

- Common knowledge is activated, e.g., how tall is an average person, do body parts exist in certain proportions, etc. If necessary, information from other sources will be obtained.
- The relationship between the chosen yardstick and the measurements obtained will be mathematically defined and refined.
- A suitable mathematical model and methodology for the solution of the problem will emerge from this process, as opposed to students being handed pre-conceived ones. The students may choose the model they feel is most suitable, i.e., they must choose the model themselves.
- Technology allows for solutions previously denied students until much later in their mathematical development, or not at all.
- This entire process is guided and enlightened by fundamental mathematical considerations, strategies and concepts, which make a solution possible.
- Throughout this process other questions or ideas emerge, mathematical or otherwise, which can then be expanded upon, time permitting.

A discussion of these examples highlights the essential aspects of the process of mathematical modelling. In mathematics education knowledge and skills are necessary pre-requisites which assist us in various stages of the process, but to accomplish the entire task at hand, certain central ideas or concepts are necessary, namely the concepts of measurement, approximation, and linearization.

All of the above is in accordance with Freudenthal's view of mathematics (Freudenthal, 1968) – 'mathematizing' as the activity of looking for problems and solving them, by organizing all the information you have about this problem situation and then choosing and using suitable mathematical tools.

When using these "picture mathematics" exercises, the role of the teacher in the classroom changes from being mainly the disseminator of information to becoming a moderator or facilitator of knowledge. The teacher must carefully consider the various methodologies chosen by the students, and gently guide and direct their efforts in their quest for a solution. (For example, the most common error for wide discrepancies of approximations was incorrect handling of units, e.g., incorrectly changing cm^3 into m^3, or m^3 into liters, etc.) Furthermore, the teacher should encourage discussion and reflection on the various strategies employed, and point out the central concepts and ideas contained in the various solutions. Now more than ever expertise is needed in handling information overflow, performing thorough research, discerning the important from the unimportant and the correct from the questionable. Good communication skills are required in order to make the procedures and processes accessible to others. Lastly, all of the incorporated and integrated information should lead to a higher level of knowledge.

6. MODELLING AS A CENTRAL THEME

Nowadays, tasks requiring pure technical calculation can be solved with the help of a calculator or computer software. Consequently, more demanding activities are gaining importance, e.g., the analysis of problems affecting the "real world", i.e., mathematical modelling. It is to the child's benefit to discover the "discomforts" of uncertainty in mathematics as early as possible. For students as well as for teachers the shift from problem solving (one correct answer) to mathematical modelling (multi-solution paths leading to approximate answers each with possible limitations or potential for extension) requires a new set of teaching and learning skills – symbolizing data, cleverly translating the task into the language of mathematics, i.e., into a mathematical model, the internal treatment of this problem in the field of mathematics right up to its solution(s) possibly with the aid of technology, and finally, a deliberate interpretation and critical examination of the results obtained. "Has our original question really been answered? How accurate and how reliable is the result? How applicable are my results to other (new) situations?" Thus using carefully selected examples, the typical process of mathematical modelling may become a central theme in the classroom, including both modelling methods and the accuracy of the mathematics involved, of course without losing sight of the discrepancy between the mathematical model and reality.

7. ASSESSMENT OF MODELLING TASKS

We all know that in good classroom practice assessment is a natural by-product of the classroom experience and should be a celebration of achievement. In using such activities for assessment purposes we recommend, therefore, a criteria-based assessment that is defined from the sub-tasks themselves. Since in these open-ended tasks "many roads lead to Rome", the focus of the assessment should be uppermost on the selection of the path and the markers that the student has set up along the way to ensure a secure journey toward the goal. Considerations as criteria should therefore be – *Communication:* To what degree has the student used tables, diagrams, graphs, etc. as aids in defining the problem? Has the student used correct mathematical notation and terminology throughout the activity? (*Yes*, this is still very important, perhaps now more than ever, the need for correct and clear communication.) *The Model:* What has the student used as a yardstick, and is it well justified? Has the student taken into account all vital variables to the problem? Does the model suit the problem – how *well* does it fit the problem? *Mathematical Content:* Are the calculations correct and justified? Were

formulas correctly applied? *Evaluation:* To what extent has the student evaluated the meaningfulness of the approximations obtained from the model in light of the real life problem? Has the student considered limitations of the model, or possible extensions and applications of it?
The assessment of such activities can provide students with opportunities and rewards for carrying out mathematics without time limitations of tests and exams and their accompanying stresses. Furthermore it can provide the student who has difficulty performing well on traditional exams a sense of success and achievement in this subject. Hence, the emphasis in this kind of assessment should be on good mathematical writing and thoughtful reflection.

8. CARRY ON COLLECTING AND COMPILING

Additional information and many further examples are available under Herget (2000, 2002), Herget, Jahnke, & Kroll (2001) and Herget & Klika (2003). But, dear colleagues, you will surely find up-to-date pictures, perhaps even in your local newspaper, featuring events which are of interest to the pupils, and are closely related to *their* world.

REFERENCES

Freudenthal, H. (1968). Why to teach mathematics so as to be useful? *Educational Studies in Mathematics 1*, 3-8.

Herget, W. (2000). Wie groß? Wie hoch? Wie schwer? Wie viele? – Mathe-Welt. *mathematik lehren 101*, 23-46.

Herget, W. (2002). "Pictorial Problems". One Question, but Many Ways, and Many Different Answers. In Weigand, H.-G., Neill, N., Peter-Koop, A., Reiss, K., Törner, G., & Wollring, B. (Eds.), *Developments in Mathematics Education in German-speaking Countries. Selected Papers from the Annual Conference on Didactics of Mathematics, Potsdam 2000,* (pp. 76-87). Hildesheim, Berlin: Franzbecker.

Herget, W., Jahnke, T., & Kroll, W. (2001). *Produktive Aufgaben für den Mathematikunterricht in der Sekundarstufe I.* Berlin: Cornelsen.

Herget, W., & Klika, M. (2003). Fotos und Fragen. Messen, Schätzen, Überlegen – viele Wege, viele Ideen, viele Antworten. *mathematik lehren 119*, 14-19.

International Baccalaureate Organization (1998). Internal Assessment: the Portfolio. In *Teachers Support Material: Mathematics Higher Level,* (pp. 47-58).

Chapter 3.6.3

LEARNING MATHEMATICAL MODELLING – FROM THE PERSPECTIVE OF PROBABILITY AND STATISTICS EDUCATION

Ewa Lakoma
Institute of Mathematics, Military University of Technology, Warsaw, Poland,
Email: e.lakoma@ita.wat.edu.pl

Abstract: The competency of *mathematical modelling of real phenomena* is a necessary component of *mathematical literacy for all*, strongly needed in education of today's and future society. This article considers the main aspects of a process of learning mathematical modelling, according to student's cognitive development, discussed from the perspective of probability and statistics education.

1. INTRODUCTION

Recent transformations in all fields of our lives, especially wide use of information and communication technology, cause mathematical knowledge to be necessary in every domain to an extent as never before. It implies urgent need of a new *mathematical literacy for all* (Noss, 1997). One of its indispensable components is the competency of *mathematical modelling of real phenomena*. The process of developing this competency in students' minds is complex and requires recognition of its nature from *epistemological perspective*. In this article I will present some remarks to the main research questions concerning the process of learning mathematical modelling: *How to teach mathematical modelling for all? How to help students to learn mathematical modelling? How to recognise symptoms of students' competencies in this field? What are natural ways of thinking accompanying a process of mathematical modelling?* I will consider these problems from the perspective of probability and statistics (for brevity called stochastics) education, a field of mathematics that seems to be natural for learning mathematical modelling. In discussion of the questions above I will refer to my research studies on the nature of a process of stochastics learning (Lakoma, 1990, 2000, 2002, 2003).

2. MATHEMATICAL EDUCATION IN THE DIGITAL ERA

Great progress in science and technology, rapid extension of information technology, widespread access to information, changes of standards of everyday life – all these transformations imply strong demand of people who are able to handle new circumstances rather than behave typically in routine situations. Thus, the main goal of today's education is to equip students with an operative knowledge and key competencies, which will create a base for further learning and acting, according to actual social and professional demands and potentialities. Regarding mathematical literacy for all, instead of exposure to formal structure of mathematics, which used to be the educational standard in the past, it is necessary today to treat mathematics as *a language for communication and as a tool for predictions and explanations of reality* (Freudenthal, 1983). In order to understand real phenomena, students should be able to describe, simplify, and distinguish their essential features – this is *mathematical modelling.* From this base students can make predictions and conclusions, generalise and justify them *(mathematical reasoning)*, and apply them to practice or to present and explain to other people *(mathematical communication).* It is important to enable students to develop a deep understanding of mathematical concepts and to create their own mathematics in these three aspects. Learning mathematics means developing these aspects, which involve students' creativity and are embodied into real contexts, which are interesting to students (Freudenthal, 1983; Sierpinska & Kilpatrick, 1998). Instead of transmitting definitions of concepts of mathematical theory, and showing their simple applications, *forming student's ability to create mathematical models of real phenomena, posing hypotheses and verifying them by means of mathematical tools* will be essential aims of mathematical education. These goals are possible to achieve when there is an opportunity to stimulate students' initiative and to let them learn *in accordance with their actual cognitive development.* The activating style of teaching: promoting didactical methods which involve interactions among students and co-operation in small groups, is appropriate at every stage of education and seems to be especially effective for students who are going to be *users of mathematics* in their professional lives where relations of mathematics to reality are natural and it is therefore necessary to be exposed to this in the learning process (Lakoma, 1990, 2000). Usually in the process of problem solving it is necessary to make some trials and to gain experience, which can then be a point of departure for mathematical reasoning. Using new technology can enhance this important stage of the process of mathematics learning (Laughbaum, 2000).

3. MODELLING AS A COMPONENT OF MATHEMATICAL LITERACY FOR ALL

Davis and Hersh (1981) distinguish two principal elements of functional use of mathematics: *dealing with data*, which characterises the power of mathematical symbols, and *applying mathematical models*, which reflects the formatting power of mathematics. Although the notion of a model occurs in mathematics in various connotations, one definition may be common to all meanings: *a model of a given object is another object, which is not identical with it, but it in some respect is similar, to such a degree that it can be used for some purposes in place of what it stands for.* This indicates two basic features of a model: ability to imitate an original object and helpfulness in predicting results of acting on this object. Investigations in a model should provide information about the original object. Thus, the notion of a model can be described even more concisely: *M is a model of an object O, since observing M makes possible to get to know something about O.* It stresses that a model is not the same as what it models and that usually there are many models for any given object; which of these models is suitable, very much depends on what we intend to do with it. In the process of teaching, a sense of creating and exploring a model is best motivated in situations that do not come from ready-made mathematical tasks, but are connected with real phenomenon. Observations of a learning process show that modelling is naturally developed when there is a need to solve a problem.

What is a natural way to lead from observation of a phenomenon to solution of a problem connected with that phenomenon?

Observations of a phenomenon lead us to questions and to formulate a problem. In order to find answers to these questions, we present the phenomenon in an idealized way, as simple as possible, regarding its most important aspects, abandoning everything insignificant from the viewpoint of the posed questions. From this we build a model of the phenomenon, in order to solve the problem. When analysing students' explorations of real phenomena, we can distinguish four main steps of the process of students' thinking and acting: *discovering and formulating of a problem; constructing a model of 'real' phenomenon; analysing the model; interpreting results obtained from the model with the 'real' situation.* This process arises directly from student's common sense thinking while considering simple problems as well as more advanced ones. Mathematical models used by students are usually as simple as possible, have a local character and a strong *explanatory value*. These are *the local models*. This observation can be the groundwork for a holistic approach to mathematics teaching. What is essential is the general way of reasoning and acting, which can be developed in unique ways at every level of mathematics learning (Lakoma, 1990, 2000, 2003). Classes of problems, and suitable tools useful for solving them, change depending on the level of education. Information technology becomes helpful here, shifting an emphasis from calculation into reasoning, and the development of mathematical thinking (Lakoma, 2002, 2003).

4. MODELLING IN STOCHASTICS EDUCATION

Previous research on students' stochastic reasoning (Lakoma, 1990, 2000) indicates that natural ways of developing main stochastic concepts are closely connected with considering real random situations and modelling concrete random phenomena. We can recognise the following steps of a process of a students' natural reasoning: *a). exploring a situation involving randomness; b). formulating a problem; c). creating a local model of the phenomenon under consideration; d). analysing the mathematical model in order to solve the problem; e). comparing solutions obtained using the model with results of observations of the random phenomenon.* In order to present students' natural ways of modelling and reasoning during the process of solving probabilistic problems, the following situation will be analysed. Students are asked to consider the following game:

> Two boys – Jack and Mark – try to score at a basketball target. They both have the same frequencies of success: 50%. They decided to play a game: each will throw the ball until he fails to score. When one fails, the other takes over the throwing. The first who scores a hit is a winner. Jack always starts first. What are chances for winning for these boys? Do you think the game is fair?

Recording individual scores and communicating this in percentage form is a common way of presenting a "free throw percentage" of a basketball player. Thus, the situation described above seems to be "a real situation" for students.

a).-b). Exploring situations involving randomness and formulating a problem

Initially, students' behaviour is to try to understand what is going on, what the rules of this game are. They find it helpful to make experiments and to observe their results. How to get a good simulation of this game and under what conditions? What does "frequency 50%" mean? How do the rules of this game influence the chances of players?

c). Creating a local model of the real phenomenon under consideration

Students usually try to understand this game by making a simple, rough model: a Monte Carlo simulation – throwing a ball to the basket is replaced by throwing for example a coin by turns and recording results. They realise that they are not able to predict "for sure" results of this game. They find it useful to confront empirical results with theoretical considerations and by recognising characteristic features of the game and considering the random mechanism theoretically, they build a more sophisticated model.

Students realise that the situation is not symmetrical, the game is not fair: – *"This game is not fair because the first has a better chance. If he hits the other has not even got a chance to try! Let them throw at the same time."* – *"If the 1st player scores, the 2nd player has no chance so the game is unfair. The game is fair if the 1st misses then the 2nd gets an equal chance."* – *"This game is not fair because the 1st player always takes from the whole stake and the 2nd always from the rest. The 2nd player always has to wait for his turn in order to throw a ball."*– *"The 1st guy who starts is directing the probability. But mathematically it will maybe turn out that their chances are equal, showing that our thinking is wrong. This game may never end."*

For students, "a fair game" means a game with symmetrical situations for both players and with even chances of winning. Students often refer to the concept of symmetry. Drawing a game-board for the simulation usually is accompanied by this reasoning. Fig. 3.6.3-1 shows game-boards created by 13 year old students. Such constructions are developed at the stage which precedes drawing the graph presented in Fig. 3.6.3-2. They turn out to be even more natural in students' reasoning than this graph.

Drawing graphical representations of random mechanisms is usually accompanied by expressing its rules.

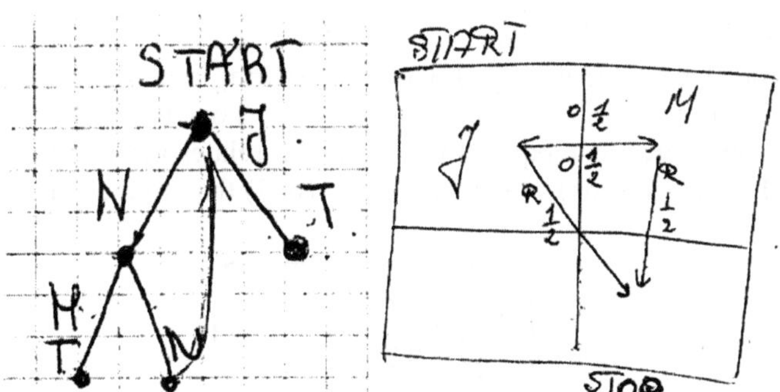

Figure 3.6.3-1. Game-boards

One of the students pointed out the very essence of the rules of this game as follows:

–"Both players throw a ball with chance 50%, which is 1/2. From start – when Jack hits, he obtains one point, if not, he transmits the ball to his opponent's hands. And the opponent – Mark – when he hits he obtains one point, if not, he transmits the ball to his opponent's hands, and so on – by turns – until... it never ends!?"

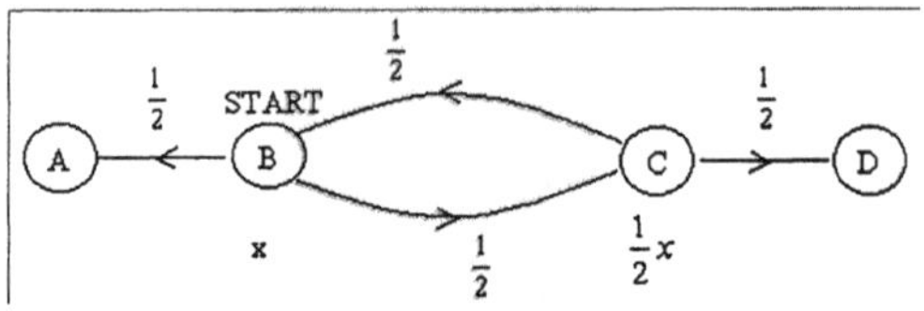

Figure 3.6.3-2. Graph of the game

Many students draw a tree diagram of the game. They quickly realise that such a tree is infinite, and they easily transform this model to the graph shown in Fig. 3.6.3-2, if they find it useful for further explorations.

d). Analysing the mathematical model in order to solve the problem

In order to calculate chances of winning the game, students try to discover some regularities of the constructed model:

– "The tree can continue theoretically to the infinity. Here Mark tries, he fails – with a chance one half, Jack tries again .... and this is as if the game begins again! One half of one half of a chance of the second boy – this is as if he starts to play again."

This student was able to grasp the fact of infinite number of possible outcomes in recursive way. He also solved the equation:

$$x = \frac{1}{2} + \frac{1}{2}\left(\frac{1}{2}x\right) \qquad \frac{3}{4}x = \frac{1}{2} \qquad x = \frac{2}{3}$$

Another local model (Fig. 3.6.3-3) – "Sharing a pizza" step by step according to the chances of the players, also allows one to find recursion and to compose an appropriate equation. This model relates more to the concept of expected value (in sense of Ch. Huygens, 1657) than to the concept of probability.

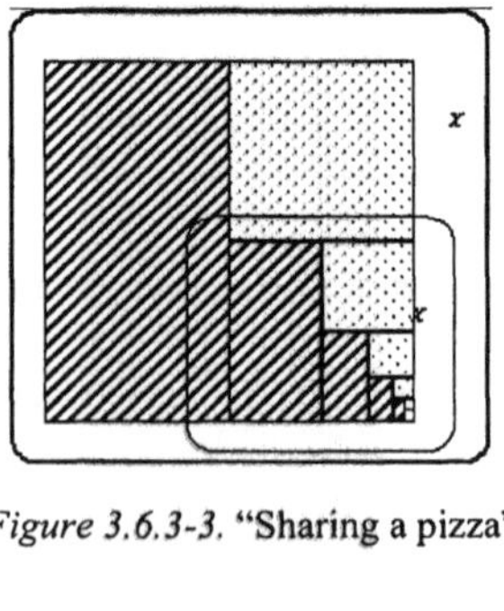

Figure 3.6.3-3. "Sharing a pizza"

Students are able to notice regularity in this model:

–*"Area for the first player is at each stage of this game twice bigger than area for the second player. Thus the whole area of Jack is twice bigger than the area of Mark. Jack has chance $^2/_3$ and Mark $^1/_3$."*

During analysis of this reasoning, we find two parallel ways of thinking: comparing chances of both players and, as result of these comparisons, calculating chances. Many students, trying to formulate an equation, use simultaneously both geometrical representations. When calculating chances of winning, they use geometrical and algebraic (an equation) models. Reasoning simultaneously in two or more models turns out to be more convincing for students. It seems that they not only convince each other, working together in small groups, but they also convince themselves during their reasoning.

e). Comparing solutions obtained using the model with results of observations of the random phenomenon.

After an analysis of a model, students gather experimental frequencies of results again, repeating some experiments, and compare them with theoretical chances. They check whether their model is appropriate.

In spite of models based on the concept of expected value, students also make considerations on the basis of comparing empirical and theoretical frequencies. It seems to be sufficient to replace real throws of a coin by distributing a number of trials exactly according to chances of results.

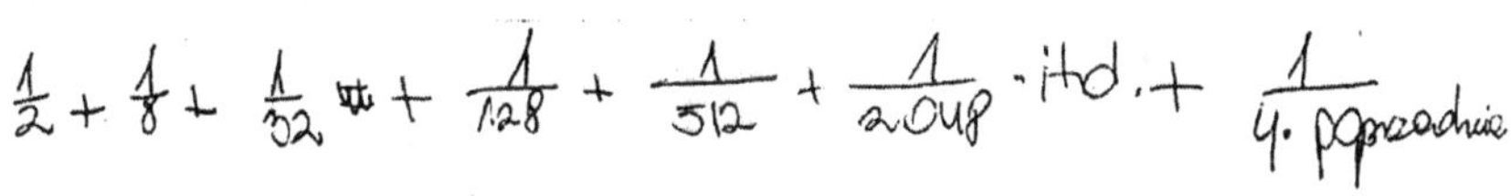

Figure 3.6.3-4. Observing regularities

This model we call idealized simulation, it bases on the idea of Arthur Engel's Probabilistic Abacus. However, it is not often created originally by students. They prefer just to share a unit among states in the graph or the tree, and to approximate successively theoretical frequencies of winning (Fig. 3.6.3-4: author's age 15; Polish text "etc.+1 over [4 times previous one]").

Students calculate it easily "by hand" or by preparing simple program in order to obtain theoretical frequencies at successive stages of the game:

$$1 \to B:0 \to A:0 \to D:A+0.5B \to A:0.5B \to C:D+0.5C \to D:0.5C \to B:D$$

Another local model – presenting the random mechanisms by matrix of probabilities of transitions in the graph, belongs to models having for students a strong explanatory value. Raising the transition matrix to successive powers leads to probabilities of winning. In this case information technology allows learners to focus on developing their conceptual thinking instead of carrying out difficult calculations. Reasoning, presented here, is common for learners independently of their age, and is connected with their actual stage of understanding stochastic concepts.

5. FINAL REMARKS

Analysis of students' reasoning became a base for developing a holistic approach to stochastics teaching: the Local Models Approach (Lakoma, 2000). Its ideas are implemented in school practice for students of age 10 – 19 (Lakoma a.o., 1996 – 2001, 2002 – 2004). Observing learning shows that stochastics gives students natural motivation to develop their competency of mathematical modelling. During the process of mathematics learning students should be equipped with various local models, which are available to them and are not too abstract. When solving concrete problems, students do not often understand the need for generalization and are often not able to appreciate the beauty of a global model. It will come later. What is fundamental is learning the unique style of thinking and acting necessary in the process of mathematical modelling.

REFERENCES

Davis, P., & Hersh, R. (1981). *The Mathematical Experience.* Boston: Birkhaeuser.

Freudenthal, H. (1983). *Didactical Phenomenology of Mathematical Structures.* Dordrecht: Reidel.

Lakoma, E. (1990). *The Local Models in Probability Teaching* (in Polish). Unpublished doctoral dissertation, Warsaw University, Department of Mathematics, Informatics and Mechanics, Warsaw.

Lakoma, E. (2000). Stochastics teaching and cognitive development. In Fauvel. J., & van Maanen J. (Eds.), *History in Mathematics Education. An ICMI Study.* Dordrecht: Kluwer.

Lakoma, E. (2002). On the impact of hand-held technology on mathematics learning – from the epistemological point of view. In Borovcnik, M., & Kautschitsch, H. (Eds.), *Technology in Mathematics Education, Proceedings of ICTMT5,* (pp. 461-46). Vienna: Hoelder–Pichler–Tempsky & Oesterreichischer Bundesverlag.

Lakoma, E. (2003). On the role of information technology in the process of teaching mathematical modelling. In Triandafillidis, T., & Hatzikiriakou, K. (Eds), *Technology in Mathematics Teaching, ICTMT6,* (pp. 75-81). Dardanos, Volos: University of Thesally.

Lakoma, E., & Zawadowski, W. a.o. (1996-2001). *Mathematics 2001* (in Polish). Series of textbooks, teachers' guides, films for students (10-16). Warsaw: WSiP.

Lakoma, E., & Zawadowski, W. a.o. (2002-2004). *Mathematics counts* (in Polish). Series of textbooks, internet service – for lycee students (17-19). Warsaw: WSiP.

Laughbaum, E. (Ed.). (2000). *Hand-held technology in mathematics and science education: a collection of papers.* Columbus, Ohio: The Ohio State University.

Noss, R. (1997). *New Cultures, New Numeracies, Inaugural Professorial Lecture.* London: University of London, Institute of Education.

Sierpinska, A., & Kilpatrick, J. (Eds.). (1998). Mathematics Education as a Research Domain: A Search for Identity. An ICMI Study. Dordrecht: Kluwer.

Chapter 3.6.4

CONSIDERING WORKPLACE ACTIVITY FROM A MATHEMATICAL MODELLING PERSPECTIVE

Geoff Wake
University of Manchester, U.K., Email: geoff.wake@manchester.ac.uk

Abstract: In recent years researchers investigating the mathematical activity of workers have highlighted how this is historically and culturally situated and consequently school mathematics often seems to provide inadequate preparation for the workplace. Here I draw on two case studies to illustrate how worker activity might be (re-)interpreted from a mathematical modelling perspective and suggest that such an approach may usefully be developed to inform future curriculum development in terms of mathematical modelling.

1. INTRODUCTION

Mathematics educators often turn to the world of work to attempt to understand how the mathematics used by workers might inform design of future curricula. Some, particularly early, studies sought to identify appropriate mathematical content by noting its use, or required understanding, across a range of different workplaces, and, perhaps by necessity, used frameworks that rely on what might be considered 'formal' or 'academic' mathematical content to organise findings. Other, particularly more recent, studies have used theoretical frameworks, such as Cultural Historical Activity Theory (see for example Engestrom & Cole, 1997), with researchers adopting the role of ethnographers, to explore more fully the richness of workplace mathematical activity and how this may be better understood by taking into account how it is culturally and historically situated as part of a community of practice (see for example Lave, 1988).

As might be expected mathematical models, modelling and applications are at the centre of much of the activity of workers as often this is focused on solving problems, but perhaps more often on monitoring and measuring workplace routines and output. However, although mathematical models loom large, implicitly if not always explicitly, in accounts of workers' activities there has, as yet, been little emphasis on what we can learn about mathematical modelling from workplace based research and the implications we can draw for the teaching and learning of mathematics in schools have not been well developed.

2. MATHEMATICAL ACTIVITY IN WORKPLACES

Much recent research in the field of the use of mathematics in the workplace has been strongly influenced by ideas of situated cognition which at its most extreme suggests that mathematical understanding and competence cannot be separated from the socio-cultural setting in which it is constructed. At first sight this seems an attractive proposition: for example, it perhaps allows one to take account, to some extent, of the apparent lack of visibility of mathematics to workers. The more comprehensive theoretical framework of Cultural Historical Activity Theory (CHAT) draws attention to other aspects of the workplace and how these mediate the activity of workers. Perhaps most easily visible to researchers, and often significant in coming to understand the activity of workers, are 'instruments', including often idiosyncratically developed artefacts, which mediate the actions of workers so that they, as individuals and teams, successfully achieve the outcomes required of them. Often these artefacts can be used successfully by workers without recourse to mathematical thought or understanding. However, CHAT also draws our attention to other influences that mediate the activities of workers: the 'rules', 'community' and 'division of labour' of workplaces are often most pertinent in shaping and forming day-to-day workplace relationships and activity.

Other analyses have led researchers to develop constructs that take account of the situated nature of the mathematics they have observed in workplaces whilst bridging to the perhaps more familiar mathematics of education and academic communities. These include:

- the idea of *situated abstraction* (Pozzi et al, 1998), which allows one to understand how workers may develop a generalised mathematical understanding, but within the situational context of their work, using a discourse other than that of standard/formal mathematics but which may be mapped to this;

- *general mathematical competences,* (Williams et al, 1999, Wake & Williams, 2000) which were developed from a mathematics education stand point to attempt to take account of common ways that workers might bring together coherent bodies of mathematical knowledge, skills and models, for example when "handling experimental data graphically";
- *techno-mathematical literacies* (Kent et al, 2004) which are currently being developed to assist understanding of how mathematics in workplaces is not only very much grounded in day-to-day workplace activity but is also often highly integrated and dependent on the use of modern technologies.

These constructs whilst not always referring explicitly to ideas associated with modelling do, often by implication, suggest that mathematical modelling and mathematical models are central to the activity of workers. To inform discussion of what such activity might look like I will illustrate, by necessity very briefly, mathematical activity that was investigated as part of a research project 'Using College Mathematics in Understanding Workplace Practice'[1]. The project and research methodology is described in detail in the final project report (Wake & Williams, 2001).

3. MATHEMATICAL MODELS AND MODELLING IN WORKPLACES

In one case study we investigated the work of a finance office worker, Alice, whose activity included computing performance data in a medium-sized retail company selling power tools to customers operating in construction (CON) and engineering (ENG) sectors. One industry-standard measure, "debtor days", that Alice calculates is used as an indicator of how long customers take to settle their accounts. Fig. 3.6.4-1 shows a screenshot of Alice's spreadsheet, which she built from scratch and which she uses to calculate "debtors days", for the two sectors into which customers have been classified.

During a workplace visit by a researcher and two students undertaking a pre-vocational course in Business Studies, Alice explained that she takes as data the most recent three month's sales. She totals these (row 12) and divides by the number of working weeks (in this case she takes this to be eleven and a half which takes into account breaks for the Christmas and New Year holidays) and multiplies this by fifty-two (row 14), "Because there are fifty-two weeks in a year". At this point Alice is not concerned to give meaning to the figure calculated as this is just a sub-step on the way to calculating the final indicator. The researcher and students, however, tried to give meaning to this figure as this turned out to be important to them in allowing them to eventually come to an understanding of the final measure, "debtor days".

	A	B	C	D	E	F	G	H
1	**DEBTOR DAYS FEBRUARY**							
2								
3								
4	Number of working weeks			11.5				
5								
6								
7								
8					**CON**	**ENG**	**DES**	**DH**
9	3 Months Sales			FEB	7,440.24	4,155.02	97,031.20	60,536.46
10				JAN	3,776.98	1,226.75	57,164.49	80,697.46
11				DEC	2,985.64	1,755.66	74,853.72	42,977.12
12					**14,202.86**	**7,137.43**	**229,049.41**	**184,211.04**
13								
14	3Mths sales/No weeks x 52				**64,221.63**	**32,273.60**	**1,035,701.68**	**832,954.27**
15								
16	DEBTORS				14,952.46	4,493.16	245,820.50	208,239.45
17								
18	LESS 17.5%				**12,725.50**	**3,823.97**	**209,208.94**	**177,225.06**
19								
20	**DEBTOR DAYS**		Debtors/Salesx365		**72.32**	**43.25**	**73.73**	**77.66**

Figure 3.6.4-1. Spreadsheet developed to calculate "debtor days"

Agreement was reached that this "gives you what that [sector] would have sold in the whole year, if things stayed the same"; in effect then this figure gives a measure of "sales" for the year. Attention now turned to the measure "debtors" (row 16): this is the total outstanding debt for the sector and as it includes sales tax (VAT) of 17.5% this is deducted in row 18 by dividing by a factor of 1.175. Finally the measure "debtor days" can be calculated (row 20) using the figures previously found: this is given by ("debtors (less sales tax)"/sales)·365. Alice described this as "the average number of days it takes a customer to pay us", but the researcher had a problem with understanding this idea of "average" particularly when an actual statistical average could have been calculated using the raw data available. To try to make sense of the measure he suggested, "Perhaps I should try it with some numbers. So if I was owed, a hundred pounds and the total turnover for the year was two hundred pounds that would give me a half times 365, giving me half the year which is about..." Suddenly he was enlightened, "It's not actually the number of days it's taking people to pay is it, it's just a [indicator of this]".

As a second illustration we turn to just one aspect of the work of a railway signal engineer/designer. This worker, Alan, checks calculations made by other workers of where signals and speed restriction indicators should be placed along railway tracks so that train drivers can respond safely. He explained in detail to a researcher accompanied by three students on a pre-vocational engineering course this aspect of his work. Fig. 3.6.4-2 shows an illustration from a training manual he used to explain the calculations he carries out. Alan summarised the problem as, "What's the minimum distance for the approaching line speed – say 60 miles per hour – how far back do we have to position *that* (signal A) so the driver can safely stop at *this* signal (B) and not run through." Crucial to the calculations he performs is the idea of average gradient.

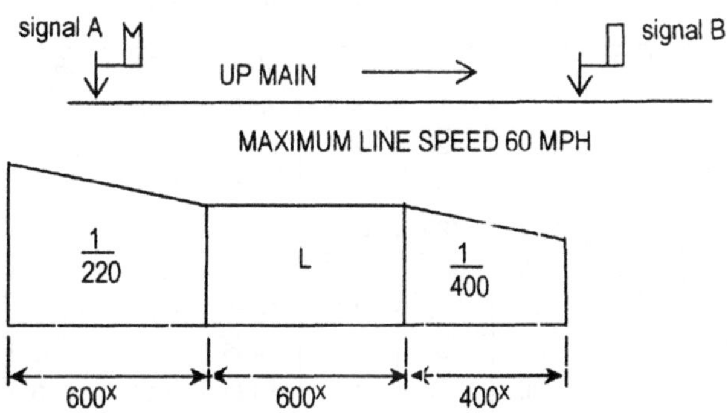

SECTION	DISTANCE IN YARDS	GRADIENT	RISE (YDS) (DIST * GRAD)	FALL (YDS) (DIST * GRAD)
X - A	600	$\frac{1}{220}$	-	2.73
A-B	600	L	-	-
B-Y	400	$\frac{1}{400}$	-	1.00
TOTALS	1600			3.73

$\frac{1600}{3.73} = 428.95$ THERFORE AVERAGE GRADIENT= $\frac{1}{429}$ FALLING

Figure 3.6.4-2. Training manual example calculation of average gradient.

At a later stage the researcher and students re-created the calculations that would be carried out to find the spacing between these signals. Although the training manual example (Fig. 3.6.4-2) shows considerable detail of how to calculate an average gradient the students struggled to make sense of this with one student suggesting, "you start adding them together ... adding the gradients together and divide by two". Having calculated the average gradient Alan then uses a Table such as that illustrated in Fig. 3.6.4-3 to look up the distance required between warning and stop signals. For the training example of 1 in 429 falling for trains with a maximum speed of 60 mph a distance of 1435 (yards) would be used. Alan explained that he always errs on the side of safety and as a gradient of 1 in 429 falling lies between the level and 1 in 200 falling, it is safer to take the value associated with the gradient of 1 in 200 falling.

	GRADIENT								
	Rising					Falling			
INITIAL SPEED (Mph)	1 in 50 2.0%	1 in 67 1.5%	1 in 100 1.0%	1 in 200 0.5%	Level Level	1 in 200 0.5%	1 in 100 1.0%	1 in 67 1.5%	1 in 50 2.0%
20	175	180	195	215	240	275	320	395	520
25	240	255	280	315	355	410	485	625	840
30	320	340	380	425	485	575	700	895	1425
35	405	440	485	550	635	780	1010	1380	2237
40	495	550	620	720	865	1080	1420	1903	2237
45	630	710	805	935	1130	1435	1660	1903	2237
50	688	748	816	935	1130	1435	1660	1903	2237
55	770	831	901	984	1130	1435	1660	1903	2237
60	849	911	980	1061	1165	1435	1660	1903	2237

Figure 3.6.4-3. Table used to give stopping distances for trains travelling at various speeds for a range of typical average gradients

4. DISCUSSION AND CONCLUSION

It is clear that the brief descriptions of the mathematical activity of two different workers outlined here exhibit features that might be understood by reference to the constructs of situated abstraction, general mathematical competences or techno-mathematical literacies referred to previously. It is equally clear that they also exhibit features of mathematical modelling, which I would like to explore in a little greater detail. To assist with this it is useful to have an overview of aspects of the modelling process (see, for example, Blum, 1992) as illustrated by the schema in Fig. 3.6.4-4. This depicts the various constituent processes of mathematical modelling whilst also indicating that to develop an appropriate model it is often necessary to go round the modelling loop more than once. In each case described here the worker is clearly working with a mathematical model of reality: in each case this is "industry standard" with important meaning in the workplace. Alice, the finance worker built her spreadsheet (Fig. 3.6.4-1) to calculate the agreed measure "debtor days" which has meaning to her and others including the company's directors, who use it to make decisions about how the company is performing. In developing the spreadsheet, Alice carefully took account of factors pertinent to her real world ensuring that the resulting mathematical model / measure was able to be ultimately interpreted in the standard way.

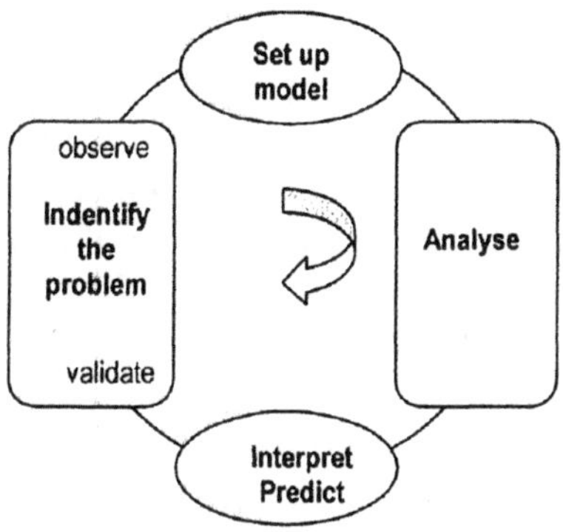

Figure 3.6.4-4. Mathematical modelling cycle

Alice's day-to-day activity, however, does not usually require her to undertake such modelling activity; rather she (and others) receive the outputs from her spreadsheet and are required to interpret these, taking into account their vast wealth of work process knowledge (for example, why variation between sectors, or at different times of the year, might be expected). In the case of the railway worker, Alan, he too is working with an agreed "industry standard" model of average gradient. Although his calculations could also be automated by the programming of a spreadsheet, he chooses each time to carry out the calculation from scratch using the agreed procedure (Fig. 3.6.4-2). To reach the final answer to his problem he uses the results of his calculation of average gradient to look up the spacing between signals in a table (Fig. 3.6.4-3). To Alan this has become procedural: he knows that if values for the gradient he has calculated are not given in the table he uses the closest value given that is a safer case (that is, he chooses the value "to the right" of where the calculated gradient lies). Alan knows that in reality the signals may be placed further apart, but never closer, than his calculated value, as the warning signal has to be placed in a position that allows train drivers good visibility of it.

These two examples of workers' mathematical activity suggest, therefore, that workers may be expected to use standard or accepted mathematical models and are required to be able to interpret data that these produce drawing on a rich contextual background that is part of their everyday life at work. Even when Alice developed her spreadsheet to calculate "debtor days", with all the flexibility that a blank spreadsheet allowed, she worked towards the standard model drawing on the data that she had collected for this purpose.

These examples suggest that workers do not have freedom in how they set up their mathematical models, rather that they have to organise numerical data in such a way that they can arrive at a model that is familiar or standard in their industry, and that their day-to-day activity is often focused on interpretation of the results of their application of these models. Our research suggests that, when explaining to others the mathematical models that they use, or perhaps when attempting to make sense of new mathematical models themselves, workers, whilst requiring a deep understanding of the workplace context, also need to be able to draw on a range of strategies, such as considering simple cases or extreme values, to help them in effect make sense of the mathematics of others (for further discussion of this see Wake & Williams, 2003).

It would seem that there is a need for further research, perhaps accompanied by further analysis of existing data, focussed on mathematical modelling, and the application and use of mathematical models in the workplace. This needs to be supported by sensitive curriculum development that explores how such workplace activity could be better supported by curriculum specification and mathematical activities designed for use in classrooms at all levels.

REFERENCES

Blum, W. (1992). Applications and modelling I mathematics teaching. In Niss, M., Blum, W., & Huntley, I. (Eds.), *Teaching Mathematical Modelling and Applications*. Chichester: Ellis Horwood Publishing.

Engestrom, Y., & Cole, M. (1997). Situated cognition in search of an agenda. In Whitson, J.A., & Kirshner, D. (Eds.), *Situated cognition: social, semiotic, and psychological perspectives*. Hillsdale, N.J.: Lawrence Erlbaum Associates.

Kent, P., Hoyles C., Noss R., & Guile, D. (2004). Techno-mathematical Literacies in Workplace Activity, Paper presented to International Seminar on Learning and Technology at Work. London: Institute of Education. From www.lonklab.ac.uk/kscope/ltw/seminar.htm.

Lave, J. (1988). *Cognition in practice: Mind, mathematics and culture in everyday life*. Cambridge: Cambridge University Press.

Pozzi, S., Noss, R., & Hoyles, C. (1998). Tools in Practice, Mathematics in Use. *Educational Studies in Mathematics, 36*, 105-122.

Wake, G. D., & Williams, J. S. (2000). Mathematics in pre-vocational courses. In A. Bessot, & J. Ridgeway (Eds.), *Education for Mathematics in the workplace*. Dordrecht: Kluwer.

Wake, G.D., & Williams J.S. (2001). *Using College Mathematics in Understanding Workplace Practice*. Manchester: University of Manchester.

Wake, G.D., & Williams, J.S. (2003). Using Workplace Practice to Inform Curriculum Design. In S.J. Lamon, W.A. Parker, & S.K.Houston (Eds.), *Mathematical modelling a way of life*. Chichester: Ellis Horwood Publishing.

---

[1] This project was supported by a grant from the Leverhulme Trust to the University of Manchester.

Section 3.7

Assessment and Evaluation

Edited by Peter Galbraith

Chapter 3.7.0

ASSESSMENT AND EVALUATION - OVERVIEW

Peter Galbraith[1]
University of Queensland, Brisbane, Australia, Email: p.galbraith@mailbox.uq.edu.au

Abstract: This chapter samples issues that continue to challenge the theory and practice of assessment and evaluation. The contributions of selected papers are complemented by comments drawn from participants in a working group representative of nine national contexts. The contributions are significant for their diversity, but some issues of longstanding continue to emerge.

1. INTRODUCTION

Assessment continues as a major issue – because it is challenging to conduct, and because it is a potent driver of curriculum change. Internationally items with applications and modelling relevance are found within TIMSS and PISA testing programs; at the national level several countries have introduced curricula with applications and modelling as specified components; universities and colleges continue to search for richer assessment practices within modelling oriented subjects; and there is a continuing demand for new problems and more incisive means of establishing competence at all levels of education. All these issues are represented in the present section, where emphases raised within the selected papers are complemented by input from a targeted discussion involving representatives from nine countries, who addressed issues of assessment across all levels of education.

Elementary school level. In the elementary school, teaching and assessment must together reward both effort and progress. As many primary teachers are afraid of mathematics, and teacher interest is variable, large differences in practical issues emerge e.g. how much time is allocated or actually utilised for project work. There are major differences between countries and there is a continuing challenge to identify, exemplify, and communicate good practices that are both successful and feasible within a given context.

Secondary school level. Stimulating project type experiences continue to be developed. For example in one initiative, students put on "mathematics glasses" to describe a modelling situation. They work for 4 weeks, with each student producing a poster, and being allocated 5 minutes to describe their work, with the teacher asking a common set of questions about each poster. In another approach children decide what to do with situations involving a range of mathematical procedures. They present activities and discussion at a school conference (as well as submitting written work), where other children participate and ask questions, with teachers providing feedback. This process has been welcomed by the school community, to which teachers of other subjects, and the headmaster, have attributed a change of school atmosphere.

Undergraduate level. In one first year program, students work on modelling problems throughout the course, and write a report for the group on their solutions, followed by an end of term oral exam. They have 20 minutes to describe their project work, during which examiners challenge students to reflect on different parts of the modelling process - the oral exam is essential in gauging the level and quality of student reflection. The three days needed for the oral examinations, are built into the overall structure of the course. An undergraduate modelling based Physics course involves students working in groups on topics chosen on the basis of interest e.g. hydrodynamics, refrigeration. They design experiments to display essentials for their chosen topic, and report by oral presentation and poster. As no lectures/theory are presented formally students need assistance to proceed. In the second semester the students identify where and how their chosen phenomena appear (e.g. in industry), how people deal with it, where it is hidden if appropriate. This involves conducting interviews, and the final product is a 20-page report.

The above examples illustrate ongoing activity in the area of assessment of mathematical modelling. They reinforce the judgment that assessment of modelling performance needs to be part of a complete package, integrated into the teaching timetable as a central part of the whole curriculum.

The influence of politics and culture cannot be ignored in deciding what can be attempted or accomplished in regards to assessment practices, and this featured strongly in group comment. For example, in some national settings where even tertiary students are used to the teacher telling everything, it becomes important to monitor what happens when students express themselves freely between themselves. Elsewhere there are problems with availability of materials, including places where technology is unavailable. In some countries the climate and culture of education is changing, so that for example, when instead of centrally controlled curricula, content emphasis is placed in the hands of a headmaster, it depends how this freedom is used. When schools can now decide how or what to teach given that some minimal standard is reached, this on the one hand potentially opens the door for applications and modelling initiatives. However when schools are simultaneously paid according to the number of students attending, parents must be convinced of the worth of programs. So messages need to be clear about what students (in undertaking applications and modelling work) are able to achieve successfully. It is clear that great diversity continues to exist among contexts in which applications and modelling are being implemented – and the associated assessment needs imply a range of practices both established and new. A challenge is to obtain and maintain integrity and practicality among this diversity.

2. PAPER SUMMARIES

Turner identifies items within the OECD's Programme for International Student Assessment (PISA), which value the ability of students to *use* mathematical skills to meet real-life challenges. Noting that students from different countries perform disparately on such items, PISA might be instrumental in promoting increasing interest in how modelling-related learning tasks influence achievement. A second issue concerns the level of complexity and difficulty of modelling activities. Within the item set (for PISA 2003) those with no modelling demand proved markedly easier than those that involved modelling, items involving only interpretation or reflection phases of the modelling cycle were of intermediate difficulty, while items requiring students to construct or manipulate a model were the most difficult. We need to ask, what kinds of instructional activities promote facility with such items, and ultimately with more extensive modelling activities?

Antonius notes that while modelling has had a central position in Danish curricula at upper secondary level for 15 years, the final examination still consists of traditional written and oral components. He describes a project examination trial in which part of the centrally designed project is based on an open modelling problem. Students work on the project in class and at home, can ask advice from their teacher, discuss work with peers, and use technology. Each student must write an individual report. Teachers note that 'less able' students obtain better results in the project examination compared with the traditional written examination, whereas 'gifted' students receive approximately the same mark. However eight required competences were assessed at a higher level in the project examination than in the traditional written examination. Questions that arise concerning the authenticity of the assessment include: Has the student written the report herself/himself, and does the student understand what is written? Constant 'monitoring' by teachers and an oral defence of project work by students are seen as of central importance here in authenticating the assessment.

Vos notes that within the Netherlands RME-based curriculum, assessment has proved problematic, as it does not require students to apply their skills practically. The author researched whether grade 8 students were improving their abilities to apply mathematics in practical situations, noting that in designing alternative assessment for applied mathematics and modelling, formats such as observation, interviews and portfolios have proved labour and cost-intensive, with reliability in coding responses also an issue. Results on TIMSS items with application content suggested that Dutch students in 2000 had not gained practical competencies in mathematics relative to those attained in 1995, and the author suggests the null result may be attributed to continuing conservative teaching and assessment practices.

Haines and Crouch seek to create items that provide information on component skills of modelling. Tasks cannot always be identified with specific stages of the modelling cycle, but rather are often concerned with the *transition* between two or more stages: for example *moving from the real world to the model, specifying the model, choosing variables, constructing equations, moving from mathematics to the real world.* Processes involving transitions between the real and mathematical worlds are demonstrably difficult for students, and to explore these dimensions a selection of multiple choice items have been constructed, tested, and analysed. Robust items have been identified and further development is taking place.

Lege provides the only paper concerned with the evaluation of modelling programs as distinct from issues associated with assessing student performance. Two sets of junior secondary school students were introduced to mathematical modelling over a three week period using contrasting instructional approaches: (a) having students examine examples of previously-constructed models, and (b) actively engaging them in modelling a problem for themselves. Assessment involved a new contextual situation, but reflected the same structure as the curricular approach in how it was presented to the respective groups. The group that learned about modelling by "doing" outperformed the group that learned by studying examples on four performance goals – two related to modelling, and two related to model structures.

In summary this section offers a selection of snapshots covering a spectrum of issues that are persistent and significant within the domain of assessment. A challenge is to distil the substantial experience and expertise available, as a basis for the progression of this crucial element in enhancing the successful implementation of applications and modelling in education.

---

[1] Valued contributions to the discussion that formed a basis for this introduction to the chapter were provided by: Salett Biembengut, Morten Blomhoj, George Ekol, Djordje Kadijevik, Akio Matsuzaki, Susan McNab, Jarmila Novotna, Torulf Palm, Jacques Treiner.

Chapter 3.7.1

MODELLING BASED PROJECT EXAMINATION

Søren Antonius
Danish Institute for Upper Secondary Education, University of Southern Denmark,
Email: sa@ceus.dk

Abstract: In this article I present a case study of a project examination trial in one of the upper secondary education programmes in Denmark. I shall give some empirical support to the argument that this project examination can be seen as valid not only in relation to modelling but in relation to all mathematical competences.

1. INTRODUCTION

For 15 years modelling has had a central position in the Danish curricula in mathematics on upper secondary level. But teaching does not yet reflect goals and intentions. One of the reasons is probably that the final examination still consists of a traditional written examination with a number of independent, standardised, pre-structured and rather closed tasks, and an oral examination in which the student is supposed to define and explain certain concepts, prove certain theorems etc. Much can be said of these tests, but they are definitely not suitable for assessing open modelling, and what is not assessed tends to disappear from teaching.

A project examination trial was launched for a general course in mathematics on upper secondary level in 1999. The projects are prepared centrally and distributed to students on a CD. The first part of the projects consists of fairly traditional tasks, but the second part is based on an open modelling problem. Students work for 3 weeks with the projects – not full time, but in 15 math-lessons. They can use the teacher as a guide, and they can discuss problems with peers. They are also allowed to work on the project at home. All kinds of information technologies are allowed, including CAS. Each student must write his own report.

The teacher and an external examiner assess the report, but before deciding on a mark the student has to 'defend' the report in an oral test of 10 minutes. In that respect this trial is different from the Victoria-projects in Australia (Barnes et al., 2000). The oral defence is supposed to eliminate or reduce authentication problems. During the oral defence the teacher and the external examiner ask questions to certain parts of the report, which the student must be able to make an account for. Finally the teacher and the external examiner decide on a mark.

I should add that project examination substitutes the traditional written examination. There is no additional written test, and this also distinguishes the Danish trial from the Victoria-projects. In this respect the trial is – as far as I know – unique for a general course on upper secondary level, which prepares students for university studies[1].

2. RESEARCH DESIGN

I have used quantitative as well as qualitative approaches in my research. The main focus has been on students' and teachers' (examiners') attitude to the project trial, but in order to make a comparison I have also collected information on the traditional written examination. Part of my empirical investigations is a case study of one specific class of students working with the project examination in May-June 2002. I have made classroom observations during the 3 weeks of project work, and I have been present at the oral defence. Additionally, I have conducted a student inquiry based on a questionnaire for the class after having finished the reports. Finally, I have interviewed four students and the teacher after the oral defence.

3. PROJECT MARKS VERSUS MARKS IN TRADITIONAL WRITTEN EXAM

Students in project examination make considerably better results than students in the traditional written examination. The average mark in the project trial was for the May – June 2002 examination 7.9, and the corresponding average for the traditional written examination was 6.9, cf. Tab. 3.7.1-1. According to teachers particularly 'less able' students make better results in project examination than they would have done in the traditional written examination, whereas 'gifted' students make approximately the same (top) mark, as they would have made in the traditional examination.

But this does not automatically imply that project examination promotes learning to a higher degree than the traditional examination. However, I do think that this is the case. The reason for this position is not the marks, but the fact that students work with meaningful problems and that they use much more time than they would have done in the traditional written examination. According to the questionnaire students work 15 lessons at school and in average 20 hours at home with the reports. But when observing students work I do not find that 'less able' students are those, who profit the most from project examination as regards to learning. Some 'less able' students work very isolated and do not use the teacher as a guide, and my observations suggest that 'gifted' students are better at handling the chaotic project work in a constructive way. Therefore the question of who is profiting most from project examination may have 'opposite' answers if seen from a *learning* perspective than if seen from the perspective of *marks*.

Table 3.7.1-1. Marks in project examination, in traditional written examination, and in the school based written assessment, May – June 2002 , average and standard deviation

Number of students / average mark / standard deviation	N	$\bar{x}$	s
Project examination	208	7.9	2.05
Traditional written exam.	1549	6.9	2.50
School based written mark	165	7.7	1.69

4. PROJECT MARKS VERSUS WRITTEN SCHOOL BASED MARKS

The distribution of the project marks is different from the distribution of the marks in the traditional written examination. But it appears that there is not much difference between the average of project marks and the average of written marks given in the school based assessment, cf. Tab. 3.7.1-1. The latter is given by the teacher alone, based on student activity in the classroom, homework assignments and small diagnostic tests from the very beginning to the end of the course.

The fact that all students are given a written school based mark makes it possible to do a simple linear regression analysis of the relation between the project mark (y) and the written school based mark (x). The estimated regression line is $y = 1.366 + 0.839x$, which implies the following expected project marks given a specific written school based mark.

Table 3.7.1-2. Expected project marks given a specific written school based mark

Written school based mark	3	5	6	7	8	9	10	11
Expected project mark	3.9	5.6	6.4	7.2	8.1	8.9	9.8	10.6

It appears from Tab. 3.7.1-2 that a student can expect almost the same project mark as he has attained in the school based assessment. The regression line is not far from $y = x$. The t-value of a test of the hypothesis 'intersection = 0' is 2.347 and the p-value is 0.02, which indicates moderate significance. The t-value of the hypothesis 'slope = 1' is 2.176 and the p-value is 0.03, which also indicates moderate significance.

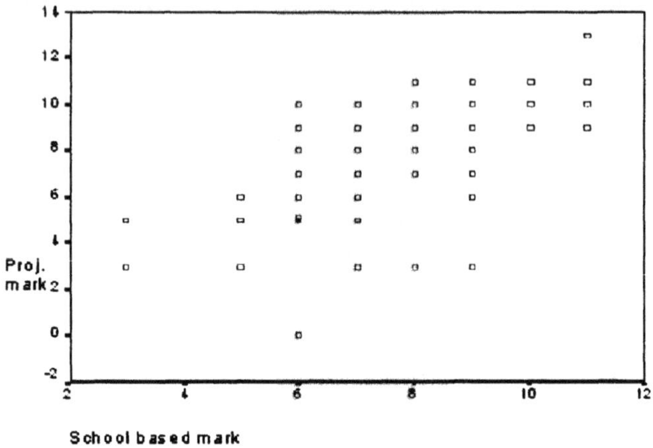

Figure 3.7.1-1. Scatter plot: Project marks versus written school based mark, May–June 2002

The consistency between the written school based mark and the project mark can easily be explained. First of all, project marks as well as school based marks are based on the teacher's assessment. Secondly these marks are based on the same activities, e.g. students collaborating with peers, using CAS and using the teacher as a guide. They do so in the ongoing teaching practice and they do so in project examination.

The consistency can be seen as a token of validity. If we assume that the teacher is the best to know what the student has deserved (Watson 2001), and that this knowledge is expressed in the teacher's school based mark, the fact that the student can expect almost the same project mark as he has attained in the school based assessment indicates that the project assessment is fair.

However, it appears from Fig. 3.7.1-1 that the actual project mark may differ from the expected mark by ± 3. A student who has been given the school based mark 8 may attain the project mark 5. A similar student may attain the project mark 11. How are these residuals to be interpreted? Part of the variation is due to random errors. Marking is *ultimately (radically) subjective* according to Matos (2000), who also characterises the whole assessment process as *arbitrary.* In that respect the residuals can be seen as lack of reliability: *The actual mark is to some extent random.*

But the residuals may be due to other explanatory variables. Teachers and students explain the residuals by referring to students' commitment to the project work. Some students put a lot of effort into the project examination.

They work concentrated at school and at home, and they make use of all legal support from peers, teacher and family. On the other hand, some students work very un-concentrated – at least in the beginning of the project period – and some lose concentration during the 3 week long project period. In that respect the residuals indicate validity of the project assessment: *It pays to make an effort.* It would be interesting to test a regression model, which includes an *effort*-variable. Unfortunately my research was not designed to take this question into account.

5. COMPETENCES

Being a final examination the question of educational goals becomes crucial. It is a question of validity: do we value educational values? The question can be analysed in relation to the Danish KOM-project and the competences: 1) *mathematical thinking*, 2) *mathematical argumentation*, 3) *modelling*, 4) *problem posing and solving*, 5) *representation*, 6) *symbols and formalism*, 7) *communication*, and 8) *aids and tools* (Niss & Jensen 2002). An appropriate final examination form should give students the possibility to demonstrate all competences, and examiners must be able to detect and assess all competences.

To address this problem I asked this question to examiners: *To what extent have the competences been made the basis of your assessment?* The question was given to examiners at the project examination and to examiners at the traditional written examination as well. I split the 'aids and tools'-competence in three technology competences: a competence in using a graphic calculator, a competence in using advanced math-IT (e.g. CAS), and a competence in using other sorts of IT (e.g. Word), and I added two non-mathematical competences: *personal competence* (independence, initiative, commitment, responsibility etc.) and *social competence* (openness, sociability, ability to collaborate etc.). The question could for each of the competences be answered on a 4-points scale from 'no importance' with numerical value 0 to 'much importance' with numerical value 3. The average values are shown in Tab. 3.7.1-3.

Table 3.7.1-3. [To what extent have the following competences been made the basis of your assessment? 0: not at all, 1: little importance, 2: some importance, and 3: much importance. Average values, May – June 2002 examinations]

Competences	Project exam. $n = 20$ examin.	Traditional exam. $n = 11$ examin.	p-values for test of equal expectations
Mathematical thinking	2.35	2.00	0.13
Problem posing and solving	2.30	2.55	0.25
Modelling	2.35	1.70	0.01

Competences	Project exam. n = 20 examin.	Traditional exam. n = 11 examin.	p-values for test of equal expectations
Mathematical argumentation	1.68	1.38	0.42
Representation	2.05	1.89	0.51
Symbols and formalism	1.85	1.60	0.35
Communication	2.10	1.56	0.02
Graphic calculator	1.70	1.78	0.83
Advanced math-IT	2.05	0.75	0.00
Other sorts of IT	1.55	0.50	0.00
Personal competence	2.15	1.78	0.18
Social competence	1.00	-	-

I find the results interesting in two ways. They suggest that the competences generally are made the basis of the assessment at *a higher level in project examination* than in the traditional written examination, and the differences are for some competences statistically significant (modelling, communication, advanced math-IT and other sorts of IT). The different competences seem to be more visible in project examination. This could be due to the fact that the examination product is more extensive in project examination than in the traditional examination, since it includes a written report and an oral presentation. In the traditional written examination there is only written answers with no possibility of exercising 'control of understanding'.

The second interesting result is that the average competence-values for project examination vary less than the values for the traditional written examination. The different competences seem to be visible on *a more uniform level in project examination* than in traditional written examination. This gives some empirical evidence for the analytical statement in the KOM-report: "Projects can (...) be used to assess the complete spectrum of mathematical competences (...)." (Niss & Jensen, 2002, p. 128f, my translation) In that respect project examination has a high degree of validity. Project examination seems to reflect all educational goals, not only modelling goals. I am fully aware of the fact that this does not exhaust the question of validity; there reminds at least the question of how a student's profile of competences is being detected, characterised and assessed in a project examination – in absolute terms or relative to goals and intentions.

6. AUTHENTICATION PROBLEMS

A central issue is the question of authentication. As stated by Stephens & McCrae (1995) this question can be divided into two: Has the student written the report himself, and does the student understand what is written? These two questions represent two forms of cheating.

In principle a student can order a project report from outside, or he can copy the report written by a student from a different school doing the same examination. Technology makes it possible to transmit project reports easily from one part of the country to another part. It happens that a student, who has been progressing very slowly and with great difficult, suddenly ends up with a complete and brilliant report. Such incidents cause suspicion, but it is almost impossible to prove this kind of crib. However, I believe that crib is a small problem quantitatively speaking. The teacher follows the project work as a guide during the whole period, he is constantly in dialogue with students, and this 'monitoring' has definitely a preventing effect.

Not understanding what is written in the report is a different way of cheating. It is not illegal to cheat this way, but it represents in my opinion a more serious problem, partly because it happens more often, partly because it is the student himself who is being cheated to believe that he understands the meaning of what he has written, and partly because the oral defence of 10 minutes is not enough to check whether the student really understands. The conversation during the oral defence is fragmented into small bits of a short 'teacher question' followed by an even shorter 'student answer', sometimes by just repeating a sentence in the report. The student response can be characterised as 'guess what the teacher is thinking'. There is no time for a genuine dialogue[2].

7. CONCLUSION

I find that the statistical analysis gives some evidence of validity of project examination. The assessment is valid because the student can expect a project mark on the same level as the school based mark given by the teacher, assuming that the teacher is the best to know. The validity would be improved if the 'residuals' could be explained by an *effort*-variable.

Project examination can also be seen as valid in regards to competences. The results indicate that the competences are made the basis of the assessment on a rather uniform level. Project examination can be used to assess modelling, but modelling competence will not dominate the other competences.

Authentication problems are being taken care of. 10 minutes of oral defence is not enough to reveal whether the student understands what is written in the report or not. Consequently the oral test will be extended to 30 minutes from 2005.

Of course, this does not close the question of assessment. I have not addressed problems such as assessment criteria (are they transparent and do they ensure reliability), and fairness (concerning sex and 'less able'/'gifted' students). But the results so far are promising.

REFERENCES

Antonius, S. (2003). Autenticitet og nye evalueringsformer (Authenticity and new forms of assessment). *Forum for Matematikkens Didaktik, 7*(1), 7-10

Barnes, M., Clarke, D.J., & Stephens, M. (2000). Assessment: The engine of systemic curricular reform? *Journal of Curriculum Studies, 32*(5), 623-650

Clarke, D. (1996). Assessment. In A.J. Bishop, K. Clements, C. Keitel, J. Kilpatric, & C. Laborde (Eds.), *International Handbook of Mathematics Education*. Dortrecht: Kluwer Academic Publishers.

Cumming, J.J., & Maxwell, G.S. (1999). Contextualising Authentic Assessment. *Assessment in Education, 6*(2), 177-194

De Lange, J. (1995). Assessment: No Change without Problems. In Romberg (1995)

Lajoie, S.P. (1995). A Framework for Authentic Assessment in Mathematics. In Romberg (1995)

Matos, J.M. (2000). Dilemmas in assessment – dilemmas in mathematics education. In *Proceedings of the 2nd International Mathematics Education and society Conference,* from http://correio.cc.fc.ul.pt/~jflm/mes2/matos.doc.

NCTM (1995). *Assessment Standards for School Mathematics*. Virginia: The National Council of Teachers of Mathematics.

Niss, M. (1993). Assessment of Mathematical Applications and Modelling in Mathematics. In de Lange, J., Huntley, I, Keitel, C., & Niss, M. (Eds.), *Innovation in maths education by modelling and applications*. Chichester: Ellis Horwood.

Niss, M., & Jensen, T.H. (2002). *Kompetencer og matematiklæring – Ideer og inspiration til udvikling af matematikundervisning i Danmark* (Competences and Learning in Mathematics). The Danish Ministry of Education, Uddannelsesstyrelsen, Report no. 18

Oke, K.H., & Bajpai, A.C. (1986). Assessement in Mathematical Modelling. In J.S. Berry, D.N. Burghes, I.D. Huntley, D.J.G. James, & A.O. Moscardini (Eds.), *Mathematial Modelling, Methodology, Models and Micros*. Ellis Horwood, Chichester

Palm, T. (2002). *The Realism of Mathematical School Tasks – Features and Consequences.* Umeå University: Department of Mathematics,

Romberg, T.A. (Ed.) (1995). *Reform in School Mathematics and Authentic Assessment*. Albany: SUNY Press.

Stephens, M., & McCrae, B. (1995). Assessing Problem Solving in a School System: Principles to Practice. *Australian Senior Mathematics Journal, 9*(1), 11-25

Van den Heuvel-Panhuizen, M. (1996). *Assessment and realistic mathematics education.* Utrecht: Freudenthal Institute

VCE (1999). *Assessment advice for school-assessed CATs for 1999, Mathematics*. Victoria, Australia: Board of Studies.

Watson, A. (2001). Making judgements about pupils' mathematics. In P. Gates (Ed.), *Issues in mathematics teaching*. London: RoutledgeFalmer.

Wiliam, D. (1994). Assessing authentic tasks: alternatives to mark-schemes. *NOMAD, Nordic Studies in Mathematics Education, 2*(1), 48-68

---

[1] The project examination – with minor modifications – will be the general examination for this course from 2005.

[2] The Ministry of Education has recognized this problem and consequently extended the oral defence to 30 minutes from 2005.

Chapter 3.7.2

MATHEMATICAL MODELLING AND APPLICATIONS: ABILITY AND COMPETENCE FRAMEWORKS

Christopher Haines and Rosalind Crouch
Education and Lifelong Learning, City University, London, UK, Email: c.r.haines@city.ac.uk / Department of Computer Science, University of Hertfordshire, UK, Email: r.m.crouch@herts.ac.uk

Abstract: There is much reported research on experiences of learners of mathematical modelling and applications. In this paper we discuss contextual frameworks for such learning and associated teaching together with developmental research concerned with establishing a widely applicable mathematical modelling continuum.

1. PROCESS FRAMEWORKS

Mathematical modelling and applications in an undergraduate context are a feature of a wide range of disciplines apart from mathematics and, given a wider definition, occur at all levels and in all sectors of education and beyond. This wider definition would include posing and solving open-ended questions, quantitative tasks linked to real problems, and engaging in applied problem solving generally. There are dedicated courses in mathematical modelling, but wherever modelling and applications occur, activities associated with them have the potential to enhance students' performance in mathematics, for *mathematical modelling seems to provide a promising territory to explore the question of learning as an activity where students' give meaning to ideas, problems [and] mathematical and non-mathematical concepts* (Matos, 1998, p. 26).

Various teaching and learning paradigms for mathematical modelling have been developed, including:

A Holistic approaches in which students learn through experience of complete case studies. They might be through modelling problems that admit simple and straightforward outcomes and progressing through to more difficult situations. They could also be single instances of complex situations requiring extensive in-depth analysis. This approach focuses on specific applications and relies on experience of process and expertise being transferred from one application to the next.

B The teaching of mathematical modelling by defining, and examining in detail, processes and stages through which the modeller passes. This approach focuses on process and could distance the learner from specific applications.

This paper focuses on B; both A and B emphasise engagement in the mathematical modelling activity in contrast to the common practice in higher education where modelling may involve the presentation and critical analysis of particular models within the normal lecture programme.

Mathematical modelling may be described and defined as a cyclic process in which real world problems are abstracted, mathematised, solved and evaluated in order passing through six stages: real world problem statement; formulating a model; solving mathematics; interpreting solutions; evaluating a solution; refining the model, before reconsidering the real world problem statement again and repeating the cycle. Often a seventh stage is included: the writing of an appropriate report after stage five. That modelling activities can be viewed in terms of staged cyclic processes does not mean that successful mathematical modellers inevitably exhibit this behaviour. Galbraith & Stillman (2001) report that during the modelling activity in schools, there is continual referencing back to the real world context at several stages of the modelling cycle and not simply following stage 5: evaluating a solution, or stage 6: refining the model.

The many different elements and characterisations of modelling and applications also include: open ended questions, mathematising situations, simulations, word problems and applied problem solving, wherever they occur. Processes used in these characterisations may also be stages in a cyclic process.

Sometimes tasks in mathematical modelling and applications, whilst clearly located within a modelling cycle, cannot always be identified with specific stages of the cycle but rather are concerned with the *transition* between two or more stages. In their study of modelling behaviours amongst new undergraduates, Haines, Crouch & Fitzharris (2003) used short modelling tasks in multiple choice format concerned with: *moving from the real world to the model, specifying the model, specifying variables, constructing equations, moving from mathematics to the real world, graphical information interpreted in the real world, using mathematics linked to the real world.*

Noting the importance of linking knowledge and moving freely between the real world and the mathematical world, and the continual referencing between them, descriptors of students' modelling behaviour can be classified (Tab. 3.7.2-1). Teaching and learning paradigms such as A and B above, stages within a modelling cycle and behavioural process descriptors (Tab. 3.7.2-1) could be appropriately applied to mathematical modelling and applications at different educational stages and levels (Fig. 3.7.2-1).

Table 3.7.2-1. Broad descriptors of mathematical modelling behaviour

Process a	Evidence of taking into account the relationship between the mathematical world and the real world input to the model
Process b	Some limited evidence of the above, such as (i) mentions having thought about the model, but little evidence that this has been done, or (ii) has obviously thought about the model, but lacks knowledge of the real world and/or mathematics to solve the problem effectively
Process c	(i) No evidence that the relationship between the mathematical world and the real world input to the model has been taken into account, nor that a modelling perspective has been adopted, or (ii) The problem has been looked at simply in real world terms, or entirely in terms of reasoning or maths (according to its position in a modelling cycle) without reference to the needs of the model nor to the interface between maths and the real world

2. MODELLING EXPERTISE CONTINUUM

The development in students, of expertise and associated skills in mathematical modelling implies the existence of a continuum. One would expect that, in the context of investigations, open ended questions and projects in schools, a level of competence to be achieved would be at a certain level that reflects both the complexity of the situation considered and the processes through which the pupil passes in reaching a solution to the problem. That level for school applications would appear in other sectors, so that for example, the processes could be considered to be the same for pre-university students (Fig. 3.7.2-1) but the level and the complexity involved would be higher. These several overlapping continua, one for each level or sector, would collectively contribute to a single overall continuum (Fig. 3.7.2-2). It is reasonable to postulate the existence of a continuum of expertise in general terms, as there is anecdotal and experiential evidence of a range of achievement amongst modellers.

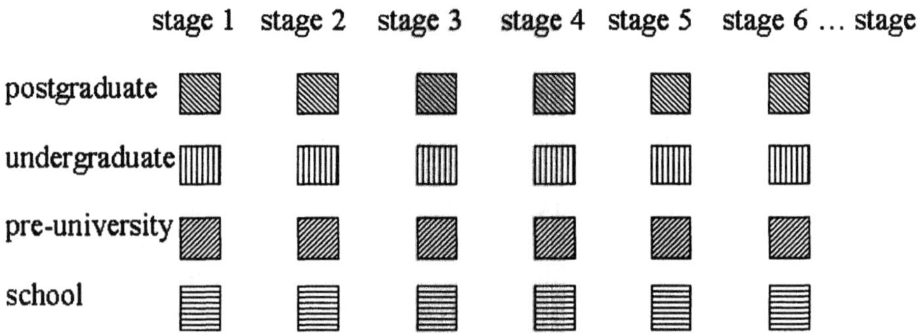

Figure 3.7.2-1. Mathematical modelling at most levels can be described as a staged process. The stage descriptor is common across all levels. Some project or investigational activities might not cover all stages.

To evaluate a modeller, or related learning, in applications of mathematics it would be necessary to establish an achievement scale with descriptors of behaviour appropriate to modelling and/or applied problem solving.

In their research Haines, Crouch & Fitzharris (2003) used twelve multiple-choice questions previously devised for base-level assessment of modelling skills. It is clear that composite questionnaires can provide a snapshot of students' skills at key developmental stages without carrying out a complete modelling exercise. Houston & Neill (2003) constructed six more, parallel, questions and these, with an additional four questions were subject to extensive testing in the United Kingdom. Present analysis (Izard et al., 2003), indicates that the 22 questions are robust and that they could contribute to an overall modelling rating scale of achievement (Fig. 3.7.2-2). They cover: *real world to the model*, *specifying the model*, *specifying variables*, *constructing equations*, *mathematics to the real world*, *graphs to the real world*, *using mathematics*. These items do not address the full range of modelling skills, for they do not, as yet, cover *solving mathematics*, *refining a model* and *reporting*. More research is needed on the applicability of items such as these at different levels. Mostly novice undergraduates in various disciplines in the United Kingdom provided the data in these studies, but some data from more experienced modellers, final year students and research students, is included. Schools in Europe and in Asia are interested in applying these questionnaires.

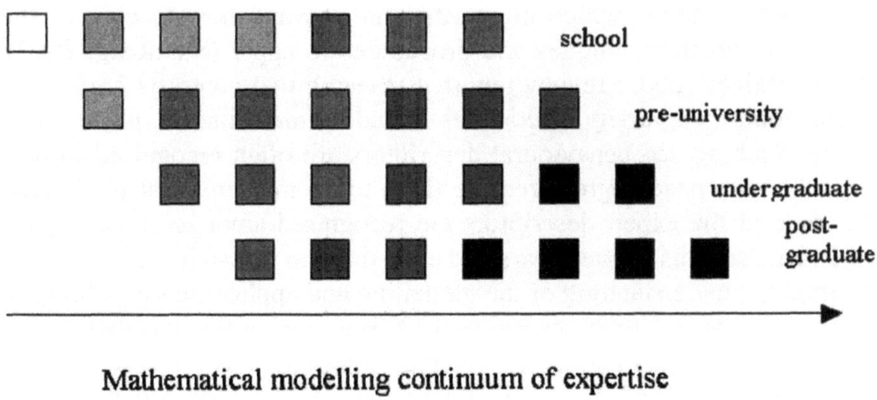

Figure 3.7.2-2. Structure of a mathematical modelling expertise continuum. Behaviours in school for example, might be qualitatively the same as in other sectors, but differ substantially in terms of acquired expertise

3. NOVICE AND EXPERT BEHAVIOURS IN MATHEMATICAL MODELLING

Successful mathematical modelling and applications problem solving involves an ability to move between the real world and the mathematical world, bearing both in mind. The modeller-problem solver needs to consider the real world problem and decide how to mathematise it, deciding which aspects of the real world are relevant and which not – a process of abstraction – and deciding what mathematical principles and techniques to bring to bear, even when technology is used to apply them (Kent & Noss, 2000). The solution also needs to be checked against the reality provided by the applications context and modified if necessary. These processes involving transitions, between the real world and the mathematical world are demonstrably difficult for students and similar experiences are reported in a classroom context (Christiansen, 2001). These difficulties could arise because, despite the fact that formal education gives practice in thinking about topics in a decontextualised way leading to a greater ability to abstract (Galotti, 1994), in abstraction and decontextualisation the rich connections that often provide motivation for the subject are lost.

Some difficulties are undoubtedly due to students being new to mathematical modelling or applications problem solving. In this sense, such novices will perform poorly when compared to experts. This seems to be due to novices having a restricted and poorly structured knowledge base, making it hard for them to know which information is relevant, how to classify the problem, and which techniques and procedures to apply (Sternberg, 1997). Crouch & Haines (2003) review reported research that contrasts novice expert behaviours in a variety of contexts including mathematics, physics and medicine. Such novice behavioural descriptors are often recognised in new undergraduates whose degrees require them to do mathematical modelling, whilst some of the expert descriptors are recognised amongst more experienced undergraduates in later years and even more in research students.

To improve understanding of the modelling and applications capability of new undergraduates, Crouch & Haines (2003) looked at the correlations between nine descriptors of student behaviour: *processes used in solving (Tab. 3.7.2-1), credit attracted, ease of understanding the question, whether problem was regarded as real world, whether problem was interesting, whether problem was located in mathematics as a discipline, time taken to choose answer, ease of making the choice, confidence in their choice.* Amongst their reported research outcomes are:

- Given contextual differences in applications and diversity amongst students it is imperative that they understand and identify with the problems, especially as novices think they have understood them when they have not
- When faced with practical real world problems students are less likely to succeed
- Anxiety amongst students faced with problems regarded as located in mathematics
- Weak knowledge base and a lack of experience in abstraction cause difficulties in the transition from real world to mathematical world.
- Expert behaviours are more in evidence when solving mathematics and interpreting solutions
- Some novices exhibit aspects of expert behaviours at particular stages of the modelling cycle
- There is more evidence of engagement and interest at the transition from the mathematical model to the real world than vice versa.

The development of expertise, including a structured and organized knowledge base, takes some time to develop, needing strong teaching and learning structures with motivated practice on suitable tasks with defined and appropriate learning goals. Students also need to be engaged in the tasks and to get appropriate feedback. Moving to expertise in modelling involves building on this basis to develop appropriate knowledge and skills (Niss, 2001).

4. CONCLUDING REMARKS AND FURTHER RESEARCH

The foregoing discussion indicates that further research is required on the defining processes employed in mathematical modelling and applications, not simply in an abstract sense but also from the point of view of those new to such tasks when compared to those whose expertise is better developed; differential processes might also be present. Useful preliminary work has been done on the establishment of a rating scale for mathematical modelling which appears to meet some requirements for evaluating and monitoring achievement at the transition stages between the real world and the mathematical world. A comprehensive scale should be achievable. In order to understand more fully, how novices develop into experts it would be helpful to define concise descriptors of such behaviours specifically focused on mathematical modelling and applications.

REFERENCES

Christiansen, I.M. (2001). The effect of task organisation on classroom modelling activities. In J.F.Matos, W.Blum, K.Houston, & S.P.Carreira (Eds.), *Modelling and Mathematics Education: ICTMA9 Applications in Science and Technology,* (pp. 311-334). Chichester: Ellis Horwood.

Crouch, R.M., & Haines, C.R. (to appear). Mathematical Modelling: transitions between the real world and the mathematical model. *International Journal of Mathematics Education in Science and Technology, 35*(2), 197-206 (to appear).

Galbraith, P.L., & Stillman, G. (2001). Assumptions and context: Pursuing their role in modelling activity. In J.F. Matos, W. Blum, K. Houston & S.P. Carreira (Eds.), *Modelling and Mathematics Education: ICTMA9 Applications in Science and Technology,* (pp. 300-310). Chichester: Ellis Horwood.

Galotti, K.M. (1994). *Cognitive Psychology In and Out of the Laboratory.* Pacific Grove, California: Brooks/Cole.

Haines, C.R., & Crouch, R.M. (2001). Recognising Constructs within Mathematical Modelling. *Teaching Mathematics and its Applications, 20*(3), 129-138.

Haines, C.R., Crouch, R.M., & Fitzharris, A. (2003). Deconstructing Mathematical Modelling: Approaches to Problem Solving. In Q-X. Ye, W. Blum, K. Houston, & Q-Y. Jiang (Eds.), *Mathematical Modelling in Education and Culture: ICTMA10,* (pp. 41-53). Chichester: Ellis Horwood.

Houston, K., & Neill, N. (2003). Investigating students' modelling skills. In Q-X Ye, W.Blum, K.Houston & Q-Y. Jiang (Eds.), *Mathematical Modelling in Education and Culture: ICTMA10.* Chichester: Ellis Horwood, 54-66.

Izard, J., Haines, C.R., Crouch, R.M., Houston, S.K., & Neill, N. (2003). Assessing the impact of the teaching of modelling. In S. Lamon, W. Parker & K. Houston (Eds.), *Mathematical modelling: A way of life* (pp. 165-178). Chichester: Ellis Horwood.

Matos, J.F. (1998). Mathematics Learning and Modelling: Theory and Practice. In S.K. Houston, W. Blum, I. Huntley, & N. Neill (Eds.), *Teaching and Learning Mathematical Modelling,* (pp. 21-27). Chichester: Albion.

Niss, M. (2001). Issues and Problems of Research on the Teaching and Learning of Applications and Modelling. In J.F. Matos, W. Blum, K. Houston, & S.P. Carreira (Eds.), *Modelling and Mathematics Education: ICTMA9 Applications in Science and Technology,* (pp. 73-88). Chichester: Ellis Horwood.

Sternberg, R.J. (1997). Cognitive Conceptions of Expertise. In P.J. Feltovich, K.Ford, & R.R. Hoffman (Eds.), *Expertise in Context,* (pp. 149-162). Menlo Park, California: AAAI/MIT Press.

Chapter 3.7.3

"TO MODEL, OR TO LET THEM MODEL?" THAT IS THE QUESTION!

Jerry Legé
Department of Mathematics, California State University, Fullerton, USA,
Email: glege@fullerton.edu

Abstract: Two sets of students with weak content skills and no prior experience were introduced to mathematical modelling. Contrasting instructional approaches were used – exposure to a variety of models, versus constructing the model themselves. Both groups demonstrated understanding of aspects of modelling and model structures. The students actively modelling the situation statistically out-performed the other group on four evaluation parameters.

1. INTRODUCTION

In introducing students to mathematical modelling for the first time, which teaching approach would more effectively convey the complexities and subleties of the work involved – having students examine instances of previously-constructed models, or actively engaging them in modelling a situation for themselves? This research question established the basis for a comparative case study, and is the focus of attention for this paper.

The term 'mathematical modelling' is used interchangeably to describe or interpret two related types of activity: translating the real world into mathematical terms (Gravemeijer, 1997) for the purpose of solving a problem or analyzing a situation (Dossey, 1996), and the various steps associated with accomplishing that goal (often called the 'Modelling Process'). The following operational definition for a mathematical model was used: "...a mathematical construct designed to study a particular real-world system or phenomenon. We include graphical, symbolic, simulation and experimental constructs." (Giardano et al., 1997, p. 34)

However, because these terms are familiar enough to mathematics educators, and for the sake of simplicity with respect to the participants in the study, the statements that "Modelling is the process of making a mathematical model." and "A model is the product formed by engaging in mathematical modelling." were not seen as circular reasoning, but rather a way of clearly distinguishing between the two activities – one as process, the other as product. Additionally, the method of evaluation assumes that the learning that takes place from those two activities is disjoint, although the author makes no such claim.

2. EXPERIMENTAL DESIGN

Two high schools having similar economic and ethnic make-ups were found in an inner-city school district near New York City, with each site containing one group of participating students. A course called Foundation Math, offered to all 9th-grade students deemed "not-ready" for Algebra 1, was identified, standardizing the age and ability level of the participating students. Two teachers and three class sections of students from one school, and two teachers and two class sections from the other school, agreed to participate in the study. The study took a total of three weeks to complete both curriculum and assessment phases, with students predominantly working in pairs and having access to four-function calculators at all times.

The curriculum phase did not actively teach students about modelling. Rather, it created an environment in which students explored the contextual problem of planning a vacation trip. The two groups received very different exposures to how mathematical modelling could be used to understand and optimize the situation. The dissimilar curricular approaches, to be detailed in 3., were the only experimental differences between the two groups of students. When the curriculum phase was finished, students in both groups took two identical assessments that were based on a different situation – that of determining the best rap artist. The first assessment had a task structure similar to the curriculum which one group of students experienced, while the second was similar to the curriculum which the other group experienced. Details about those assessments are also provided in 3.

Both student groups were evaluated with respect to two sets of performance goals that were developed for the study. The performance goals were drafted with consideration for what might be accomplished by the participating students in the anticipated exposure time, but also reflected the learning that ideally would take place in each curricular approach. Twenty "Performance Goals for the Assessment of Modelling" were used to evaluate the first assessment. A brief description of each of those performance goals is provided in Table 3.7.3-1:

Table 3.7.3-1. Description of performance goals for assessing modelling

No.	Description	No.	Description
1	Approach modelling systematically	11	Define objects and relationships clearly
2	Show modelling as dynamic activity	12	Stay with the plan previously made
3	Use creative/unusual approaches	13	Use appropriate mathematics to describe
4	Define the problem	14	Perform calculations correctly
5	Separate useful/irrelevant information	15	Report how assumptions, results related
6	Use contextual knowledge not given	16	Interpret results within the context
7	Use planning behaviors	17	Check whether the model makes sense
8	Clarifying/simplifying assumptions	18	Test validity of predictions made
9	Use an organizing representation	19	Refine/extend the original situation
10	Use sub-models & links	20	Identify strengths/limitations of model

They reflected the open-ended modelling approach that one group of students had received. The modelling performance goals included overall considerations for modelling in a procedural way, as a dynamic activity, and with a potential for applying creative or unusual approaches. There were also specific goals related to stages of the Modelling Process, including the problem definition, the organization and planning, the execution (building the model), and evaluation.

Twenty "Performance Goals for the Assessment of Model Structures" were used to evaluate the second assessment. They reflected the curricular approach taken by the group of students that learned about modelling in a more structured environment. Those goals involved general themes of problem identification, simplification, model construction and calculations, and evaluation and revision, but focused on understanding when, and why, something was done. A brief description of each performance goal related to assessing model structures is provided in Table 3.7.3-2:

Table 3.7.3-2. Description of performance goals for assessing model structures

No.	Description	No.	Description
1	Identify the main features	11	Know how mathematics supports model
2	Know if/when clarifying is needed	12	Explain connection between model parts

No.	Description	No.	Description
3	Connect problem to real situation	13	Vary quantity through a range of values
4	Identify assumptions that were made	14	Identify model refinement stages
5	Distinguish types of assumptions	15	Explain changes made in refinement
6	Spot inappropriate/unnecessary info.	16	Determine inherent limitations
7	Know when given information is insufficient	17	Determine whether model makes sense
8	Explain how assumption can simplify or when one should be made	18	Know whether model is sensitive to a given quantity or initial condition
9	Recognize real features from model	19	Interpret solution in context
10	Verify calculations inherent in model	20	Identify when an inference is valid

For each performance goal, a [0..5] rubric scale was developed so that individual students could be monitored on their relative abilities in the specific performance goals, as well as report a cumulative score that would reflect overall performance, even if not interpretable on a normative scale. Three judges were employed to score the student work. Initially, they discussed how to interpret the rubrics, reached an understanding of what the descriptors meant to them collectively, and then assigned scores on each goal for every student separately. In the case of variability among the scores, the median was reported when the range of scores differed was one, and the mean (rounded to the nearest integer) was reported when the range was greater than that.

The statistical analysis of the data concentrated on several specific areas. First, the mean scores would identify those performance goals which could be demonstrated easily (or not), and skills either present at the beginning of the study or promoted as a result of participation in the study. Second, the total score would be used to indicate a relative degree of difficulty in demonstrating overall performance in modelling versus an understanding of model structures. Third, a comparison-of-means statistical analysis would reveal if either student group performed significantly better overall, and would identify plausible knowledge and skills that are better supported by each curricular approach.

3. CURRICULUM AND ASSESSMENT TOOLS

The curricular problem of planning a vacation gave students eleven cities to choose as destinations, with activities, lodging and dining options, all with associated costs. They had an allocated amount of money to budget, potential schedule conflicts with certain activities, and a constraint of having to visit one particular city for two days and one night. Both groups of students were provided an overview page which contained the problem description, information about the cost for renting a car, the fuel economy rating and expected cost for gasoline purchases, and the posted speeds along various roads. They were also given other handouts that contained a description of the Modelling Process, a map showing the major roads in the region, a chart of driving distances along various roads, the fees to be charged along specific toll roads, bridges and tunnels, and details about activities, lodging and restaurants that were available by city. The stated problem was to: "Build a model for this situation – in this case, a description of exactly where you're going, what you're going to do, how much it's going to cost you, and how you arrived at your decision." (Legé, 2003, p. 239)

Students in the first group explored the problem by examining a sequence of five activities, each containing several models that described facets of the situation. The activities were designed to explore the following themes (in order): *Is it possible to build such a model that is conflict-free? What constraints exist on the problem? How might a 'best' model be determined? What features will a reasonably good model have? Can the reasonably good model be improved further (and if so, how)?* The models all had the form of tables of information, containing brief descriptions of the thinking that motivated the development of the model, the approach taken in response to that thinking, clearly stated assumptions made for the model, and all calculations completed. Students would have to answer questions that served a variety of purposes, including reproducing the calculations that were provided, explaining why decisions were made or things done in a particular way, comparing current models to previous ones encountered, and opportunities for the students to make decisions independently. The content of those questions, the student-student and group-teacher interactions, and the level to which students were actively engaged in the study were the factors which shaped the students' understanding of modelling via a critical examination of the structure of completed models.

Students in the second group explored the same problem by actively working on it as a task. They were provided several organizers for structuring the task, which helped students monitor their progress in completing the work and structured the thinking and decisions made in modelling the situation. One organizer was a blank schedule to record the city to be visited each day of the vacation period, the desired activities to do, and the costs for those activities, meals and lodging. A second organizer was to give students an additional copy of the map, a highlighter pen, and the suggestion that they mark the route along which they intended to travel. A third organizer recorded the miles traveled along the various route segments, and the costs incurred from tolls – either roadways, bridges or tunnels. The final organizer categorized the costs on a daily basis, so that students would have a record of all costs in one location, and could quickly determine if they were over budget. The efforts made in building their own model of this situation, combined with the same types of classroom dynamics and their level of engagement, gave students their understanding of model structures via actively (and successfully) modelling a real-world situation.

The assessment involved a new contextual situation, but contained the same structure as the curricular problem in how it was initially presented. Students were introduced to a mild argument among some friends about whom the best rap artist was. They were provided musical industry charts (both single hits and albums) that were current, a year old, and even five years old. They were also given lists of all the albums that each artist had released, and the year in which the first and last albums were made. All of this given information was reviewed with the students, before the first assessment was actually begun, as a way of familiarizing them with the contextual situation. The first assessment simply asked students to make a mathematical model of the situation for themselves, consistent with how the second group of students had explored their curriculum. Students worked on the problem over two consecutive class periods, and were encouraged to be as detailed as they could in their work. The models that were produced were holistically evaluated with respect to the "Performance Goals for the Assessment of Modelling" – each group's model was examined through twenty different lenses for evidence of goals being met, and to what degree.

After that task was completed, students were then provided a second assessment which was consistent with how the first group of students had explored their curriculum. A series of models were presented, and specific questions were asked to which students needed to respond. The questions were written in such a way that the primary evidence for evaluating specific "Performance Goals for the Assessment of Model Structures" came from the elicited response to particular questions. Additionally, each set of responses was reviewed again to see if evidence of having met any goals could be found in unexpected locations.

4. RESULTS

Based on the means of the reported scores, a large percentage of students were able to demonstrate understanding of aspects of modelling on about half of the performance goals. This suggests that many of the "students could define a problem, discriminate between useful and irrelevant information, utilize a representation structure, work with defined objects, adhere to some kind of plan, mathematize the situation, perform calculations correctly and/or interpret the results in a manner consistent with their model." (Legé, 2003, p. 157) Similarly, a large percentage of students were able to demonstrate an understanding of specific details about model structures related to thirteen of the twenty performance goals. Summarily, this means that "students could identify main features, recognize when a problem reflects a situation, identify assumptions, distinguish between types of assumptions, determine when information is unnecessary or insufficient, explain how an assumption can simplify a situation, recognize features of a context from a model, explain the relationship between the mathematics used and assumptions made, explain the connection between disparate parts of a model, recognize refinement stages, determine limitations, and/or recognize valid inferences." (Legé, 2003, p. 157)

It was also the case that students' mean cumulative score for the model structure goals was approximately 12.5 points higher than the mean cumulative score for the modelling goals, whereas the standard deviations were approximately the same. It is uncertain whether that difference in performance is due to the type of assessment used (specific questions versus holistic evaluation of modelling work), the demand for detail required by the rubrics, the degree of difficulty between the two sets of performance goals, or whether students do acquire understanding of model structures more quickly (perhaps because it *is* structured). More research is needed to provide a plausible explanation for this result, assuming that it is not simply a coincidence.

In comparing the overall performance, there was no statistical difference between the two student groups. However, the second group (the one that learned about modelling by just "doing it") significantly outperformed the first group (the one that had learned by looking at examples) on four performance goals – two related to modelling, and two related to model structures. One of the modelling goals was concerned with the use of sub-models, especially for organizing thinking about the model. Most students in the first group primarily constructed models based on a single consideration, whereas students in the second group tended to cycle through the Modelling Process more than once by varying the key assumption, and in an attempt to incorporate all their efforts, would link those models together with some kind of selection criteria. The other modelling goal that produced significantly different results was related to accuracy on mathematical calculations. Students in the first group tended to use the given information on chart rankings to create formulas like "Total of All Current and Previous Weeks' Rankings" – formulas which required calculations. Students in the second group tended toward models like "Most Albums Released" or "Most Years in the Business", which only required counting.

The first of the model structure goals in which a significant difference existed between the two student groups was the ability to recognize the contextual features that are inherent in a model. The assessment required students to interpret a score of '0', which reflected how many times the artist appeared in the singles charts on the first three handout pages. The first group tended to focus on the artist – not popular, didn't get many votes, or not ranked as high – while the second group tended to concentrate on the fact that the songs were not popular or there hadn't been any released prior to the date the chart was made. The second model structure goal in which there was a significant difference between the two groups related to determining limitations that are inherent in a model. Students in the first group contained many answers that addressed ancillary issues – "lack of given information, perceived errors in having scores of '0', calls for conclusions that agreed with their personal opinions, and excellent answers to some other question which was not asked." (Legé, 2003, p. 162) Students in the second group used answers that targeted the restrictive information used, suggested better sources of information upon which to base the prediction, and critiqued the process used in the original model.

In all four of the performance goals in which the second group of students significantly outperformed the first group, it may be the case that those students had more experience within the contextual situation used. But it could also be argued that the additional sense-making exhibited by them, and the quality of the responses provided by them, was supported by being actively engaged in modelling the curricular problem.

REFERENCES

Dossey, J. (1996). Modelling with functions. In T. Cooney, E. Wittmann, G. Schrage, J. Dossey, & S. Brown (Eds.), *Mathematics, pedagogy, and secondary teacher education.* (Ch. 5) Portsmouth, NH: Heinemann.

Giardano, F., Weir, M., & Fox, W. (1997). *A first course in mathematical modelling*. 2nd Edition. Pacific Grove, CA: Brooks/Cole Publishing Company.

Gravemeijer, K. (1997). Commentary. Solving word problems: A case of modelling? *Learning and Instruction*, 7(4), 389-397.

Legé, G. (2003). A comparative case study of contrasting instructional approaches applied to the introduction of mathematical modelling. Ed.D. dissertation thesis, Teachers College, Columbia University.

Chapter 3.7.4

MODELLING AND APPLICATIONS IN PISA

Ross Turner
Australian Council for Educational Research, Camberwell, Victoria, Australia
Email: turner@acer.edu.au

Abstract: Several test items from the OECD's Programme for International Student Assessment (PISA) are now in the public domain. Some of these incorporate elements of modelling. This paper highlights a selection of those items, shows some interesting student outcomes such as gender differences and item difficulty, and poses some questions about the implications of these items for school mathematics.

1. BACKGROUND

The OECD's Programme for International Student Assessment is an ongoing collaborative project among OECD member countries and an increasing number of non-OECD countries, designed to measure how well 15-year-old students nearing the end of compulsory schooling are prepared to meet the challenges of today's knowledge societies. The resulting data will be of interest to educational policy makers in participating countries, and have the potential to promote debate among educational researchers and others and to contribute useful information to educational reform processes.

The nature of the assessment material used and the constructs being tested through PISA are quite broad. For details the reader is referred to the most recent framework document (OECD, 2003). PISA uses a 'literacy' orientation that is intended to extend well beyond simple curricular knowledge, to the ability of students to *use* their knowledge and skills to meet real-life challenges. This provides the possibility of a strong link with ideas about applications and modelling in mathematics.

The PISA Mathematics tests used in both 2000 and 2003 were pencil and paper tests. The mathematics items have been presented in a mixture of formats including multiple-choice, short constructed-response (for example requiring students to carry out a brief calculation), and more open constructed-response items (involving, for example, students writing an explanation of some kind). A number of PISA Mathematics items have been released into the public domain and published by the OECD. Several of these released items involve mathematical modelling, or at least elements of the modelling process.

2. SOME SELECTED PISA ITEMS

PISA Mathematics items can be found in three OECD publications. The first set (OECD, 2000) comprised 14 items (including coding guides and some commentary) from the field trial conducted in 1999 in preparation for the first round of testing in 2000. The second set (OECD, 2002) comprised 11 items from the PISA 2000 Main Study. A further set comprising 27 items from the field trial conducted in 2002, were published to accompany the PISA 2003 Assessment Framework (OECD, 2003). The OECD will release 39 PISA 2003 items in December 2004.

Sample questions coming from four PISA units, each with obvious modelling elements, are presented here. The first is from the PISA 2000 test. The other four are from the PISA 2003 field trial, items that were not included in the final item selection for PISA 2003. Each of the items is reproduced, and some data are presented showing certain aspects of student performance. Note, however, that this is in no sense a random selection of items representing a broader class, so any generalisations can only be very limited. The purpose, rather, is to highlight item features and to facilitate discussion of some important issues.

2.1 Continent Area

This item demands some spatial insight, and requires students to come up with a geometric model and to use that model together with suitable calculations involving scale to estimate an area.

Responses to this item (see Fig. 3.7.4-1, next page) were coded according to the geometric model students used to estimate the area (modelling the shape as a square, circle, using multiple simple shapes, or by any other method). Full credit was awarded for any method that resulted in an approximation in a specified range (12 000 000 – 18 000 000 sq km). Partial credit was awarded to responses that used an appropriate method, but where the result was outside the specified range.

This proved to be a relatively difficult item, indeed an average across OECD countries of only about 9.6% of students obtained full credit for the item, placing it as the third most difficult item in the PISA 2000 set. An average of 19.5% of students obtained partial credit, and about 20% of students obtained no credit. On average, students from Canada and UK did best on this item.

Below is a map of Antarctica

Estimate the area of Antarctica using the map scale.

Show your working out and explain how you made your estimate. (You can draw over the map if it helps you with your estimation).

Figure 3.7.4-1. Sample item – Continent Area

About half of students given this item omitted it. Open constructed-response items such as this typically have relatively high omission rates, and the very high omission rates for this item are another indication of item difficulty. OECD countries with particularly high omission rates for the item (above 60%) were Greece, Italy, Korea, Mexico, Poland and Spain. Those with relatively low rates of omission (below 40%) were Australia, Austria, Belgium, Canada, Finland, New Zealand, Switzerland, UK and the lowest of all USA. Boys found the item easier than girls in all OECD countries except New Zealand. In several countries the margin was quite large.

2.2 Rock Concert

The second sample item, shown in Fig. 3.7.4-2, also involves spatial insight. For this item, students need to decide on a suitable model to quantify the amount of space occupied by a human, then perform an appropriate calculation to estimate how many people would fit into a given space.

For a rock concert a rectangular field of size 100 m by 50m was reserved for the audience. The concert was completely sold out and the field was full with all the fans standing.
Which one of the following is likely to be the best estimate of the total number of people attending the concert?
A 2000
B 5000
C 20 000
D 50 000
E 100 000

Figure 3.7.4-2. Sample item – Rock Concert

The difficulty of this item is close to the average of items used in the field trial. An average of about 26% of students answered the item correctly (response 'C'). Students from Japan and the Netherlands did relatively well, and students from USA, Mexico and Greece did relatively poorly. Response 'B' was chosen by 49% of students.

Omission rates were very low (2% on average), which is typical for multiple-choice items such as this. Boys found the item easier than girls in about two-thirds of countries. The international average percent correct was 24% for girls and 28% for boys.

2.3 Twisted Building

The item 'Twisted Building', shown in Fig. 3.7.4-3 (next page), also involves spatial insight and geometric modelling. It invites students to make some realistic assumptions about the height of the ground floor of the building described, and the height of each floor above, leading to an estimate of the total height of the building.

Responses to this item were coded according to the explanation given, which exposed the modelling the student had used to estimate the total height of the building. Full credit was awarded for a method that led to an approximation in a specified range (50 m to 90 m). Partial credit was awarded to responses that assumed 20 floors rather than 21, but used an otherwise appropriate method.

This item is somewhat more difficult than the average of items used in the field trial. An average of about 15.7% of students obtained full credit for the item, about 9.5% of students obtained partial credit, and about 28.6% of students obtained no credit. On average, students from the Netherlands did best on this item (44% full credit, 21% partial credit). Students in the USA and Mexico did particularly poorly (relative to other OECD countries).

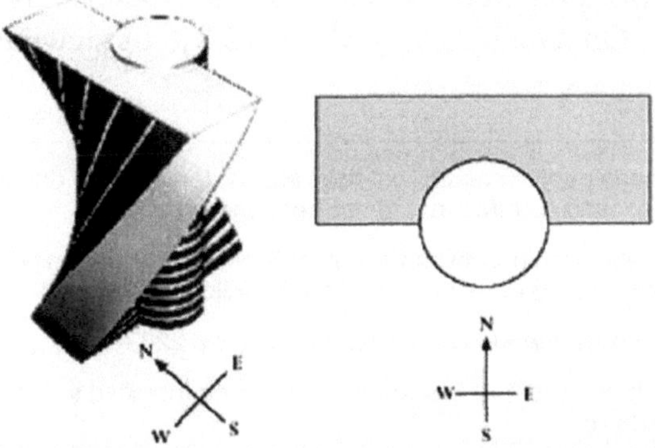

In modern architecture, buildings often have unusual shapes. The picture shows a computer model of a 'twisted building' and a plan of the ground floor. The compass points show the orientation of the building.

The ground floor of the building contains the main entrance and has room for shops. Above the ground floor there are 20 storeys containing apartments.

The plan of each storey is similar to the plan of the ground floor, but each has a slightly different orientation from the storey below. The cylinder contains the elevator shaft and a landing on each floor.

Estimate the total height of the building, in metres.

Explain how you found your answer.

Figure 3.7.4-3. Sample item – Twisted Building

About 46.3% of students omitted this item (an open constructed-response item). OECD countries with particularly high omission rates for the item (above 60%) were Greece, Italy, Japan, Portugal and Spain. Those with relatively low rates (below 30%) were Austria, Belgium, Canada, the Netherlands and Switzerland. Boys found the item easier than girls in all but seven participating countries. The difference was extreme in Japan and Italy. In New Zealand the item strongly favoured girls.

2.4 Heartbeat

In this question (see Fig. 3.7.4-4, next page), students are given a model that combines algebra and words. Students need to understand and interpret the model and the surrounding text. The interpretation required proved to be very difficult indeed, in fact this question was one of the most difficult in the set of field trial items. Full credit was awarded to students giving either 40 or 41 as their answer. On average, only about 12.8% of students answered the item correctly.

> For health reasons people should limit their efforts, for instance during sports, in order not to exceed a certain heartbeat frequency.
>
> For years the relationship between a person's recommended maximum heart rate and the person's age was described by the following formula:
>
> *Recommended maximum heart rate* = 220 – *age*.
>
> Recent research showed that this formula should be modified slightly. The new formula is as follows:
>
> *Recommended maximum heart rate* = 208 – (0.7 × *age*).

A newspaper article states: "A result of using the new formula instead of the old one is that the recommended maximum number of heartbeats per minute for young people decreases slightly and for old people it increases slightly".

From with age onwards does the recommended maximum heart rate increase as a result of the introduction of the new formula? Show your work.

Figure 3.7.4-4. Sample item – Heartbeat, Question 1

Omission rates were very high (53.6% on average). OECD countries with relatively low rates (below 40%) were Canada, Korea, the Netherlands and USA. Hong Kong also had a very low omission rate. Countries with relatively high omission rates (above 70%) were Italy and the Slovak Republic. The international average percent correct was 11.9% for girls and 13.8% for boys. In other words, a large overall gender gap was not apparent for this item, though there was a large gap in favour of boys in Canada, Japan and Korea.

A second question, shown in Fig. 3.7.4-5 (next page), was asked in relation to the same stimulus. In this case, students are required to modify the given algebraic model so that it satisfies an additional condition.

Full credit was awarded to responses giving a formula that is the equivalent of multiplying the formula for *recommended maximum heart rate* by 80%. On average, only about 13% of students answered this item correctly, making it of similar difficulty to the previous question. Countries in which students tended to do relatively well on this item (above 20% correct) were

Belgium, Denmark, Hong Kong, New Zealand, the Russian Federation, Switzerland and Scotland.

> The formula *recommended maximum heart rate* = 208 – (0.7 × *age*) is also used to determine when physical training is most effective. Research has shown that physical training is most effective when the heartbeat is at 80% of the recommended maximum heart rate.
>
> Write down a formula for calculating the heart rate for most effective physical training, expressed in terms of age.

Figure 3.7.4-5. Sample item – Heartbeat, Question 2

Omission rates were again very high (60% on average). Countries with relatively low omission rates (below 40%) were Hong Kong, Indonesia, the Netherlands and Scotland. Italy had an extremely high omission rate (79.8%). Boys found the item easier than girls in about two-thirds of countries. The international average percent correct was 11.9% for girls and 14.2% for boys. In three countries (Czech Republic, Italy and Scotland) the item strongly favoured boys.

3. DISCUSSION

A number of interesting issues arise from these examples. First is the extent to which those responsible for curriculum and instruction in different OECD countries value the kinds of mathematical thinking that underpins these tasks. Tasks involving modelling certainly reflect the importance PISA places on applying one's knowledge to solving the problems confronted in various lifetime contexts. One strand of thought would say that these are the most highly valued kinds of tasks, and that there should be more of them in the PISA tests. This might have the effect of encouraging participating countries to incorporate such activities more extensively into school curriculum, but perhaps would happen only if the benefits of doing so were evident. To what extent has empirical evidence demonstrating the benefit of inclusion of modelling and applications activities in the mathematics classroom been produced and disseminated? What are the implications for national curriculum among participating countries of including items such as these in the PISA test instruments? It is clear that students from some countries routinely do relatively well on such items, and others do relatively poorly. It remains to be seen to what extent PISA might be instrumental in promoting increasing interest in modelling-related learning tasks.

A second issue is the level of complexity of the mathematical modelling activities that 15-year-old students can cope with. It seems to be rather low. What is it that makes these items so difficult? Which elements relate to gender differences? The nature of the item context can have an influence on difficulty, and this can impact differentially on girls and boys. The kind of thinking demanded by the different phases of the modelling process certainly has an impact on difficulty. When looking at the full PISA 2003 item set, items with no modelling demand were on average markedly easier than those that involved modelling. Items involving only the interpretation or reflection phases of the modelling cycle were of intermediate difficulty. Items requiring students to come up with a model (such as Continent Area, Rock Concert, or Twisted Building), or to manipulate a given model (such as Heartbeat) were, as would be expected, the most difficult. Breaking the modelling process down into its component parts seems a strategy that at least can make modelling accessible to a wider range of students.

Third, what kinds of instructional activities might promote facility with items such as these, and in the longer term with more extensive modelling activities? How can teachers be more effectively empowered to explore and promote the mathematical thinking underlying these tasks, and what kinds of teaching and learning activities will be most effective in facilitating this kind of mathematical thinking among 15-year-old students? These are empirical questions, however it might be reasonable to expect the following: first, that good modelling behaviour can itself be modelled by teachers – students can be taught to model by example; second, that structured modelling experiences, including exposure to the separate components of the modelling process as well as guided exposure to 'full-blown' modelling activities, would be part of an effective teaching process for mathematical modelling; and third, including applications and modelling tasks in assessment activities, especially in high-stakes assessment, will encourage both teachers and students to take modelling more seriously.

REFERENCES

OECD (2000). *Measuring Student Knowledge and Skills – The PISA 2000 Assessment of Reading, Mathematical, and Scientific Literacy*. Paris, France: OECD.

OECD (2002). *Sample Tasks from the PISA 2000 Assessment: Reading, Mathematical, and Scientific Literacy*. Paris, France: OECD.

OECD (2003). *The PISA 2003 Assessment Framework – Mathematics, Reading, Science and Problem Solving Knowledge and Skills*. Paris, France: OECD.

Chapter 3.7.5

ASSESSMENT OF APPLIED MATHEMATICS AND MODELLING: USING A LABORATORY-LIKE ENVIRONMENT

Pauline Vos
University of Groningen, Groningen, Netherlands, Email: f.p.vos@fwn.rug.nl

Abstract: In the Netherlands, since 1993 the mathematics core curriculum for junior secondary schools states that students should develop skills for using and applying mathematics in practical situations. For monitoring purposes, a trend study was carried out using mathematical hands-on tasks in a laboratory-like environment. The study revealed that this kind of alternative, practical assessment can have a satisfying curricular validity, higher than written tests based on the same curriculum. However, comparability of test results (between students, schools, etc) depends on the uniformity of test circumstances.

1. INTRODUCTION

Three decades ago Hans Freudenthal and his colleagues started to transform the mathematics curriculum in the Netherlands with a treatise, known as Realistic Mathematics Education (RME). In 1993, a common core curriculum based on RME for Dutch junior secondary schools was legislated. This curriculum emphasized data modelling and interpreting, visual 3D Geometry, approximation and rules of thumb, the use of calculators and computers. The approach to the subject was more practical and investigational, being described by three keywords: *applications*, *skills* and *coherence*. As a result, chapters of Dutch mathematics textbooks start with real life situations, in which mathematics is used and applied, instead of ending with these.

National assessment was adjusted to the new content approach. Generally, test items in the RME-based curriculum describe an appealing daily life situation, often with authentic photos to enliven imagination. The test items contain modelling activities, requiring students to mathematize the context (e.g. into a graph), apply mathematical skills to use the model adequately (e.g. derive a solution from the graph), interpret the mathematical answer in its context, and reflect upon the methodology used.

The new, RME-based curriculum was considerably different from the prior curriculum. A large exercise was undertaken to introduce secondary school mathematics teachers to the new content and its approach. Many workshops on the new curriculum were organized by the curriculum developers, and by teacher training institutes. Also, the national exams were adapted to the new intended curriculum. Nevertheless, the introduction went hand-in-hand with a dilution of the initial ideals. One of the observed weak points in the dissemination of the curriculum was that the assessment practice remained of the written form and did not require students to apply their skills practically, as in small investigation projects. Therefore, a study was designed in which students were tested on their skills to use mathematics in practice, in a laboratory environment. Its objective was to investigate whether, in the years following the introduction of the new curriculum, grade 8 students were indeed improving their abilities to apply mathematics in practical situations. The study also served as an empirical study to investigate what valid and reliable alternative assessment methods can be used to monitor the implementation of a RME-based curriculum.

2. INNOVATING MATHEMATICS ASSESSMENT

When measuring student achievement in mathematics for a large population, in many cases, paper-and pencil tests have been used. However, these tests have come under debate, as they cannot evaluate all practical competencies from an intended mathematics curriculum. Attempts to alter assessment methods have been made, defining criteria for alternative assessment, such as: (a) testing through open questions and for higher order skills, (b) being open to a range of methods or approaches, (c) making students disclose their own understanding, (d) allowing students to undertake practical work, (e) asking for performances and products, (f) being as an activity worthwhile for students' learning, and (g) integrating real-life situations and several subjects (Niss, 1993).

In this section, I will concentrate on alternative assessment for applied mathematics and modelling at a nation-wide scale, for example, to monitor curriculum developments. In this area of study, a number of issues have emerged. First, formats such as observation, interviews and portfolio have shown to be labor- and cost-intensive. Second, the interpretation of students' answers can result in unreliable data because of inconsistencies between examiners (Kitchen & Williams, 1993) Especially the coding of borderline answers (which are neither totally correct nor totally incorrect) is conditional to the coders' background (e.g. coding experience, subject matter knowledge, teaching experience, etc). Despite obvious disadvantages, nation-wide alternative assessments of mathematics have been carried out. For example, countries participating within the Third International Mathematics and Science Study (TIMSS) had to administer a standard written test at grade 8 level (students at the age of approx. 14 year). However, participating countries could opt to administer additionally an alternative assessment, as a complement to the written test. This TIMSS Performance Assessment consisted of practical investigative tasks in science and mathematics (for a full description, see Harmon et al., 1997). The test was developed from the educational vision that seeks coherence between procedural, declarational and conditional cognition. Students were expected to investigate systematically, being provided with a practical context (manipulatives and instruments). They were tested through open-ended tasks like: designing and executing an experiment, observing and describing observations, looking for regularities, explaining and predicting measurements, etc. The test provided students with a worksheet that guided them through the tasks. Students had to record their answers on the sheet, and hand in products (lumps of plasticine, cut-out models, etc.). The use of manipulatives was considered very appropriate as these help students to better understand the context of the question. Instead of describing real life situation in words, the equipment offered the context directly into students' hands. Especially second language learners and students with lower reading abilities were expected to gain from these circumstances.

In 1995, the test was administered in 21 countries, amongst which the Netherlands. The test raised questions on reliability and international comparability of its results; in the international report a league table of countries was avoided. However, in the Netherlands, the test was judged as being very valid in light of the new RME-based curriculum, to such an extent that the test was replicated in 2000 (Vos, 2002). Trend results would allow to monitor the implementation of the RME-based curriculum. Moreover, a replication could give experience in analyzing issues on validity, reliability and comparability of alternative assessments.

In 1995, the TIMSS Performance Assessment was administered to a random sample of Dutch grade 8 students (n = 437 from 49 secondary schools). In 2000, the test was replicated at a slightly smaller scale because of financial constraints (n = 234 from 27 secondary schools). The research questions were 1. *To what extent is there a trend between 1995 and 2000 in the mathematics achievement of Dutch grade 8 students on the TIMSS Performance Assessment?* and 2. *What opportunities and obstacles do exist in large-scale mathematics assessment using hands-on tasks?*

3. VALIDITY

Validity of a test can be established in various ways. For the TIMSS Performance Assessment in the Netherlands, an expert appraisal was carried out to establish the curricular validity of the test with respect to the Dutch RME-based intended mathematics curriculum for junior secondary schools. Six experts from a variety of mathematics education institutes were invited to assess the test items. Their appraisal showed that eight out of twelve tasks matched well with the intended mathematics curriculum (Vos, 2002). The other four tasks were from biology, physics and chemistry, or a hybrid of disciplines. These tasks were maintained in the test, but were not considered relevant for the measurement of mathematics achievement. Below are the eight tasks, which were considered to match well with the intended RME-based curriculum for grade 8.

- The task *Dice* is related to probability.
- The task *Calculator* is related to discovering patterns in numbers.
- The task *Folding* is related to symmetry and spatial abilities.
- The task *Around the Bend* is related to scale drawing and finding geometrical rules: which rectangles can go around a cardboard bend?
- The task *Packaging* is related to measuring and the design of nets: how to pack four ping-pong balls in different ways?
- The task *Rubberband* covers the topic of tables, graphs and extrapolation: with given ten washers to hang on a rubber band, can you estimate how far the rubber band will stretch if you would have twelve?
- The task *Shadows* is related to geometrical transformations.
- The task *Plasticine* asks for problem solving heuristics in combinatorics.

In hindsight, we should have consulted a sample of students on the validity of the test. Anecdotal evidence said that students especially loved the tasks *Around the Bend*, *Folding* and *Plasticine.* One observer noted that students, walking out of the testing session, said they did not feel as having completed a mathematics test; instead, they had produced something worthwhile.

Validity of a test can also be checked through an assessment grid, for example the grid designed for assessment of modelling and applying mathematics, as in Kitchen & Williams (1993). The grid contained the following assessment categories: mathematizing, rewriting (generalizing and simplifying), interpreting, and reflecting. All test items were allocated to one of these categories. If appropriate, an item could be fitted into two categories, but then the weight of that item would be spread. Two curriculum experts were asked to categorize the test items independently. Their inter-rater score was 87% and their average results are reported in Table 3.7.5-1. For comparison, a standard written RME-based test for the same level of schooling was analyzed through the same procedure: the *Afsluitingstoets Basisvorming 1999* (Final test for the core curriculum 1999), developed by the *Centraal Instituut Toetstontwikkeling* (National Institute for Educational Measurement). The TIMSS Performance Assessment showed a better spread over the grid, with a stronger emphasis on mathematizing than the RME-based control test. Often, in written tests, an item already readily states the mathematical formula, which models the context (and thus, these items do not require mathematizing). Also, the skill to reflect is better covered in the TIMSS Performance Assessment. As a result, the TIMSS Performance Assessment can be considered valid on its spread of required modelling activities.

Table 3.7.5-1. Percentage of test items in each modelling category, comparison between the TIMSS Performance Assessment and a standard, written RME-based test.

		Rewrite			
	Mathematize	Generalize	Simplify	Interpret	Reflect
TIMSS Perform. Assessment	35	20	14	16	15
Standard RME-based test	19	25	13	38	6

4. RELIABILITY

Reliability of test data depends on a number of issues. First, uniform test conditions must be created. In the TIMSS Performance Assessment, tests administrators traveled from the testing center to the schools with a large box containing all test materials and an abundance of supplies. To ensure uniform procedures throughout the measurement, the administrators were trained in how to set up the laboratory environment in an ordinary classroom (even if the school had no laboratory), how to introduce the test to the students, how to communicate with students during the testing session, etc.

Alternative assessments contain open-ended questions, and when testing at a large scale, students' answers need to be interpreted and transformed into a code, which can be entered into a database. The reliability of these data depends largely on the evaluation of students' answers. Students' answers must be interpreted in such a way, that the resulting code is independent of the coder. In the TIMSS Performance Assessment, coders were trained during a one-day workshop on the application of the codes. To verify interpretation differences between coders, two different coders coded a systematic sub-sample of 10% of students' responses independently. In this way, the inter-coder agreement was an indicator of the reliability of coding. This agreement was calculated as the percentage of items on which the two coders agreed with their codes. In the 1995 administration, the agreement on the correctness code ranged between 52% and 100%. The lowest percentage was reached on one item (from the task *Shadows*), where the coders only agreed on 52% of students' answers. In the international protocols, no limits were set for the inter-coder agreement, but in hindsight, 52% should have been considered as too low to yield reliable results.

At the onset of the repeat study in the Netherlands in 2000, it was clear that if ever a trend was to be measured, the data needed to be consistent throughout the measurement, but also comparable in time. Therefore, it was decided to check through other means than the inter-coder agreement. Comparability could be affected, for example because the 2000 measurement was carried with slightly different laboratory equipment. One example will illustrate how small equipment differences can have a multiplier effect on students' performance. In the task *Shadows* a torch is used. The torch used in 1995 gave a vague shadow, while the torch of 2000 gave a sharper edge to the shadow. The latter made student's measurements easier giving them more time for remaining items.

Table 3.7.5-2. TIMSS Performance Assessment Mathematics tasks, 1995 and 2000 in the Netherlands: reliability, comparability between years, and results.

Task (number of items)	Reliability Cronbach Alpha 1995 (n=437)	2000 (n=234)	1995-1999 Trend comparability $p(\chi^2)$	Students' achievement results 1995 (n=437)	2000 (n=234)
Dice (6)	0.50	0.64	0.77	77 (3)	74 (4)
Calculator (7)	0.71	0.68	0.99	62 (4)	60 (5)
Folding (4)	0.83	0.76	0.53	73 (4)	77 (5)
Around the bend (8)	0.59	0.62	1.00	68 (3)	70 (4)
Packaging (3)	0.61	0.65	0.28	52 (4)	58 (5)
Rubberband (7)	0.58	0.39	0.00	---	---
Shadows (6)	0.64	0.61	0.01	---	---
Plasticine (8)	0.85	0.78	0.00	---	---

Note. --- Dashes indicate omitted results, which did not satisfy the comparability test. Standard deviations are shown in parentheses.

To detect unreliable and incomparable results, two statistical tests were carried out (Vos, 2002). The results are shown in Table 3.7.5-2. First, for each task, Cronbach's alpha was calculated for 1995 and 2000 separately. Results higher than 0.5 were considered acceptable. On this test, the task *Rubberband* failed. Second, a χ^2-test was carried out. The outcomes, indicated by their significance $p(\chi^2)$, indicated the probability that answer patterns were comparable. Values lower than 0.05 were considered as indicators of unequal testing circumstances in 1995 and 2000. As a result of the tests, and to avoid distortions of the trend measurement, tasks with questionable data were eliminated: *Rubberband, Shadows* and *Plasticine.*

5. RESULTS

Five mathematics tasks (28 items) remained suitable for analysis. The achievement of Dutch students on the Performance Assessment in 1995 and its repeat in 2000 is given in Table 3.7.5-2. For each task, the average percentage of correct scores on the items is calculated. Compared to 1995, the achievement results did not show significant changes on these tasks. On each mathematics task, the shifts were statistically insignificant. The average percentage correct on all five mathematics tasks in 1995 was 66 (not included in Table 3.7.5-2), which did not differ significantly from the average score correct of 68 in 2000. The results showed that Dutch students in 2000 had not gained practical competencies in mathematics since 1995, despite the increased emphasis on these competencies in the RME-based curriculum. This answers the first research question. The null-trend could be caused by the classroom practice, in which students never encounter tasks in a laboratory environment. In classroom practice, hands-on tasks are lacking, as the assessment practice stuck with a paper-and-pencil format, in which students only read texts about real-life contexts. Despite curricular intentions, tests offering students tangible real-life contexts (through projects, or through manipulatives) are still rare in the Netherlands.

The project also showed, that testing conditions need to be well controlled, for example by minimizing differences between schools and between measurements from different years. Small changes in equipment can destroy valuable data. Also, it is important to have different coders, who can code and re-code students' answers at different stages in time. However, provided these conditions, alternative mathematics assessment in a laboratory environment is feasible at a large scale. Reliability of data is to be scrutinized closely, but the high validity of the test will compensate for this. This answers the second research question.

Finally, anecdotal evidence showed that the TIMSS Performance Assessment was an eye-opener to many mathematics teachers. During the testing sessions, they observed the tasks and how their students coped with these.

Some teachers admitted that they had never thought mathematics could be tested in a laboratory environment. They associated manipulatives with 'fun mathematics' as used on the day before holidays. Now, the assessment context created a serious atmosphere. As such, the TIMSS Performance Assessment could be used as exemplary curriculum material, not only in the Netherlands. If laboratory-based tests are part of national exams, then teachers who 'teach to the test' might better implement a mathematics curriculum based on modelling and applications.

REFERENCES

Harmon, M., Smith, T.A., Martin, M.O., Kelly, D.L., Beaton, A.E., Mullis, I.V.S., Gonzales, E.J. & Orpwood, G. (1997). *Performance Assessment in IEA's Third International Mathematics and Science Study.* Boston, MA: Boston College.

Kitchen, A. & Williams, J. (1993). Implementing and Assessing Mathematics Modelling in the Academic 16-19 Curriculum. In T.Breiteig, I.Huntley & G.Kaiser-Messmer, *Teaching and Learning Mathematis in Context* (pp. 138-150). New York: Ellis Horwood.

Niss, M. (Ed.) (1993). *Investigations into Assessment in Mathematics Education, an ICMI Study.* Dordrecht, NL: Kluwer.

Vos, F.P. (2002). *Like an Ocean Liner Changing Course; the Grade 8 Mathematics Curriculum in the Netherlands 1995-2000.* Enschede, NL: University of Twente.

Part 4

EDUCATIONAL LEVELS

Chapter 4.1

MODELLING AND APPLICATIONS IN PRIMARY EDUCATION

Maria Salett Biembengut
Universidade Regional de Blumenau, FURB, Brasil, Email: salett@furb.br

Abstract: This chapter is the result of the reflections on contributions presented by the participants of the Primary Education Modelling Group during the Modelling and Application Study conference held in Dortmund, Germany in February, 2004. The material presented by the group participants gives opportunity for some brief considerations regarding the day-to-day school life of children and for the presentation of modelling procedures for use in primary education, as well as it invites reflections on the possibility of these procedures becoming school practices in light of the level of training of the majority of teachers.

1. DAY-TO-DAY SCHOOLING

In everyday life, children perceive their environment, obtain information and select from it, compare it to what they already know, and after assimilation, confer significance to the various scenarios that surround them. The child is always interactively researching everything within his grasp. His imagination surpasses the limits of the image, leading him to create symbols or objects and to form ideas, giving form, color and sense to the world in which he lives.

This complex process peculiar to the human mind passes basically through three stages, which can be described as those of *perception*, *comprehension* and *signification*. This means that each *sensation* or *perception* that the child takes in from his environment generates imagination and ideas in his mind that, starting from the *comprehension* that he already has, may be transformed into *significance*, a mental model that results into understanding (Kovacs, 1997). Mental models, or representations of the world of which he belongs, have an ever- increasing capacity to express and reproduce, in various ways. This means that the child creates and re-creates models in his mind that can allow him to establish ways of being and acting (Sacks,1995).

In most cases, children are inserted into the knowing and doing of things. However, when they end up going to a formal schooling, concerns about rules, convention and program curricula result in the fact that there is not enough time left to stimulate creative and imaginative talents. The teaching of mathematics, for example, frequently leads children to respond to specific questions (generally arithmetic questions) in a certain standard way, without considering the amount of information that they have already received from the outside world, much less their particular capacities. This contributes to passivity and inhibition on the part of the child in his resolution of significant questions, becoming an obstacle especially when learning mathematics.

Several studies show that during the grade school years children tend to apply strategies in a superficial way when solving problems, leaving out their knowledge of the real world. Among the reasons for this, Bonotto (2004) points to textual factors related to stereotypical problems in the textbooks associated with classroom practices which contribute to this dissociation between school mathematics and the mathematics applied to various, day-to-day situations. A lot of empirical data points to the fact that teacher education courses in various countries do not foster consistent or sufficiently encompassing education in the future teacher, one which would enable the use of alternative practices in the classroom in accordance with the sociocultural reality in which they perform (Palm, 2004).

Currently, in most countries, curriculum reforms and their accompanying documentation more or less explicitly assume that one of the most important objectives in mathematical education is to help students acquire the ability to develop and use mathematical models as a means of making sense out of day-to-day situations, which leads to making sense of the complex systems that make up modern society (Blum, 2002). The purpose is not only to motivate students by daily 'contextualizing', but also to create conditions in which they can learn to research, and come to comprehend the significance of what they are studying. Research shows that the use of applications and mathematical modelling in teaching can enable students to learn and develop abilities for making use of mathematics outside of the classroom and further, it provides motivation for studies that are relevant to mathematics (Biembengut, 2004).

Both curricula reform and research confirm that although mathematical modelling has been shown to be an advantageous strategy for the student' academic formation, it is hardly adopted in most countries, and even less in the primary school, The justification of many teachers is that they do not feel able to deal with the situations or questions posed by children in the daily education practice, or that they do not feel able to explore the links between curricular and extra-curricular knowledge.

2. PROCEDURES IN PRIMARY EDUCATION

In the process of perceiving a real context, understanding and explaining by means of a language or system of signals, followed by external description or representation, one can recognize the same mental processes that are used to construct what has been perceived. That is to say, in making a model of an observed phenomenon or in using a model for understanding or solving something, one can identify the three phases of the cognitive process: *perception*, *comprehension* and *model-signification*. In these terms, modelling procedures in primary education are synthesized into three phases that will be called *perception and apprehension*, *comprehension and explanation* and *signification and modelling*. These procedures can be adapted at any level of teaching in the course of the academic year, with some or all curriculum subjects. Furthermore, these procedures can be realized in flexible phases in a circular, give and take process.

1st phase: Perception and Apprehension

This first phase seeks to stimulate the perception and interest of children with material and artifacts that illustrate the environment. The idea is to promote activities that involve them with nature (beauty, harmony) and with other participants and symbols that they already know in this context, as well as to sharpen observation and attention towards things that are as yet unperceived. This means that this context has value as a model or something that motivates them, at another time, in the learning of mathematics. This is the phase in which children seek to inform themselves from the context in question and obtain the greatest number of facts (Gravenmejer; Winter apud Schwarzkopf, 2004). Although perception is not the only source of knowledge, it is without doubt essential to the first description of the environment that surrounds them, allowing children to decode, effect representations and furthermore, to deal with new situations, visualizing the occurrence of phenomena, and judging and comprehending something in this respect.

2nd phase: Comprehension and Explanation

In this phase, we seek to promote activities that allow children to go beyond images already learned, leading them to conceive other images and to delineate symbols, stimulating association of ideas and comprehension. It consists of teaching children to understand the real world in a quantitative sense and leading them towards using mathematical symbols as a means of representing things that they observe and find interesting. Based on ideas that they already have about comparison or measurements, for example, it comes to be a means of teaching mathematical concepts symbols that are as yet unknown. What is important is that it is a 'back and forth' process among the materials and artifacts that surround the students and that they can handle or observe the mathematical symbols. Mathematics needs to be learned and understood as another language, another way of representing, visualizing, comprehending and communicating. If mathematics is learned as a language, or rather if the materials or artifacts can be described in mathematical language, and vice-versa, then there is a better chance that the children will not reject them, especially during the later phases of teaching (van den Heuvel-Panhuizen, 2004). In accordance with the level of education of the child, activities that integrate other areas of knowledge may contribute. In this way, children do not disconnect mathematics from reality, while comprehension of unknown facts is facilitated, by means of a process that assimilates such knowledge or reduces it to already familiar facts (Bonotto, 2004).

3rd phase: Signification and Modelling

At this phase, the child should recognize both the materials that surround him and accumulate mathematical symbols and concepts, based on previous knowledge and available references (mathematical or otherwise). According to Steinbring (1999), symbols are necessary to the process of knowledge but a referent context is required in order for these symbols to be understood and interpreted. Learning is a circular process of construction of relations between these functional components of knowledge. Building relationships between symbols and a referent context requires the creation of a sub-adjacent conception (mathematical), which provides integration of the knowledge within a theoretical structure (Schwarzkopf, 2004). Thus this 3rd phase, the most challenging, consists of sharpening the child's creative sense in solving questions or making representations of some material in terms of a model. The goal here is that children be encouraged to reorganize a variety of situations, capable of being translated into mathematical language, which permits them to inform themselves in detail about mathematics and the possibilities of using it to learn more about the complexities of the real world outside of a school context.

3. POSSIBILITIES OF MODELLING AND APPLICATIONS

In an ongoing research (cf. Biembengut, 2005) the above procedures were applied for two consecutive years, in an experimental phase, with 2 classes (70 children) from the 2st grade in 2001, and with the same children in the 3nd grade (64 children) in 2002, in primary education. One of the activities developed with children from the 3rd grade was *the growth of plants*. The mathematical content to be taught was the *system of linear measurement*. As such, in the 1st phase the children were taken to the school's garden to observe the plants around them to explain what they perceived, knew and felt. Next, a container of earth was distributed to each group of two children in order to plant corn or beans. The containers were kept in a place suitable for growing plants, to which the children had easy access in order to care for them and accompany the development of their respective plants. In the 2nd phase, during the germination period, the children were taught, among other programmatic contents, linear measurement. As soon as the plants started to grow, each group of children took daily measurements of their plant and recorded data in table form. In the 3rd phase, they represented their respective data on graph paper, obtaining graphic representation of the linear growth of the plant in relation to time. Next, it was proposed to the children that they compare their data and graphic representations with each other. Most of the children verified that the growth data and graphic representations of each plant did not coincide, but that the graphic representations seemed alike – in the form of the letter 's'. Even without formalizing a 'logistical growth model', the activities developed allowed the children: to observe and interpret symbols and their significance; to relate, integrate and represent data from external means and to comprehend the environment conceptually.

4. CONCLUSION

In the primary school, where the mathematical syllabus (elementary arithmetic and geometry) can be richly and lively applied to the universe of children, it is not difficult to plan activities that make children to understand mathematical contexts and to play with the mathematical language. Empirical research has shown that curiosity and comprehension in children in regard to the environment in which they live can be strongly stimulated. By formalizing or representing different events or information perceived and by elaborating particular categories such as, for example, symbols and messages, most of the children exhibited gradual advancement in their ability to understand and respond to the activities proposed. This affected both evaluation of what they know and what they do not know. Thus, children gifted with a sharpened sense of imagination can dare to look for solutions and may find effective means for predicting the course of events that occur around them (Bonotto, 2004).

It is worth noting here that knowledge flourishes to the degree in which different events or perceived information can be represented by means of symbols and messages. Thus mathematical modelling in primary education can contribute to this 'flourishing' since the activities involved in the process can lead the child to understand a situation or context and get to know the mathematical language that allows him or her to *describe*, *represent* and *solve* a real-life situation or context and to interpret/validate the result within this same context.

This means that the teacher is in control of various areas that make up the school curriculum, and has the means of facilitating various levels of expression (linguistics, mathematics, technological artistic) and feels capable of modifying objectives and classroom content along the way. It is also positive that teacher is going to deal with a significant number of children from distinctly different sociological and cultural realities, all of who require the general education that is necessary and sufficient for taking part in the environment in which they live. But the teacher has to be aware of this: changing the beliefs, conceptions and attitudes of teachers is essential to turn mathematical modelling a natural practice of teaching.

REFERENCES

Biembengut, M.S. (2005). *Creativity in the Early Grades (Technical Report FAPESC)*. Florianópolis.

Biembengut, M.S. (2004). *Mathematical Modelling & Implications in Teaching and Learning*. 2ª ed. Blumenau: Edifurb.

Blum, W. et al. (2002). ICMI Study 14: Applications and Modelling in Mathematics Education – Discussion document. *Educational Studies in Mathematics, 51*, 149-171

Bonotto, C. (2004). How to replace the word problem with activities of realistic mathematical modelling. In H.-W. Henn, & W. Blum (Eds.), *ICMI Study 14: Applications Modelling in Mathematics Education. Pre-Conference Volume*, (pp. 41-46). Dormund, Germany: University of Dortmund.

Kovacs, Z. L. (1997). *The Brain and the Mind: an introduction to computational neuroscience.* São Paulo: Acadêmica

Palm, Torulf. (2004). Features and Impact of the Authenticity of Mathematical School Tasks. In H.-W. Henn, & W. Blum (Eds.), *ICMI Study 14: Applications Modelling in Mathematics Education. Pre-Conference Volume*, (pp. 205-210). Dormund, Germany: University of Dortmund.

Sacks, O. (1995). *An Anthropologist on Mars: 7 paradoxical tales.* São Paulo: Cia das Letras.

Schwarzkopf, Ralph (2004). Elementary Modelling in Mathematics Lessons: The Interplay between "Real-World" Knowlwdge and "Mathematical Structures". In H.-W. Henn, & W. Blum (Eds.), *ICMI Study 14: Applications Modelling in Mathematics Education. Pre-Conference Volume*, (pp. 241-248). Dormund, Germany: University of Dortmund.

Steinbring, H. (1999). Mathematical Interaction as an Autopoietic System Social and Epistemological Interrelations. In I. Schwank (Ed.). *Proceedings of the First Conference of the European Society for Research in Mathematics Education Vol. 1+2.* Osnabrück, Germany: University of Osnabrück, Forschungsinstitut für Mathematikdidaktik.

Van den Heuvel-Panhuizen (2004). Shifts in underwstanding – The didactical use of models in mathematics education. In H.-W. Henn, & W. Blum (Eds.), *ICMI Study 14: Applications Modelling in Mathematics Education. Pre-Conference Volume*, (pp. 285-290). Dormund, Germany: University of Dortmund.

Chapter 4.2

POSSIBILITIES FOR, AND OBSTACLES TO TEACHING APPLICATIONS AND MODELLING IN THE LOWER SECONDARY LEVELS

Toshikazu Ikeda
Yokohama National University, Yokohama, Japan, Email: ikeda@ed.ynu.ac.jp

Abstract: Reports on teaching applications and modelling in eight countries dealt with similar arguments regarding obstacles, each with a different emphasis. The following four obstacles are seem to be common across eight countries in which applications and modelling are located within the national curriculum; "Teachers' perceptions of mathematics", "Teachers' understanding of modelling", "A lack of adequate textbooks and curricular modelling tasks", "A lack of adequate assessment, and of modelling tasks in central examinations".

1. INTRODUCTION

Many issues emerge in the context of applications and modelling at the lower secondary level. For example, reasons for teaching applications and modelling; approaches to teaching applications and modeling; obstacles to teaching applications and modelling. Because of space restrictions, this document focuses on the issue of obstacles.

It seems there exist a variety of obstacles to teaching applications and modelling in different countries around the world. This document samples experiences within eight countries that exemplify general aspects of the issue of obstacles, and summarizes some general features of the reports from these countries.

2. EIGHT CASE STUDIES

Obstacles to teaching applications and modelling in the eight countries are described as follows.

2.1 Gabriele Kaiser (Germany)

Being able to apply mathematics competently is an accepted overall aim of German mathematics teaching. The question as to why applications and modelling examples do not gain the same importance in other domains as they have within didactic discussions, can be approached from various points of view:

(1) The system: Mathematics teaching is dominated by a subject-based understanding of theory, and this implies a partial and predominantly subject-based implicit didactic. The lesson structure follows the subject structure, which means, if put into concrete terms, that the lessons start from general concepts and phrases and then continue with general conclusions.

(2) The teacher: Empirical research has made it clear that the perceptions of mathematics, held by mathematics teachers, are dominated by an understanding of mathematics as a logical and consistent construction of thinking. The notions of usefulness and applications of mathematics play a minor role in the perceptions of mathematics and in the learning of mathematics as well.

(3) The student: Empirical studies have demonstrated that the mathematical beliefs of most students are dominated by an understanding of mathematics as an accumulation of knowledge. The higher the age-group, the lower the importance of application-based mathematical beliefs.

2.2 Hugh Burkhardt (England)

In England, everyone talks of the importance of being able to use mathematics. However, in most lower secondary school classroooms, few applications are actually taught, and there is no modelling.

Reasons for this include: (1) An inward-looking view of mathematics on the part of specialist mathematics teachers, focussing on the concepts and procedural skills of pure mathematics. Applications, where they exist, are mainly seen as concept reinforcement. (2) The political emphasis on 'basic skills' for weak students and schools. While the recent focus on "functional mathematics" may change this, it could again degenerate into just 'basic skills'. (3) The belief that you have to learn skills before you can apply them leads to indefinite postponement of work on solving non-routine problems of any kind. Change will be difficult because many teachers have no experience of the teaching skills needed, and regard non-routine problems as 'unfair'. (4) The 1989 National Curriculum for mathematics and its tests introduced a view of mathematics as a checklist of narrowly-defined skills. (5) 'High stakes' testing at ages 7, 11, 14, 16 and 18 affects the future of every teacher and student. It leads to teaching that focusses on those fragments of mathematical performance that are tested – you cannot do applications, let alone modelling, this way. (6) TIMSS has focussed on pure mathematics – this may be mitigated by PISA.

All of these factors reflect a traditional approach, nationally and internationally. Indeed, over the last 20 years, things have become worse, rather than better, in England.

2.3 Florence Mihaela Singer (Romania)

The most pre-eminent obstacle is the theoretical orientation of the teacher training pre-service and in-service programs. Their content is focused on a highly theoretical level of mathematics, with a very low emphasis on aspects connected with teaching methodology, and even lower on applications and modelling.

With regard to learning, there is a discontinuity at the transition from primary to lower secondary education. This discontinuity is generated, on the one hand, by different teaching expectations: the primary teachers teach a number of subjects, while the lower secondary teachers teach only mathematics. On the other hand, it is compromised by the dual system of initial teacher training: primary teachers are trained in colleges/high schools and are more pedagogically oriented than subject oriented; while the lower secondary teachers graduate at university level, with a very low emphasis on the psychological and sociological aspects of teaching and learning.

This discontinuity is also manifested at the level of curriculum interpretation. While the written curriculum recommends a progressive development following a unifying system of objectives, content, and learning activities, the teaching practice lags far behind this; due to insufficient training programs being provided to support the curriculum reform process.

2.4 Christine Suurtamm (Canada)

In Canada, although each province is responsible for its own educational system, there are many similarities in the mathematics curricula. Most provinces have a curriculum and curriculum resources that would support applications and modelling in the lower secondary level. Mathematics educators in Canada realize that students at this age level often need to see the relevance of mathematics and mathematical modelling helps them to make connections.

Teachers are encouraged to present students with a mathematical problem first and then to develop mathematical ideas through the process of problem solving. However, in practice, many teachers tend to use applications, modelling, and problem solving as examples of uses of mathematics once the mathematical concepts are taught. Some of the obstacles to full implementation of mathematical modelling are teachers' understandings of modelling, their view of mathematics, and their inexperience in doing mathematical modelling themselves. However, teachers are moving along the continuum as further professional development engages teachers in mathematical modelling activities.

2.5 Jarmila Novotná (Czech Republic)

The main reason for using applications and modelling in mathematics education in my country is always teaching mathematics, not teaching applications and modelling only. Two problematic issues are as follows.

Difficulties in choosing appropriate problems

Characteristics of problems suitable for applications and modelling at the lower secondary level include: (1) Minimal mathematical background required. (2) Tasks should stimulate both manual and intellectual activities. (3) Tasks need to provide for the modelling of situations either in reality or in the minds of the students. (4) Tasks should challenge students to create their own models or introduce new interesting situations to be solved.

Difficulties with the language appropriate for applications and modelling

There is still much work to be done to better understand the role of language in the theoretical-experimental domain of modelling for identifying epistemological obstacles. For example, relations between language and point of view, and between mathematical language and change of strategy.

2.6 Pauline Vos (Mozambique)

In Mozambique (as in many other countries, especially in lesser developed countries) applications and modelling are not perceived as an important part of the mathematics curriculum. The first obstacle is the existing curriculum, which has a strong enforcing role on what is taught in Mozambican mathematics education. Secondly, the national exams (at the end of grades 5, 7, 10, and 12) are an obstacle, as they do not contain any applications or modelling items. Another obstacle is the predominantly deductive approach used in mathematics education, whereby students are trained to memorize definitions and algorithms, resulting in little understanding, short-term retention, and low motivation. A further obstacle is the large number of un(der)qualified teachers. More than 80% of the mathematics teachers are un(der)qualified, and their insecurity makes them hold firmly to old routines.

From a political perspective, a general belief seems to be that if students know the definitions and can carry out the algorithms, they will be able to apply them. There are some thoughts on making the curriculum more practical, but few people have an idea as to what this means, and what the curriculum would consequently look like.

2.7 Koeno Gravemeijer (Netherlands)

Experiences in the Netherlands suggest that a lack of adequate teacher enhancement, textbooks, and assessment can be a serious obstacle, even if applications and modeling are mandated, and exemplary tasks are developed. Applications and – to some lesser extent – modelling were central traits of the curriculum reform of the early 1990's. However, the government did not facilitate much in-service teacher enhancement, and as a consequence, teachers seem to have developed a rather limited image of the innovation. The adage, 'learning mathematics by doing mathematics', seems to have been translated into, 'independently working on textbook problems'. In addition, the new textbooks were not innovative enough to alter this view. In such a setting, challenging problems interrupt the smooth flow of the lessons, so textbook authors started to make those tasks less demanding at the request of teachers. This eventually has resulted in textbooks full of contextual problems that are divided into a series of simple sub questions that in fact obscure the intended applications or modelling for the students. In addition, our conjecture is that the individual help that teachers offer tends to adjust to the expressed needs of the students, who ask for instrumental directions – a tendency that seems to be in tune with the instrumental character of customary forms of assessment.

2.8 Toshikazu Ikeda (Japan)

Most teachers use applications and modelling as examples of uses of mathematics once mathematical concepts have been taught. More process-oriented modelling tasks need to be developed, and mandated for implementation in classroom teaching.

In terms of practice, there seem to be three dominant obstacles. First is the influence of entrance examinations, in which modelling tasks hardly appear. The second is concerned with the development of modelling tasks and the use of technology – there are still too few modelling tasks that students really want to attempt to solve, and technology is not popular in Japan. The third is concerned with the belief and confidence of teachers. Modelling makes teaching more open and complex, and teachers have little experience in modelling. Further, principals of schools don't encourage teachers to undertake modelling in their classroom teaching, and therefore, teachers don't want to tackle modelling and applications.

3. SUMMARY

All of the eight reports raised similar issues regarding obstacles, with different emphases. Focussing on the treatment of obstacles to applications and modelling in mathematics curriculum, three situations can be identified. First is that applications and modelling are not perceived as an important part of the mathematics curriculum. For example in Mozambique, the following belief was expressed, "a general thought seems to be: if students know the definitions and can carry out the algorithms, they will be able to apply these."

Second is the opposite situation in which applications and modelling are (officially) a central attribute of a curriculum. Canada and the Netherlands (curriculum reform of the early 1990's) seem to belong to this category. However it is suggested that implementation falls somewhat short of the intention for a variety of reasons associated principally with teacher practices and priorities. Reports from other countries suggest a third (intermediate) category, in which applications and modelling are located in the national curriculum, but their role is not central. Within the latter two categories, the following four obstacles seem to be common.

1. Teachers' perceptions of mathematics. 2.Teachers' understanding of modelling. 3. A lack of adequate textbooks, and adequate modelling tasks. 4. A lack of adequate assessment methods, and modelling tasks suitable for central examinations.

The following suggestion from the Netherlands is especially meaningful. "Experiences in the Netherlands suggest that a lack of adequate teacher enhancement, textbooks, and assessment can be a serious obstacle, even if applications and modeling are mandated, and exemplary tasks are developed." Teacher education appears to be the central, and most commonly recognized, issue to address in the future.

Chapter 4.3

UPPER SECONDARY PERSPECTIVES ON APPLICATIONS AND MODELLING

Gloria Stillman[1]
Address: University of Melbourne, Melbourne, Australia, Email: g.stillman@unimelb.edu.au

Abstract: Several issues raised in this study are of heightened importance at the upper secondary level. There are many tensions at this level of schooling contributing to a reluctance by teachers to teach mathematics by modelling and a scepticism by many students that modelling is central to their mathematical learning. Several of these tensions are raised as the issues are discussed in this chapter. The challenge is for modelling to be seen as an essential embedded element of mathematics, mathematics teaching and assessment whether students are in academic, technical or general education courses.

1. INTRODUCTION

Upper secondary perspectives are pertinent to the issues raised in this study because the final schooling years are in most instances the last chance mathematics teachers have any influence on students' mathematical competencies or attitudes towards mathematics. Students' beliefs are the source of their attitudes and it is difficult once students pass through the formation ages of 13 or 14 to influence these. There is also some scepticism by students about the relevance of modelling activities due to too many teachers at this level placing too much emphasis on the mathematical, rather than the modelling, aspects. Furthermore, the senior secondary level is often the first time tertiary mathematics educators really care about what secondary teachers teach in school and what students learn.

At the upper secondary level, especially in some European countries (Artaud, 2004), there is a strong bias against mathematical modelling as the prevalent attitude is that high level mathematics is what is important. However, the upper secondary curriculum is ripe for modelling as it is function driven, as aptly demonstrated by the work of Azarello, Pezzi, and Robutti (2004). Pre-calculus in the United States of America, for example, is about functions. The curricula in other countries are also ripe for including modelling but when this is attempted or suggested it can be seen by some as messing up functions. In Italy modelling is present in technical schools not lycées. Modelling is not seen as culturally noble. As mathematics educators we need to convince people of the cultural aspects of mathematical modelling – it is part of mathematics and therefore culturally relevant otherwise there will be modelling only in technical schools. In other countries such as Canada (Roulet & Suurtamm, 2004) and Australia (Stillman, 2001, 2004) secondary school mathematics curriculum documents in particular provinces and states have called for classroom practices involving both modelling and applications for some time. Actual practice varies. The new Ontario mathematics curriculum (Ontario Ministry of Education, 2000), for example, calls for modelling activities and the development of knowledge via investigations and modelling. In general there has been a move to more active learning but very limited investigation and modelling except for one course, Year 12 Mathematics for Data Management. This has been a relative success with modelling (usually with the aid of technology) taking a significant place in the classroom in contrast to the implementation of the new curriculum in the remainder of the secondary curriculum. In Alberta, applied mathematics is unique to the upper secondary school. Modelling is incorporated into the curriculum which is project based. In contrast to Artaud's findings in Europe, Chapman sees the key to success as teacher training to take teachers past their own fear to see mathematical modelling for its potential use as a pedagogical approach. Chapman (2004) "suggests that teachers' conceptions of mathematics, word problems and problem solving are important factors in creating a classroom culture to support modelling and application" (p. 70).

2. TASK AUTHENTICITY

The issue of task authenticity is difficult. For a task to be authentic for a student it must be part of the culture of the student who is not necessarily scholastic or even rational at this age. Two perspectives to authenticity form the basis of the teacher's dilemma of how to address this. Firstly, there is objective authenticity that comes from the real world. Secondly, there is subjective authenticity that comes from the situation being modelled being authentic to the modeller. For learning processes in classrooms this dilemma is complicated by the tasks being done in an institution. They are usually, though not always (see e.g., Clatworthy & Galbraith, 1989), imported into the classroom. Transformation into a classroom problem then becomes an issue for authenticity. Palm and Burman (2004) have demonstrated that it is possible to analyse the authenticity of tasks used in national assessments at this level using a framework (Palm, 2002) for developing, analysing, and reflecting on task authenticity. The degree of task authenticity needed is dependent on mathematical goals. Teachers at this level might feel more constrained in this regard as the heightened sense of duty to prepare students for final assessments and tertiary entrance influences the extent of task authenticity. For those teachers who break from this, designing of the didactical experience involves choosing suitable modelling situations. Not every situation has to be real nor are all real situations good models for this level. There is also the risk that students may come to believe that all that is real is good and all that is theoretical is bad. Furthermore, with real world models it is important that we sensitise students to the assumptions we make in real world situations – a point taken up by Jablonka (2004).

3. MODELLING COMPETENCY

Modelling is a life skill (Lakoma, 2004), if not "a way of life" (Lamon, Parker & Houston, 2003, p. ix), and at the upper secondary level the sense of responsibility to focus on this is heightened. There is a strong possibility that students will need to use this life skill sooner rather than later. At this level, social phenomena can be viewed from a more global perspective and the issues dealt with broader. However, the goal of development of a skill-set for life conflicts with preparation for tertiary mathematics. It is difficult for teachers to negotiate this. In general rather than academic education courses, mathematics is taught by modelling. This raises the dichotomy for mainstream classes: Are we teaching mathematics or are we teaching modelling? If the focus is said to be teaching modelling, then the argument is that too much time is spent on modelling at the expense of mathematics. We have to have strategies for teaching mathematics *by* modelling. We need to think about these two positions in a different way – a symbolic constructive way. When teaching mathematics by modelling we need to use a multidimensional approach by beginning with many ways that are initially parallel but then converge. In the curriculum there are many critical mathematical points along the way. We need to build a network around these points.

4. APPROPRIATE BALANCE

Questions about the appropriate balance in the curriculum between mathematical activities and applications and modelling activities stem from an incorrect metaphor. Instead, applications and modelling activities should be seen as embedded within mathematics and then together the two become an effective lens for describing and analysing the real world. Teaching function, for example, as a set definition has to be linked to something very useful. This in turn provides a better understanding of reality as exemplified by the work of Azarello, Pezzi, and Robutti (2004). Otherwise, the teaching of applications and modelling becomes the teaching of something foreign or impure in mathematics. However, teaching upper secondary mathematics without modelling and applications is just the same as visiting a museum where the exhibits are mathematical objects such as a group or a matrix.

It is important that applications and modelling not be seen merely as illustrating mathematical concepts but also as a means of deriving them. Mathematical concepts can be derived from the real world and then students are able to re-invest what they have learnt in the classroom back into the real world situation. This is the essence of a conceptual rather than an applied modelling approach.

5. ASSESSMENT

High-stakes assessment at the upper secondary–tertiary interface is often seen as an unresolved problem for the infusion of modelling into the secondary curriculum at this level as other imperatives are uppermost in the minds of teachers and students driven by the demands and whims of an external examination regime in many education systems across the world. Alternatively, high-stakes assessment can be seen "as a rare opportunity" (Burkhardt, 2004, p. 57) as "professional development activities built around high-stakes assessment are usually powerful, in that: They are taken up by most teachers, not just the enthusiasts" (p. 58). There are, however, ongoing tensions between advocates of centrally set external examinations to grade, compare, and rank students for tertiary entrance and those advocating more innovative approaches more compatible with modelling such as project examination (e.g., Antonius, 2004). The latter approach replaces, rather than supplements, the traditional external examination. Issues not fully resolved in more innovative assessment practices include difficulty of defining modelling competencies (although some progress has been made in this area, e.g., Henning & Keune, 2004; Legé, 2004), validity of teacher judgements in such assessments if used, and difficulty of evaluating or verifying the assessment. Authentic evaluation of current upper secondary assessment practices, traditional or innovative, is advocated through the use of longitudinal studies to supplement existing studies at this level (e.g., Stillman, 2002) so future planning and policy can be based on actualities not myths.

6. TECHNOLOGY

Technology use in upper secondary mathematics classes can be instrumental in deepening rather than limiting students' mathematical understanding. Technology allows more authentic modelling situations; however, Strässer (2004) warns this will not be an enriching experience if the modelling is done only by hiding the mathematics into "black boxes" as commonly happens with sophisticated workplace instruments. He cautions against taking too optimistic a view. Enlightening students of the contents of "black boxes" whenever possible is strongly advocated (e.g., Kadijevich, 2004). Another concern is the apparent tension between the goals of students becoming more effective modellers through technology use (Kutzler, 2000) and the enrichment of mathematics by using authentic examples tractable with technology at this level. Azarello, from his research with other colleagues (Azarello, Pezzi, & Robutti, 2004), sees connecting technology to a real situation as essential allowing students to develop their "cognitive activities from bodily (e.g., perceptual, kinetic) to theoretical features" (p. 25) typical of learning through motor-perceptors. Use of communication technology such as the worldwide web presents further opportunities for promoting modelling on the world stage. It allows the possibility of worldwide communication of suitable mathematical modelling examples at this level and the professional enrichment of teachers by their being able to communicate with others about what is happening at this schooling level in other places. In addition the dissemination of research results to others researching applications and modelling at the upper secondary level is facilitated.

REFERENCES

Antonius, S. (2004). Validity and competencies in modelling based project examination. In H.-W. Henn & W. Blum (Eds.), *ICMI Study 14: Applications and Modelling in Mathematics Education,* (pp. 9-16). Dortmund: University of Dortmund.

Artaud, M. (2004). Contributions from the anthropological theory of didactics. In H.-W. Henn & W. Blum (Eds.), *ICMI Study 14: Applications and Modelling in Mathematics Education,* (pp. 17-22). Dortmund: University of Dortmund.

Arzaello, F., Pezzi, G., & Robutti, O. (2004). Modelling body motion: An approach to functions using measure instruments. In H.-W. Henn & W. Blum (Eds.), *ICMI Study 14: Applications and Modelling in Mathematics Education,* (pp. 23-28). Dortmund: University of Dortmund.

Burkhardt, H. (2004). Establishing modelling in the curriculum: Barriers and levers. In H.-W. Henn & W. Blum (Eds.), *ICMI Study 14: Applications and Modelling in Mathematics Education,* (pp. 53-58). Dortmund: University of Dortmund.

Chapman, O. (2004). Teachers' conceptions and teaching strategies that facilitate mathematical modelling. In H.-W. Henn & W. Blum (Eds.), *ICMI Study 14: Applications and Modelling in Mathematics Education,* (pp. 65-70). Dortmund: University of Dortmund.

Clatworthy, N.J. & Galbraith, P.J. (1989). Mathematical modelling: An attack on conformity. *Australian Senior Mathematics Journal 3*(2), 88-106.

Henning, H. & Keune, M. (2004). Levels of modelling competencies. In H.-W. Henn & W. Blum (Eds.), *ICMI Study 14: Applications and Modelling in Mathematics Education,* (pp. 115-120). Dortmund: University of Dortmund.

Kadijevich, D. (2004). How to attain a wider implementation of mathematical modelling in everyday mathematics education?. In H.-W. Henn & W. Blum (Eds.), *ICMI Study 14: Applications and Modelling in Mathematics Education,* (pp. 133-138). Dortmund: University of Dortmund.

Kutzler, B. (2000). The algebraic calculator as a pedagogical tool for teaching mathematics. In A. Gagatsis & G. Makrides (Eds.), *Proceedings of the 2nd Mediterranean Conference on Mathematics Education,* (pp. 142-160). Nicosia, Cyprus.

Lakoma, E. (2004). Modelling in mathematics education – The case of probability and statistics teaching. In H.-W. Henn & W. Blum (Eds.), *ICMI Study 14: Applications and Modelling in Mathematics Education,* (pp. 139-144). Dortmund: University of Dortmund.

Lamon, S., Parker, W., & Houston, K. (2003). Mathematical modelling: A way of life. In S. Lamon, W. Parker, & K. Houston (Eds.), *Mathematical Modelling: A Way of Life*, (pp. ix-x). Chichester: Horwood Publishing.

Legé, J. (2004). "To model, or let them model?" That is the question! In H.-W. Henn & W. Blum (Eds.), *ICMI Study 14: Applications and Modelling in Mathematics Education,* (pp. 145-150). Dortmund: University of Dortmund.

Ontario Ministry of Education (2000). *The Ontario curriculum: Grades 11 and 12 Mathematics 2000.* Toronto.

Palm, T. (2002). *The Realism of Mathematical School Tasks – Features and Consequences* (Research reports, No. 2, in Mathematics Education). Umeå, Sweden: Department of Mathematics.

Palm, T. & Burman, L. (2004). Reality in mathematics assessment: An analysis of task-reality concordance in Finnish and Swedish national assessments. *Nordic Studies in Mathematics Education 9*(3), 1-33.

Roulet, G. & Suurtamm, C. (2004). Modelling: Subject images and teacher practice. In H.-W. Henn & W. Blum (Eds.), *ICMI Study 14: Applications and Modelling in Mathematics Education,* (pp. 229-234). Dortmund: University of Dortmund.

Stillman, G. (2001). The impact of school-based assessment on the implementation of a modelling/applications-based curriculum: An Australian example. *Teaching Mathematics and Its Applications 20*(3), 101-107.

Stillman, G. (2002). *Assessing Higher Order Mathematical Thinking through Applications.* Unpublished doctoral dissertation, University of Queensland, Australia.

Stillman, G. (2004). Sustained curriculum change: The example of the implementation of applications and modelling curricula in two Australian states. In H.-W. Henn & W. Blum (Eds.), *ICMI Study 14: Applications and Modelling in Mathematics Education,* (pp. 261-266). Dortmund: University of Dortmund.

Strässer, R. (2004). Everyday instruments: On the use of mathematics. In H.-W. Henn & W. Blum (Eds.), *ICMI Study 14: Applications and Modelling in Mathematics Education,* (pp. 267-272). Dortmund: University of Dortmund.

---

[1] Valued contributions to a discussion provided by: Michèle Artaud, Ferdinando Arzarello, Olive Chapman, Dirk De Bock, Solomon Garfunkel, Jerry Legé, Vimolan Mudaly, Geoffrey Roulet, Akihiko Saeki, and Rudolf Strässer.

Chapter 4.4

TEACHING APPLICATIONS AND MODELLING AT TERTIARY LEVEL

Claudi Alsina[1]
E.T.S. Arquitectura de Barcelona. Technical University of Catalonia (UPC), Email: claudio.alsina@upc.edu

Abstract: This paper deals with the role of learning mathematics in context at tertiary level using modelling and applications, integrating technology and changing assessment practices. This opens interesting research issues.

1. INTRODUCTION

We are convinced of the importance of extending the applications and modelling approach to mathematics at tertiary level. This report shows why.

We concentrate our contribution on the following set of questions: the role of learning in context at tertiary level, the importance of modelling and applications in training mathematicians as well as all kind of university students, the role of technology, best assessment practices and research issues arising in this field from the educational point of view. We quote also some cases which have special interest for implementing modelling experiences in university courses and list some references which may be consulted by the readers.

Let us mention that in the previous ICMI Study on the Teaching and Learning of Mathematics at University Level the whole meeting held in Singapore was devoted to this level and some relevant contributions to our topic were already made (Holton, 2001). From this publication we will quote some papers, but we refer the reader to the proceedings for further general views and examples on university mathematics.

2. LEARNING IN CONTEXT AT TERTIARY LEVEL

All mathematization processes, and modelling in particular, start from real world problems and end in checking if the model's results answer, in a satisfactory way, the initial problem. This is why context is important for our considerations.

Learning in context at university level has shown to be a fruitful academic strategy either for providing motivation and interest or for engaging students in real world problems. There is no need to say that choosing a correct context is a critical matter since context depends on the area where students are involved (engineering, architecture, social sciences, life sciences,...). Context must have a real meaning for the students and not to require knowledge that students are not familiar with. Experiences show that, usually, one obtains better responses and achievements when dealing with modelling and applications within suitable contexts than when delivering lectures in a more abstract setting.

3. ON THE ROLE OF APPLICATIONS AND MODELLING IN THE TRAINING OF MATHEMATICIANS

University students, oriented towards mathematics, may benefit from learning modelling procedures. These are important aspects of the mathematics training. For example all graduates need to understand principles, techniques and applications of basic areas like statistics, operations research or computing. Student may choose to become involved in applied mathematics, but even in the case of a pure mathematics orientation they may find motivations and orientations when following a course on modelling. Through learning modelling, they may understand how many mathematical concepts and structures originated and about their applications outside math.

Of course, if students consider the possibility of becoming teachers (at any level) then some background on applications and modelling will be of great interest for their future role, as we will discuss later. In any case, modelling and applications may serve future mathematicians for the development of mathematical thinking, fostering creativity and intuition.

The MCM (The Mathematical Contest in Modelling) and the ICM (The Interdisciplinary Contest in Modelling) organized by COMAP (The Consortium for Maths and its Applications) have been a growing worldwide experiences where many mathematics students are involved, showing their interest in this field (see http://www.comap.com/undergraduate/contests/mcm).

4. ON THE ROLE OF APPLICATIONS AND MODELLING IN MATHEMATICS AS A SERVICE SUBJECT

When one faces the design of mathematics courses where mathematics is a service subject there is a need to connect the mathematics program with the examples and applications that belong to the university study hosting such courses. The main concern about this need of connections is firstly to provide motivation for learning mathematics, and arise the interest in these conditions. Instead of answering to the chilling question "what is all this stuff good for?" is to start topics by presenting concrete applied problems. Not to go into difficult modelling questions but explain appropriate applications (Krantz, 1993). Later, these connections may also facilitate the competency of applying mathematics to a concrete field of interest, improving the capacity of mathematization when facing specific problems.

In some university studies like engineering, biology, physics, chemistry, etc. maintain an old tradition in requiring quite a high level of mathematics having their students know the importance of mastering some mathematics for following their own training. In other fields like social sciences, design, architecture, etc., the quantity of mathematics needed or required depends on the aims of the study.

Let us conclude this part by mentioning some examples of illuminating actions of promoting the role of applied mathematics and modelling in mathematical courses for non-mathematicians. Among these are the English LTSN (Learning and Teaching Support Network) engineering case studies (see http://mathstore.ac.uk), the HELM project http://www.lboro.ac.uk/-research/helm.

5. TECHNOLOGY, MODELLING AND APPLICATIONS

Among the Information Technologies (IT) we find today a wide range of interesting programs and devices that may facilitate mathematical modelling and, in particular, its learning. Sophisticated calculators with high numerical and graphical capabilities, advanced mathematical software (Mathematica, MuPad, SciLab, Matlab, Maple, Derive,...), graphing programs (CAD, Cabri, Geometer's Sketchpad, Cinderella,...), statistics software (APSS, SASS,...), Internet,... constitute interesting materials to be used to make either more advanced modelling (without the numerical and visual limitations of the recent past) or more professional projects. IT engender a professional approach to work because while secondary level software and calculators are teaching-oriented, university level IT can be used as professional tools.

In a web environment students may find new experiences in learning, in including multimedia elements in their projects (Kadijevich, 2002b), in sharing improved interactions with experts, in accessing significant information, etc., bearing in mind critical attitudes towards Internet's contents and fighting against plagiarism of projects.

An interesting example of applied mathematics and technology is, e.g., the robots competitions made by computer science students in many different universities (Technion-Israel Institute of Technology, Universitat Politècnica de Catalunya,...) where it is necessary to integrate multidisciplinary skills developing a remarkable acquisition of spatial skills, programming techniques, use of computers algebra systems, etc. in order to build and control robots making concrete performances, Industrial-directed modelling courses and computer-based modelling activities are also interesting modelling courses related to technological issues.

There are also interesting uses of technology in quantitative literacy courses on statistics (e.g., the course "Chance" at Dartmouth initiated in the 90's. the Web site "Chance and Data in the News", The Chance Web site, etc.) as well as in many other university studies like economics, political sciences, pedagogy, linguistics, (e.g. in automatic translation).

6. ASSESSMENT BASED UPON MEASURING COMPETENCIES IN MODELLING AND APPLICATIONS

Teaching on modelling and applications opens an excellent opportunity for revising the traditional assessment of coursework and written examinations and go into the fruitful collection of good assessment practices that have been implemented at tertiary level: open tasks, project-portfolios, project development, journal writing, group project work, comprehension tests, self-assessment, interactive testing in the web, achievement tests, attitude questionnaires, interviews. There has been a lot of progress in developing and evaluating robust methods for assessing all kinds of student project work and associated communication skills, e.g., the work done in UK by the Assessment Research Group (Haines & Houston, 2001). Another tool which we can mention is FLAG (Field-Tested Learning Assessment Guide) a Web site offer developed by NISE (National Institute for Science Education at University of Wisconsin-Madison) where well designed assessment materials are available (Ridgway, Swam and Burkhardt, 2001).

To find a variety of theoretical aspects of assessment as well as many practical cases we recommend (Niss, 1993), a volume which originated from the ICMI Study dedicated to research on assessment.

7. RESEARCH AND LEARNING BY MODELLING AT THE UNIVERSITY LEVEL

There are two different aspects concerning research that we need to mention. On one side, *research in mathematics education* has shown that the success of the modelling approach in mathematics at tertiary level does exist. For example one of the finding that we can recall here is that there is scientific evidence that student learn better in context. Thus the emphasis on context, on applied problems, on mathematization of the real world, etc., may be a positive step towards learning success.

On the other side, the so-called research approach to learning has shown to be a valuable strategy. When students are involved in open questions, in exploratory tasks, on finding data, etc., they become more interested and more involved. In this direction applications and modelling at tertiary level offers an excellent framework for research oriented activities like project based education. Such projects give also the opportunity to be integrated with other disciplinary projects in architecture, engineering, social sciences, etc. Stimulating initiatives (many of them on engineering education) have been recently made (see, e.g., (Perrenet, 2000), http://anaw3.ana.anc.dk/fak-tekn/fins.htm), e.g., in Technische Universiteit Eindhoven, Universiteit Twente, Aalborg University, Roskilde University, Technion-Israel Institute of Technology, etc. To assure a high educational quality of university teachers we will need to provide adequate training to those entering the profession. Moreover we will need to explore new iniciatives in life-long-learning.

REFERENCES

Crouch, R.M., & Haines, C.R. (2003). *Do you know which students are good mathematical modellers? Some research developments.* Technical Report No.83, Department of Physics, Astronomy and Mathematics, University of Hertfordshire. Hatfield: University of Hertfordshire.

Crouch, R.M., & Haines, C.R. (2003). Mathematical modelling: transitions between the real world and the mathematical model. *International Journal of Mathematics Education in Science and Technology,* accepted for publication 16.9.03.

DeLong, M., & Winter, D. (2002). *Learning to teach & teaching to learn Mathematics. Resources for professional development.* MAA Notes #57.

Galbraith, P.L., & Haines, C.R. (1998). Disentangling the nexus: attitudes to mathematics and technology in a computer learning environment. *Educational Studies in Mathematics, 36*, 275-290.

Galbraith, P.L., & Haines, C.R. (2000). *Mathematics-Computing Attitude Scales. Monographs in Continuing Education.* London: City University.

Haines, C.R., & Crouch, R.M. (2003). Ability and competence frameworks for mathematical modelling and applications. In H.-W. Henn, & W. Blum (Eds.), *ICMI Study 14: Applica-*

tions and Modelling in Mathematics Education, (pp. 103-108). Dortmund: University of Dortmund.

Haines, C.R., & Dunthorne, S. (Eds.). (1996). *Mathematics Learning and Assessment: Sharing Innovative Practices.* London: Arnold, (with video).

Haines, C.R., & Houston, K. (2001). Assessing student project work. In (Holton, 2001, pp. 431-442).

Heuvel-Panhizen, M. van den (2003). The didactical use of models in realistic mathematics education: An example from a longitudinal trajectory on percentage. *Educational Studies in Mathematics, 54*(1), 9-36.

Holton, D. (Ed.) (2001). *The Teaching and learning of Mathematics at University Level: An ICMI Study.* Dordrecht: Kluwer Academic Publishers.

Houston, K. (2001). Assessing undergraduate mathematics students. In (Holton, 2001, pp. 407-422).

Izard, J., Haines, C.R., Crouch, R., Houston, K., & Neill, N. (2003). Assessing the impact of teaching modelling: some implications. In S.J.Lamon, W.A.Parker, & S.K.Houston (Eds.), *Mathematical Modelling: A way of life: ICTMA11,* (pp. 165-178). Chichester: Ellis Horwood.

Kadijevich, D. (2002a). Four critical issues of applying educational technology standards to professional development of mathematics teachers. *Proceedings of the 2nd ICTM.* Crete, Greece: University of Crete.

Kadijevich, D. (2002b). Developing multimedia lessons in the pre-service development of mathematics teachers. In Mendez-Vilas, A., Gonzales, J.A.M., & Solo de Zaldivar, I. (Eds.), *Proceedings ICTE2002, Vol. I* (pp. 460-463). Badajoz : CECT de la J. Extremadura.

Krantz, S.G. (1993). *How to teach mathematics. A personal perspective.* American Mathematical Society.

Niss, M. (Ed.). (1993). *Investigations into Assessment in Mathematics Education – An ICMI Study.* Dordrecht: Kluwer Academic Publishers.

Perrenet, J.C., Bonhuijs, P.A.J., & Smits, J.G.M.M. (2000)The Suitability of Problem-based Learning for Engineering Education, Theory and Practice. *Teaching in Higher Education, 5*(3), 345-358.

Ridgway, J., Swann, M., & Burkhardt, H. (2001). Assessing mathematical thinking via flag. In (Holton, 2001, pp. 423-430).

Shotsberger, P.G. (1999). The INSTRUCT Project: Web Professional Development for Mathematics Teachers. *Journal of Computers in Mathematics and Science Teaching, 18*(1), 49-60.

---

[1] I am grateful to the participants in the working group for their substantial contribution during the session we had in Dortmund and afterwards. Before we met in Dortmund the working group had already some e-mail exchanges in order to fix a collection of issues to be faced. We had a fruitful discussion about them during the ICMI Study and afterwards. In this final report I have tried to sum up our main conclusions.

Chapter 4.5

MODELLING IN TEACHER EDUCATION

Thomas Lingefjärd
University of Gothenburg, Gothenburg, Sweden, Email: thomas.lingefjard@ped.gu.se

Abstract: To discuss mathematical modelling in teacher education requires simplifications of both the concept of modelling as well as of the concept teacher education. Although there seems to be major obstacles and reluctance to implement mathematical modelling into teacher educations, there are several good examples of what can be done and of what has been done.

1. INTRODUCTION

During the Study conference in Dortmund, February 2004, several of the participants were teacher educators and a handful of the contributions were based on teacher education experiences. Consequently, there were plenty of occasions to explore different opinions on the concept of mathematical modelling and teacher education during the study conference. There was an overall agreement that although national curricula around the world might propose recommendations towards the use of mathematical modelling and applications in teacher education, the perspectives offered on how they could be realized are in general not well developed. The emphasis is mostly on the rhetoric level and does not define the program of study or the pathways to achieving mathematical modelling standards in teacher education.

The issue of mathematical modelling in teacher education calls for a definition: Who are really the "teacher educators?" Teacher education is not a coherent body of mutual principles, policies, rules and regulations, and content. Not between countries, and not even within countries. To address a group of professionals as "teacher educators" is an indistinguishable term, which very well might address people from quite different areas or branches like psychology, general pedagogy, mathematics and mathematics education.

The term teacher education also conceal a variety of levels like pre-school, elementary, middle, lower secondary, secondary, and upper secondary. In Northern Europe and possibly elsewhere, we define a school level by the name gymnasium. In Sweden this corresponds to grade 10 – 12. All together the rubric Mathematical modelling in Teacher education involves a vast variety of directions, levels, programs, content, context, and so on – all with different emphasizes on the scale from pedagogy and methods to rigor mathematics.

Consequently, just a short chapter can not address all these differences, obstacles and possibilities. Both generalizations and simplifications are necessary in order to actually arrive to some mutual agreement or understanding. The group decided that the phrase modelling normally means "real word problems" or "real world situations" both at the elementary and secondary level of teacher education, with the difference that such situations or problems often requires technology for proceeding through the modelling process in the training of prospective secondary mathematics teachers.

In the call for contributions to the ICMI *Study on Applications and Modelling in Mathematics Education,* it was stated that mathematics teacher education programmes rarely include orientations to modelling, and the use of the modelling process in mathematics courses. One obstacle towards such an implementation could be the limited mathematics background of primary school student teachers and the limited time available for mathematics in their education. But relevant preparation is rare also for prospective secondary teachers of mathematics, as demonstrated by several papers and discussions at the conference. Another obstacle could very well be that mathematical modelling often requires the use of technology, something many mathematicians and mathematics educators who are engaged in teacher preparation, might be reluctant to use.

As reported by Lingefjärd in Chapter 3.5.2 in this volume, the situation of mathematical modelling in the training of teachers is a complex situation and even if there are national curriculum texts like the Swedish compulsory and gymnasium curriculum, which emphasizes mathematical modelling and teaching of applied problems throughout grades 1 – 12, teacher education might very well be slow and resistant to follow. This fact is connected to the reality that Mathematical modelling is *not* a body of mathematical knowledge in the same way that Calculus or Differential Equations are, but rather a small collection of general principles which experience nevertheless has proved to be helpful in the process of applying mathematical know-how to analyze problems that arise in various non-mathematical disciplines.

2. THE SITUATION OF MODELLING IN TEACHER EDUCATION

A review of available literature on modelling, technology and teacher preparation, highlight the notion that effective infusion of modelling in schools will take place if some form of modelling is provided for teachers in the course of their preparation. According to Barron and Goldman (1994), the notion of technology's use as a cognitive tool which engages students in authentic and challenging modelling tasks should become one of the foci of teacher education. Knapp and Glenn (1996) argued that effective modelling with the aid of technological tools could enhance teaching and learning and therefore should be an integral constituent of teacher education programs. According to them, the key component in fostering change is for teacher educators to illustrate appropriate modelling activities and technology use in classroom and curricula, and for future teachers to have frequent opportunities to practice using mathematical modelling and technology as both the learning and teaching means. Hoffmann (1996) argued that schools of teacher education can better prepare teachers if university education faculty use modelling and technology throughout the teachers' own education.

The findings of research conducted by Thomas and colleagues (1996), as well as Wildmer and Amburgey (1994), substantiate the proposition that training of teachers in the use of technology should incorporate modelling by instructors in all their courses. Moreover, they specifically suggest that the technological training of teachers should be incorporated in content related assignments. A number of other research studies have documented the importance of modelling and demonstration of technology assisted instruction in the context of subject matter preparation of future teachers (Please see the reference list of Lingefjärd in Chapter 3.5.2 of this volume). See also Chapter 2.4 of Doerr in this volume.

Several participants at the Dortmund conference also claimed that teaching modelling and using technology in mathematics instruction requires a body of knowledge that is specialized in nature. This knowledge includes pedagogical presentations of mathematical concepts using appropriate software, and problem posing and questioning techniques that motivate productive use of modelling and technology within the learning environment. There was no doubt about the fact that use of modelling and technology in instruction with the purpose of enriching students' mathematical learning is valued by future teachers if they are convinced of their impact on their own learning of the content. Therefore, in order for modelling to become a part of a teacher's functioning and practice, experiences provided for them in the course of their own mathematics learning should assist them in constructing an image of the teaching and learning that is enhanced by modelling.

The process of education of teachers with the aim of learning mathematical modelling with use of technology should be viewed as a *gradual process achieved over a long period* of time. The relevant knowledge required of mathematics teachers can not be met as the result of exposure to modelling and different technological tools in just one generic course in instructional technology. Instead, modelling and technology should take place within both the teachers' content and methods courses. This, in turn, call for collaborative efforts, joint planning, and a shared vision of mathematical modelling as something useful, beneficial, and valuable in mathematics and mathematics methods courses.

3. GOOD EXAMPLES FROM LITERATURE

Although the progress of mathematical modelling within teacher education has been slow, it is not because of lack of resources. Within the English speaking community, the ICTMA books content a vastly rich source of good examples for different levels and programs in teacher education (for a full list of all ICTMA books, see Bibliography in Part 6 of this Volume). The books are so far always a selection of papers that has been prepared for and presented at the ICTMA conference. Nevertheless, it was not until the volume *Mathematical Modelling Courses* from 1987, when a section labelled Modelling and Schools was entered. The volume *Applications and Modelling in Learning and Teaching Mathematics* from 1989 has two interesting sections for teacher education, namely applications and modelling at the lower secondary level as well as applications and modelling at the upper secondary and the tertiary level, together with an emphasis on the fundamental reasons for teaching mathematical modelling and on assessment. The volume *Teaching of Mathematical Modelling and Applications*, published in 1991, also contains sections on how to teach modelling at the lower secondary, upper secondary and tertiary levels.

In 1993 the volume contained the sections General and Philosophy, New Topics and Tools, Case Studies, and Curriculum and Assessment. Assessment seemed to be the major theme of the sixth ICTMA conference. I consider several of the articles under each category to be useful in teacher education, especially when talking about the objectives to *why teach* mathematical modelling. The volume in 1997, *Teaching and Learning Mathematical Modelling*, has under the section Tertiary courses at least two articles on courses in mathematical modelling for teacher education. The use of technology in the modelling process also started to be a more frequently discussed topic in the ICTMA books by the end of the 1990's.

The 8th ICTMA conference lead to the volume *Mathematical Modelling, Teaching and Assessment in a Technology-Rich World*, with obviously many papers concerning the effect on learning and teaching when processing mathematical modelling with the aid of technology. I do not exaggerate when I claim that each of the papers in this volume could be discussed in teacher education, at some level, for some prospective mathematics teachers.

From this volume and on, each ICTMA volume contents articles, perspectives, and ideas that easily correspond with mathematics teacher education at different levels. Sometimes the ideas and problems naturally need to be slightly changed or altered and/or translated.

EXAMPLE 1

For the interested elementary teacher, or elementary teacher educator, there is a whole section in the 11th ICTMA book, called *Modelling in the Elementary school*, and I'm especially fond in the so called *Tractor problem* on page 26 (Lamon, 2003).

- When the framer drives this tractor from one end of the field to the other, will both wheels cover the same distance?

The text is accompanying by a sketch of a tractor with large rear wheels and small front wheels. What is especially interesting with this problem is that it essentially describes a situation and gives no particularly hints about how to proceed.

EXAMPLE 2

For the more advanced prospective elementary or middle school teacher in mathematics or for the corresponding mathematics educator, I like to recommend a modelling problem which I have used with quite some success in my own teaching. It originates from an idea initiated by João Filipe Matos in the 8th ICTMA book (pp. 21 – 27). Matos' paper is about the discussion of group of students in the 10th grade around a log glass and I have for the most occasions transformed the problem into a quite similar problem about an hourglass. Imagine an hourglass constructed by two identical conical vessels united by their vertices where a small tube is inserted and through which the sand is pouring when the hourglass is put upside down.

Given the volume and height of each conical vessel, as well as the draining velocity through the tube, teacher students can be asked many different questions that require them to set up a model of the situation:

- How fast is the "sand mountain" in the lower conical vessel growing?
- How is this speed connected to the draining velocity?
- How would you grade the walls of the hourglass two conical parts so that the elapsed time can be readable from both vessels?

After my students have done this as an exercise during class hours, or as a take-home assignment, I usually distribute Matos article and let the students discuss similarities and differences between the way they and the way Matos' students solved this modelling task.

EXAMPLE 3

The idea for this problem was presented at the 9th ICTMA conference. I must admit that John Masons plenary lecture gave me so many inspiring ideas, that I haven't used all of them yet. Nevertheless, I do have used the problem of a tilted Coca-Cola can and of the moving centre of gravity when a Coca-Cola can is emptied. I have addressed the following questions to several prospective secondary teachers of mathematics (Lingefjärd, 2002a, p. 77-78). Please note that the first thing the students have to do is to control if the given set of information is realistic:

- Imagine a regular full Coca-Cola can. The European can hold 0.33 litres of liquid, and if drilling a small hole in the bottom of the can and tear off the top flap, the Coca-Cola begins to flow out at the rate of 0.5 cm^3/sec. The can is positioned and the hole drilled in a way that enables all the Coca-Cola to flow out.
- The centre of gravity of the system (can + Coca-Cola), which at the beginning is located at the centroid of the can, gradually moves downward and returns to its central position when the can is empty.
- Construct a mathematical model that describes this movement of the centre of gravity over time. Illustrate the model with a diagram, and try to calculate the lowest possible position of the centre of gravity in the Coca-Cola can as accurately as possible.
- When containing a certain amount of Coca-Cola or a similar liquid, the can may be balanced on the bottom edge (see Fig. 4.5-1). For what amount of Coca-Cola is this performance possible?

Figure 4.5-1. The balance of a tilted Coca-Cola can.

The fact that the centre of gravity moves when a canister is emptied is a challenging and engaging concept for teacher students. I claim that this is a most engaging problem, and that students who are working on the process of creating a mathematical model will most likely be engaged in stimulating discussions, moving their mathematical knowledge forward. The problem can be given with any sort of canister, but only certain cans can be tilted. The article is available online from *Teaching mathematics and its Applications.*

EXAMPLE 4

The following example was published and discussed in the 10th ICTMA book. It covers a more realistic problem compared to many other problems, at the same time as the very situation becomes more complex and difficult to model. But the authors claim some success with using this problem, in a modelling course for prospective secondary teachers.

A modelling situation (from Edwards & Hamson, 1996, pp. 110-111, with our additions and modifications).

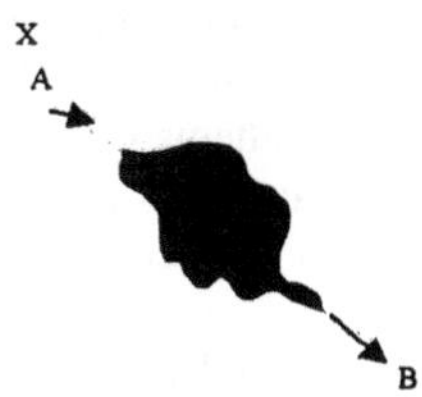

Figure 4.5-2. The pollution of a sea

- Fig. 4.5-2 represents a small lake. Although the lake is receiving and losing water in different ways, we simplify the situation and say that water flows in through stream A and out through stream B.
- At a certain time of the day, as a result of a road accident, a petrol truck overturns and spills a toxic chemical into the stream A at position X.
- Thirty minutes later the police and emergency services have brought the situation under control, and an unknown amount z (m^3) of the toxic chemical has leaked into the lake.

Develop a mathematical model that you can use to predict the concentration of the pollutant in the lake at any time and use it to estimate (for a range of possible initial pollution amounts z):

a) The maximum pollution level in the lake and the time at which the maximum is reached.
b) The time it will take for the pollution to fall below the safe level of 0,05%.
c) How will your results be affected if a constant rain starts at the same time as the accident? The rain covers the whole geographic area.

EXAMPLE 5

I will end this small survey of problems used in mathematical modelling courses in teacher education by addressing another source of modelling examples. The problem concern measuring the heart's cardiac output of a person taken in for hospital care and reveals some of the mathematics hidden behind machines and procedures in everyday life (Lingefjärd, 2002b). The measuring of cardiac output involves a lot of advanced mathematical concepts, such as integrals, differential equations, logarithms, etcetera. The cardiac output problem, and how teacher students act when modelling it, was published in *The International Journal of Computers for Mathematical Learning*. This journal can also be reached online.

So the resources in terms of problems and described experiences are there for you – you just have to reach out and get it!

REFERENCES

Barron, L., & Goldman, E.S. (1994). Integrating technology with teacher preparation. In B. Means (Ed.), *Technology and Education Reform*. San Francisco: Jossey-Bass.

Holmquist, M., & Lingefjärd, T. (2003). Mathematical modelling in teacher education. In Q. Ye, W. Blum, S. K. Houston, & Q. Jiang (Eds.), *Mathematical Modelling in Education and Culture ICTMA 10: Applications in Science and Technology* (pp. 197-208). Chichester: Ellis Horwood.

Matos, J., F. (1998). Mathematics learning and modelling: Theory and practise. In P. Galbraith, W. Blum, G. Booker, & I. Huntley (Eds.). *Mathematical Modelling, Teaching and Assessment in a Technology-Rich World.* (pp. 21-28). Chichester: Ellis Horwood.

Matos, J., F., Blum, W., Houston, S., K., Carreira, S., P. (2001). *Modelling and mathematics Education ICTMA 9: Applications in Science and Technology.* Chichester: Ellis Horwood.

Ye, Q., Blum, W., Houston, S., K., Jiang, Q. (2003). *Mathematical Modelling in Education and Culture: ICTMA 10.* Chichester: Ellis Horwood.

Lamon, S. (2003). Modelling in Elementary School: helping young students to see the world mathematically. In S: Lamon, W. Parker, K. Houston (Eds.). *Mathematical Modelling: A way of Life: ICTMA 11.* (pp. 19-33). Chichester: Ellis Horwood.

Knapp, L. R., & Glenn, A. D. (1996). *Restructuring schools with technology.* Boston, MA: Allyn and Bacon.

Lingefjärd, T. (2002a). Teaching and assessing mathematical modelling. *Teaching mathematics and its Applications 21*(2), 75-83.

Lingefjärd, T. (2002b). Mathematical modelling for preservice teachers: A problem from anesthesiology. *The International Journal of Computers for Mathematical Learning* 7(2), 117-143.

Thomas, L., Larson, A., Clift, R., & Levin, J. (1996). Integrating technology in teacher education programs: Lessons from the teaching teleapprenticeship project. *Action in Teacher Education, 17*(4), 1-8.

Wildmer, C., & Amburgey, V. (1994). Meeting technology guidelines for teacher preparation. *Journal of Computing in Teacher Education, 10*(2), 12-17.

Part 5

Cases in Applications and Modelling

Chapter 5.1

MOVING THE CONTEXT OF MODELLING TO THE FOREFRONT: PRESERVICE TEACHERS' INVESTIGATIONS OF EQUITY IN TESTING

Katie Makar and Jere Confrey
University of Queensland, Australia, Email: k.makar@uq.edu.au / Washington University in St. Louis, U.S.A., Email: jconfrey@wustl.edu

Abstract: Prospective math and science teachers were engaged in interpretation and analysis of testing data with the innovative statistical software Fathom™. Armed with a few basic statistical concepts, their ability to model complex issues was correlated with their personal engagement in the process and context at the forefront of the modelling activity.

1. INTRODUCTION

Modelling activities often neglect a major purpose of modelling: to gain insight into a problem context. Commonly, modelling instruction develops students' modelling skills first, and then gives them opportunities to apply these skills to situations. The major focus is on modelling; the context is secondary. In contrast, we chose to move the context to the forefront, assisting learners in using modelling as a tool for inquiry.

A major challenge to widespread use of modelling and applications in secondary schools is teachers' unfamiliarity with it. For teachers to use innovative practices with students, they must have experiences with such practices as learners. We designed a study where prospective secondary mathematics and science teachers could gain experience modelling in a compelling context. The study was conducted during a course on assessment and data analysis at the University of Texas. Many of these teachers plan to teach in

Texas where there is significant controversy over the effects of high states testing on educational equity.

Our purpose of bringing the context to the forefront of modelling experiences for prospective teachers was three-fold. First, we believed that learners who engaged in statistical inquiry with a compelling purpose would gain a deeper understanding of data analysis and modelling. Second, we were concerned about ways we saw schools interpret data from statewide tests and the strategies they used to raise test scores, often at the expense of underperforming students (Confrey & Makar, 2005); alternatively when testing is properly used, it can help ensure learning opportunities for all students. By heightening teachers' awareness of the context of accountability, we hoped they would be more sensitive to equity issues in testing and advocate for more equitable and systemic strategies for improving test scores in their schools. Finally, we were interested in strengthening mathematics and science teachers' knowledge of statistics, experience with modelling and inquiry, and facility with technology. Our aim was to provide them first-hand experience using a student-centered software tool to strengthen their understanding of basic statistical concepts in a context that would be valuable to them as professionals, and to help them understand the opportunities and challenges of including modelling experiences for their own students. This paper examines preservice teachers' data modelling; other aspects of the study are published elsewhere (Confrey, Makar, & Kazak, 2004; Makar & Confrey, under review).

2. BACKGROUND AND STUDY CONTEXT

The subjects in the study were eighteen prospective math and science teachers with varying backgrounds in statistics (half had no previous statistics coursework) enrolled in a course on assessment developed by the authors, and organized into three parts: an introduction to high stakes testing, classroom assessment, and concepts of central tendency and distribution using hands-on activities and explorations; association and basic linear regression including an opportunity to teach these topics at a local high school; and guided data inquiries culminating in an independent inquiry.

We assumed that the process of modelling in a relevant, complex setting would promote a richer conception of the statistical concepts of variation, distribution, and data comparisons. Content also included descriptive statistics, linear regression, and informal inferential reasoning (conditional statements, null hypothesis, sampling distributions). The purpose was for teachers to use these tools to bring to light issues of equity in testing rather than cover a complete course in statistics. The prospective teachers discussed how a particular statistical model could highlight an important issue in interpreting test scores. The concept of distribution was a key component of discussions, not as mathematical objects, but as tools for inquiry and debate.

The data analysis software used in the course, Fathom™ (Finzer, 2001), was designed for high school and tertiary students to *learn* statistics by *doing* it, and was software that the prospective teachers could later use with their own students. Fathom was central to allowing the teachers to experience statistical inquiry as learners before employing inquiry-based instruction. Unlike most statistical packages that act as a black box (data in, results out), Fathom is a dynamic statistical package that allows for a more visual, less formal approach to data analysis. Users can quickly create linked graphs, developing inferential reasoning by checking hunches without formal statistical tests. For example, if economically disadvantaged students are selected (Fig. 5.1-1, bar graph) in a graph of authentic data from a local middle school, these same students' data are highlighted in another graph (dot plot) showing their mathematics scores on the Texas Assessment of Academic Skills (TAAS). This display shows test performance widely distributed for economically disadvantaged students, countering assumptions that students from low socio-economic backgrounds do poorly in school.

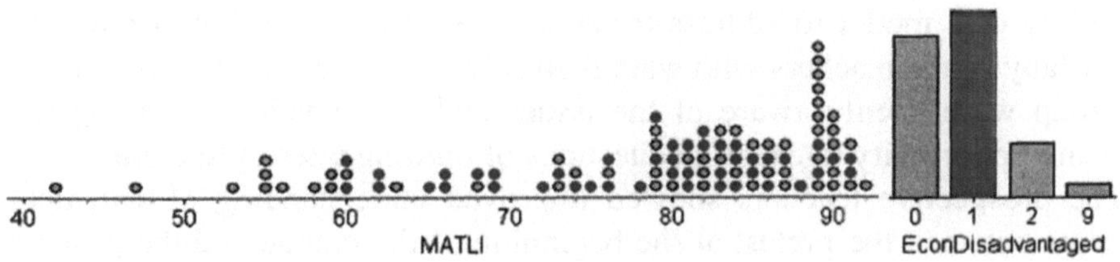

Figure 5.1-1. Student mathematics scores on TAAS with economically disadvantaged students highlighted

The prospective teachers had multiple opportunities to create conjectures, interpret data relationships, and develop facility with features of the software in shorter semi-structured inquiries throughout the semester. At the end of the course, they conducted a three-week, independent inquiry into a specific area of interest in testing and accountability related to equity. Topics of equity, diversity and assessment were major objectives in the teacher preparation program, so this task related directly to what was expected of prospective teachers. In addition to their oral presentations and written reports, research data included a prepost test of statistical concepts, video-taped data investigations, classroom discussions and observations, and artifacts collected from their self-designed data-based inquiry.

3. RESULTS AND DISCUSSION

This section will look at relevant class activities, results of pre-post tests in statistical content, and the relationship between the context of the teachers' modelling experience and their facility to create statistical models.

The context of equity in testing was uncomfortable for these prospective teachers. Since this was a major course theme, they were frequently asked to consider their own tacit beliefs through discussions and reflection papers. Initially, discussions were met with silence or safe statements repeating claims of class readings. Issues of race were especially marginalized or avoided. Many of the prospective teachers revealed strong stereotypes that they struggled to both articulate and overcome. In comparing gender performance on a mathematics test, for example, statements like "Well, boys *are* better than girls at math, right?" revealed one such stereotype. Overall, they struggled to see that these hidden stereotypes, through socialization or underestimation of ability, could lessen opportunities for girls or minorities to have access for advanced coursework in mathematics.

In this heated context, statistical modelling tools provided an analytic environment to deepen conversation about controversial issues. By diverting challenges from personal beliefs about race, gender, and economic status to the ability of a model to address these ideas, emotional tensions were lowered. Many of the teachers who were themselves members of a racial minority group were keenly aware of the issues and found voice in having the tools and opportunity to examine questions of burning interest to them.

The prospective teachers showed improved understanding of statistical concepts between the pretest at the beginning of the course and the posttest at the end ($t_{17} = 10.8$, $p < 0.01$), likely as a result of class instruction and multiple experiences interpreting data. Performance on the post-test is one measure of understanding, but the teachers' modelling ability in practice was of greater interest. We report on the teachers' project presentations and document factors showing an impact on their success in creating robust models as evidence for their findings. We also report the difficulty they had combining inquiry, data analysis, and articulating issues in equity. For example, many teachers regressed to comparing simple percentages or displaying data tables in their projects, while in structured settings (prepost test, interviews) they demonstrated greater facility with statistical concepts. This suggested to us the importance of engaging in open-ended investigations multiple times to develop confidence integrating these skills.

Three aspects of the inquiry projects were investigated to determine what factors may have correlated with the quality of the prospective teachers' models: statistical content knowledge (post-test), quality of model developed in conducting a short investigation with the software, and level of personal engagement with their inquiry topic. With regard to the first aspect, it was conjectured that the content knowledge of the teachers would be a major factor in their ability to develop a robust statistical model to support their findings. Surprisingly, there was no correlation ($r = 0.04$) between performance on the post-test and model quality. This suggests that statistical understanding, beyond a basic level, did not explain differences in their modelling ability in a complex context.

At the end of the course, the prospective teachers conducted a semi-structured investigation of test data using Fathom; all of the teachers demonstrated proficiency with using the software during these interviews. We wondered whether model quality for the open-ended inquiry project could be predicted from the quality of model developed as evidence for the semi-structured investigation. Correlation between model-quality for the semi-structured investigation in Fathom and the open-ended inquiry project, however, was not significant ($r = 0.29$, $p = 0.24$). This suggests that model quality in a well-structured problem does not necessarily transfer to an ill-structured problem context (Makar & Confrey, under review).

Finally, the choice of topic and engagement with the project was investigated as a factor in the quality of the teachers' models. Topics included the effects of small student subgroups, issues race or class, political issues (e.g. school funding), comparison studies (e.g. urban vs. suburban districts), and school case studies. Although no systematic differences were found between choice of topic and the quality of model constructed, there was a significant relationship between the level of engagement in the inquiry project and the quality of the model ($r = 0.58$, $p = 0.01$). Overall, those prospective teachers who were less personally engaged in the topic were less likely to make use of powerful statistical concepts involving variation and distribution in their models, relying instead on tables, summary statistics, and static displays in presenting their findings. The results suggest that for some teachers, regardless of previous statistical experience, the opportunity to investigate data with a compelling question stimulated stronger uses of statistical modelling. This is particularly poignant in the cases of three minority women who investigated very personal issues related to their own race, and developed some of the highest quality models in the class even though they had among the lowest scores on the statistics posttest. The evidence provided by these cases is an indication that strong personal engagement with the equity topic was a potent motivator for them to use more sophistocated statistical tools in their inquiry.

4. CONCLUSION

In this paper, we showed how the interaction of a compelling context, software-based modelling experiences with authentic data, and statistical content knowledge enabled teachers to enhance their understanding of all three of these areas. As with many projects on modelling, we struggled in creating a balance between contextual factors and technical ones. A major value of selecting the context of equity in student achievement is that the prospective teachers had many strong, tacit beliefs about the issue. In this context, the statistical tools gave the teachers a way to objectively explore some of those unspoken beliefs, make them more public, revise them, and hear the opinions of others. The complexity and relevance of the modelling context provided the mutual benefit of providing teachers a deeper understanding of statistical concepts as they developed a critical eye towards identifying equity issues in school testing.

Usually in school mathematics, the activity of modelling is a way to highlight a mathematical topic. The context being modeled is irrelevant to this process. The modeler usually has no personal stake in the context and little interest in gaining insight into it. In an environment where the context was on the forefront of the modelling, the introduction of statistics embedded in the process of better understanding the context was more natural and the context diverted attention away from teachers' fears of modelling and statistics. This kept the focus on equity while statistical concepts were being employed as tools to investigate claims of bias or difference between student subgroups. One could say that the purpose of modelling is to describe complex phenomenon, the context for which the model is meant to provide insight. We offer this study as a contrasting case to many approaches to modelling in mathematics where the use of context gives relevance, motivation and purpose to the pursuit of mathematics.

REFERENCES

Confrey, J. & Makar, K. (2005). Critiquing and improving the use of data from high-stakes tests with the aid of dynamic statistics software. In C. Dede, J. Honan, & L. Peters (Eds.), *Scaling up for success: Lessons from technology-based educational improvement* (pp. 198-226). San Francisco: Jossey-Bass.

Confrey, J., Makar, K., & Kazak, S. (2004). Undertaking data analysis of student outcomes as professional development for teachers. *Zentralblatt für Didaktik der Mathematik (International Reviews on Mathematics Education) 36* (1), 32-40.

Finzer, W. (2001). Fathom! (v1.16) [Computer Software] Emeryville, CA: Key College Press.

Makar, K. & Confrey, J. (under review). Effect of problem structure on the use of distributions as statistical evidence. *Statistics Education Research Journal.*

Chapter 5.2

MODELLING IN ONTARIO: SUCCESS IN MOVING ALONG THE CONTINUUM

Christine Suurtamm and Geoffrey Roulet
University of Ottawa, Ottawa, Canada, Email: suurtamm@uottawa.ca / Queen's University, Kingston, Canada, Email: rouletg@educ.queensu.ca

Abstract: From the research presented on mathematical modelling at ICMI-Study 14, it appears that modelling is occurring in isolated cases in classrooms around the world. In Ontario, Canada, however, modelling is embedded as a system-wide focus in secondary school mathematics education. This paper discusses the development and implementation of modelling in the province and describes the "levers" that make the development and implementation of a modelling curriculum possible and the "barriers" and challenges that are being addressed.

1. INTRODUCTION

Several research papers presented at the ICMI-Study 14 Conference on Modelling and Applications, show that, internationally, modelling is happening in isolated cases. A number of the papers report activities in particular classrooms that were sites for focussed research (Burkhardt, 2004; Kaiser & Maaß, 2006; Legé, 2004). In Ontario, Canada, we have moved beyond this with progress to a system-wide focus. Modelling, presented in the Ontario curriculum as the core activity of the inquiry process by which mathematics is to be developed, occurs, to some degree, in all secondary (ages 14 – 18 years) classrooms in the province. Modelling as a curriculum focus has had strong support from the leadership of the mathematics education community as well as the Ministry of Education through official curriculum statements and the provision of professional development and teaching resources. Many teachers embrace the idea of modelling activities in their courses, but the implementation process has some difficulties. It is important to explore the "levers" for change, the "barriers" presented, and the methods that have shown some success in overcoming barriers (Burkhardt, 2004).

2. EVOLUTION OF THE CURRICULUM

Over the past two decades, secondary school mathematics curriculum guidelines in Ontario, Canada (Ontario Ministry of Education [OME], 1985, 1999, 2000) have suggested that classroom practices should involve both modelling and applications. The introductory "process components" section of the 1985 curriculum guideline (OME) contained sections titled "Applications" and "Mathematical Models", and asserted that "applications shall be interspersed throughout the program" (OME, 1985, p. 19). Moreover the document suggested a movement away from the traditional lesson and unit sequence of "pure" mathematics study followed by applications (Hersee, 1987) to one in which "skills and concepts should be related from the beginning, to their applications" (p. 18). To that end, the guideline presented a six step modelling sequence: problem identification, production of a simplified real model, translation into mathematics, manipulation of the mathematical model, re-interpretation and testing in the real setting, and possibly refinement through a repeat of the cycle. This picture of modelling mirrored and expanded upon schemes presented by others arguing for mathematical modelling as part of the school curriculum (Blum & Niss, 1989; Galbraith, 1987; Hersee, 1987; Swetz & Hartzler, 1991).

The 1985 guideline, however, was critiqued for its failure to carry the progressive messages of its introductory pages into the list of content objectives for the courses, and it was predicted that this disconnect would lead to limited adoption of modelling and applications (Baumgart, 1985; Pravica, 1983). To some extent these fears were borne out for, ten years after the new curriculum's introduction "there was little evidence to suggest that teachers were making problem solving a focus of their lessons" (Haimes, 1996, p. 26). Applications were still appearing as end-of-unit activities, explored after isolated study of the related mathematical concepts.

Although the individual content objectives of the 1985 mathematics curriculum did not specifically call for modelling, the vision of the curriculum as a call for problem solving, as well as new paradigms of teaching and learning, inspired several creative teachers to design activities that began with rich problems and engaged students in mathematical investigation. These starting points generated the need for the creation of models, setting the stage for new mathematics learning via an emerging inquiry approach (Clark, 1995; Dewey, 1994; Montesanto & Zimmer, 1994). Coupled with this, a growing group of leading teachers elaborated on the curriculum's brief mention of technology, and began employing calculator and computer based graphing utilities to support the exploration of functions and the modelling of natural phenomena (Ontario Association for Mathematics Education, 1998; Roulet, 1994). Taken together, the innovative activities and use of technology created a growing phenomenon across the province and the 1985 – 1999 experience meant that Ontario had a group of mathematics educators with classroom experience in modelling who were committed and willing to lead the development and implementation of modelling activities.

In 1996 the Ministry of Education announced its intention to revise the secondary school curricula and as the initial step in this process commissioned the writing of a background research paper. This paper reiterated the need to move from a traditional curriculum pattern, where "applications or problems are placed at the end of a sequence or unit and are addressed after students have developed skill in the particular solution methods to be employed" (Roulet, 1997, p. 3), to a "problems-first" structure. In this proposed curriculum, units of study would begin "with an investigation of a problem setting and through a modelling activity develop" (p. 3) new (for the students) mathematical concepts and procedures. The ideas expressed in the background paper reflected the 1985-1999 work that had been done to promote applications and modelling by leaders in the mathematics education community. Further, for this round of curriculum reform the government chose a curriculum writing team largely populated by educators who had been developing classroom modelling activities and employing technology in the spirit of the 1985 guideline's introductory pages. The final version of the curriculum reflects the views held by this distinct group of mathematics educators.

The mathematics guidelines issued by the Ontario Ministry of Education (1999, 2000) present a curriculum that focuses on modelling, the use of technology, communication in mathematics, learning through inquiry, and the use of investigations in which students explore new problems in unfamiliar settings. These curriculum documents move significantly beyond the previous versions by not only addressing modelling and investigations in the introductory text, but by also reinforcing these themes within the statement of the specific course expectations. For example, students in grade 9 will "use algebraic modelling as one of several problem-solving strategies in various topics in the course" (OME, 1999, p. 11) and those in grade 11 will "determine, through investigation, the periodic properties of various models of sinusoidal functions drawn from a variety of applications" (OME, 2000, p. 24). In accomplishing such tasks, students will "collect, organize, and analyse data, using appropriate techniques and technology" (OME, 1999, p. 21). Investigations, modelling, applications, and the use of graphing calculators and computers are now required classroom experiences.

3. IMPLEMENTATION

As the new curriculum was brought forward, the Ministry of Education supported implementation through the provision of resources, professional development, and financial support for the creation of new textbooks and the purchase of the necessary technology. Significant efforts were made to assist teachers in the use of technology and the implementation of modelling activities. Yet, these supports did not necessarily address the conflict between the new curriculum's image of mathematics and the vision of many teachers in the field. Consequently, many teachers adapted the activities and incorporated them into their existing pedagogical practices. Hence, in many cases, mathematical activity was more closely aligned with "application" of previously studied concepts and procedures rather than "modelling" (Roulet & Suurtamm, 2004).

Although adoption of modelling occurred in most courses by some teachers; one setting, a new Grade 12 Mathematics for Data Management course (OME, 2000, p. 48) has witnessed teachers readily adopting and implementing modelling as a focus of instruction. In this course, students define a problem that they would like to explore; determine how they would investigate the problem; collect or find the necessary data; analyze the data and present the findings. Ontario mathematics curricula had never before presented such large tasks to students or required teachers to think about how to support their pupils as they tackle such wide-ranging projects. As well, mathematics topics such as: statistics, iteration, database organization, flowcharting, graph theory, networks, and coding appeared anew or in greater depth than in previous Ontario curricula. Because much of the course content and pedagogical approach was new for teachers, this course received extensive support through the creation of resources and professional development opportunities (Brock University; Dalrymple & Dilena, 2002; Roulet, 2002; Statistics Canada).

To meet the challenges of the new content and pedagogical approach of this course, teachers attended professional development workshops. These workshops provided experiences of exploring new content through modelling activities, reflecting what teachers could be doing with their own students. Since the content was unfamiliar to the teachers, they had the positive experience and saw the value of learning through modelling. This, in turn, had a strong impact on their own pedagogical approach in teaching the course. Furthermore, as teachers engaged in new learning with their students, and explored problems together, teaching and learning became a reciprocal process (Simon, 1995). As others have noted, when teachers, themselves, develop new mathematical knowledge through investigations, and when their epistemological views are challenged, the implementation of modelling is enhanced (Makar & Confrey, 2004; Roulet & Suurtamm, 2004).

4. CONCLUSION

There are several "levers" that helped to support the development and implementation of modelling in Ontario as well as several "barriers" to meet. One of the major challenges is that of confronting teachers' conceptions of mathematics (Roulet, 1998; Roulet & Suurtamm, 2004; Suurtamm, 2004). In fact, changing beliefs about what mathematics is and how to teach it requires new and different experiences, for changes in official curriculum documents do not necessarily result in changed teacher practice (Makar & Confrey, 2004). Since teachers often teach in ways that they were taught, they need to personally develop an understanding of new mathematical concepts through the processes of investigation and modelling for it to have meaning for them.

Several factors help to support the development and implementation of modelling in Ontario and to overcome some of the challenges. Since 1980 the vision of a modelling curriculum has been shared and nurtured by the leadership of the mathematics education community and supported by the Ontario Ministry of Education. The 1985 guideline that invited teachers to develop mathematical inquiry and modelling provided fertile ground for experimentation to occur. Further to this, the extensive period of exposure and gestation during which teachers developed and shared new ideas allowed momentum to build. As well, professional development that engaged teachers in new mathematics learning through problem solving and modelling activities helped to ground teachers in new pedagogical approaches. Thus, positive progress on both the official intended curriculum and implemented curriculum has been possible because of the synergy generated by a common vision that is imbedded in curriculum documents, is supported by the provision of technology, professional development and resources; and is experienced by teachers as they themselves engage in mathematical inquiry.

REFERENCES

Baumgart, R. (1985). Thoughts on the implementation process: A response to "Play it again, Sam – a theme from the teachers". *OAME Gazette 14*(1) 47-48.

Blum, W. & Niss, M. (1989). Mathematical problem solving, modelling, applications, and links to other subjects – state, trends and issues in mathematics instruction. In W. Blum, M. Niss, & I. Huntley (Eds.), *Modelling, applications and applied problem solving* (pp. 1-21). Chichester: Ellis Horwood.

Brock University. *Resources for mathematics for data management (MDM4U).* available: http://www.brocku.ca/cmt/mdm4u/

Burkhardt, H. (2004). Establishing modelling in the curriculum: Barriers and levers. In H-W. Henn & W. Blum (Eds.), *ICMI Study 14: Applications and modelling in mathematics education* (pp. 53-58). Dortmund: University of Dortmund.

Clark, J. (1995). Moving towards inquiry – What are the implications for curriculum? *OAME Gazette 33*(3), 7-11.

Dalrymple, S., & Dilena, S. (2002). *MDM4U - Preparing to implement*. Toronto: Fields Institute for Research in Mathematical Sciences, Mathematics Education Forum.

Dewey, R. (1994). Calculus, A student-centred approach. *OAME Gazette 33* (1), 4-8.

Galbraith, P. (1987). Modelling-teaching modelling. *Australian Mathematics Teacher, 43*(4), 6-9.

Haimes, D. (1996). Destreaming in grade 9: Teachers' strategies and sources of ideas. *Ontario Mathematics Gazette, 35*(2), 24-27.

Hersee, J. (1987). Modelling versus applications. In J. Berry, D. Burghes, I. Huntley, D. James & A. Moscardini (Eds.), *Mathematical modelling courses* (pp. 227-239). Chichester, UK: Ellis Horwood.

Kaiser, G. & Maaß, K. (2006) Modelling in Lower Secondary Mathematics Classrooms - Problems and opportunities. This volume Chapter 2.7.

Legé, J. (2004). "To model, or to let them model?" That is the question. In H-W. Henn & W. Blum (Eds.), *ICMI Study 14: Applications and modelling in mathematics education* (pp. 145-150). Dortmund: University of Dortmund

Makar, K., & Confrey, J. (2004). Modeling fairness in student achievement in mathematics using statistical software by preservice secondary teachers. In H-W. Henn & W. Blum (Eds.), *ICMI Study 14: Applications and modelling in mathematics education* (pp. 175-180). Dortmund: University of Dortmund.

Montesanto, R, & Zimmer, D. (1994). *Models and patterns: Experimenting with linear equations*. Toronto: D.C. Heath Canada.

Ontario Association for Mathematics Education [OAME]. (1998). *Teaching mathematics through technology*. Rosseau, ON: OAME.

Ontario Ministry of Education [OME]. (1985). *Curriculum guideline: Mathematics: Intermediate and senior divisions*. Toronto: Queen's Printer for Ontario.

Ontario Ministry of Education [OME]. (1999). *The Ontario Curriculum: Grades 9 and 10: Mathematics: 1999*. Toronto: Queen's Printer for Ontario.

Ontario Ministry of Education [OME]. (2000). *The Ontario Curriculum: Grades 11 and 12: Mathematics: 2000*. Toronto: Queen's Printer for Ontario.

Pravica, S. (1983). Letter to the minister. *OAME Gazette 22*(1), 31-33.

Roulet, G. (1994). Graphs to functions explorations using a computer graphing tool. In K-D. Graff, N. Malara, N. Zehavi, and J. Ziegenbalg (Eds.), *Technology in the service of the mathematics curriculum: Proceedings of Working Group 17 at ICME-7, the 7th International Congress on Mathematics Education, Quebec, 1992* (pp. 289-296). Berlin: FU.

Roulet, G. (1997). *Mathematics curriculum for Ontario secondary schools: Issues, choices and opinions*. A background research paper commissioned by the Ontario Ministry of Education and Training in support of secondary school curriculum renewal.

Roulet, G. (1998). *Exemplary mathematics teachers: Subject conceptions and instructional practices*. Unpublished doctoral thesis, University of Toronto, Ontario.

Roulet, G. (2002). Social issues and Statistics Canada resources in the mathematics classroom. *School Libraries in Canada, 22*(1), 25-27.

Roulet, G., & Suurtamm, C. (2004). Modelling: Subject images and teacher practice. In H-W. Henn & W. Blum (Eds.), *ICMI Study 14: Applications and modelling in mathematics education* (pp. 229-234). Dortmund: University of Dortmund.

Simon, M. (1995). Reconstructing mathematics pedagogy from a constructivist perspective. *Journal for Research in Mathematics Education, 26*(2), 114-145.

Statistics Canada. Learning resources. Ottawa: StatCan. available: http://www.statcan.ca/english/edu/index.htm.

Suurtamm, C. (2004). Developing authentic assessment: Case studies of secondary school mathematics teachers' experiences. *Canadian Journal of Science, Mathematics and Technology Education, 4*. 497-513.

Swetz, F. & Hartzler, J. (Eds.). (1991). *Mathematical modelling in the secondary school curriculum*. Reston, VA: National Council of Teachers of Mathematics.

Chapter 5.3

IMPLEMENTATION CASE STUDY: SUSTAINING CURRICULUM CHANGE

Gloria Stillman
University of Melbourne, Melbourne, Australia, Email: g.stillman@unimelb.edu.au

Abstract: An Australian study is being undertaken into influences on the sustaining of curriculum change designed to place real-world mathematical applications and modelling centrally within mainstream senior secondary mathematics curricula. Conditions that impeded or sustained this curriculum change in two Australian are currently being investigated. Findings from the implementation in one of the states are presented.

1. INTRODUCTION

According to Niss (2001, p. 85), "It is still the case, in general international terms, that genuine and extensive applications and modelling perspectives and activities continue to be scarce in the everyday practice of mathematical education." In Australia each state has its own education system with distinctive curriculum and assessment policies. Several Australian states have attempted to include these topics more centrally within their curricula. An on-going study described in Stillman (2004) is being conducted into the implementation of applications and mathematical modelling curricula in two states (Victoria and Queensland) where the path of curriculum change has been different with markedly different outcomes. McBeath (1995) points out that "much of the literature [on educational change] recognises the variability and liquidity of individual situations, and the difficulty of determining a single model to suit all." Even in reforms, with the same focus such as implementing mathematical modelling curricula at the upper secondary level, "it must also be [recognised that] the dynamics of each innovation" are unique; however, some of the lessons learnt in such implementations, whether successful or not, will be of benefit to planners in other systems. Some findings from the Victorian case study are presented here.

Research into, and knowledge about, applications and mathematical modelling within the informed community has reached a state of maturity where it has already been demonstrated that these approaches are possible at the senior secondary level, their inclusion is defensible and interesting case studies with teacher enthusiasts show they work. Now it is important that we move on and look more globally at why there is still a difficulty sustaining applications and mathematical modelling centrally within mainstream curricula in many countries. To do this we need to look more broadly at educational change not just within mathematics.

2. BACKGROUND

The changes to the post compulsory mathematics curriculum in Victoria came with the reform of secondary schooling at this level as a result of the Blackburn Report (1985). The Higher School Certificate Course Description for mathematics (VISE, 1985) included special emphasis on students becoming familiar with the basic steps of mathematical modelling but this was just one of a proliferation of mathematics courses and certificates existing at this level at the time. In 1986, the document, *Future Directions in Post Compulsory Schooling,* foreshadowed that in the curriculum redesign there would be only one certificate for all students and there would be a mix of school, moderated and external assessments in all subjects – a break from the traditional formal written response external examinations that existed.

In mathematics, an innovative curriculum (VCAB, 1988) incorporating "problem-solving and modelling activities…intended to provide students with experience in using their mathematical knowledge in creative ways to solve non-routine problems" (p. 24) and investigative projects which could be "an extended mathematical modelling exercise to solve a real-world problem" (p. 27) commenced formal implementation by a limited number of schools and education providers in 1989. This was "a significant attempt to promote the teaching of mathematical modelling and investigations in schools" (Stacey, 2001, p. 47) through the formal assessment of problem solving and modelling in high-stakes assessment. State-wide implementation began in 1990 at Year 11 and Year 12 in 1991.

Four centrally designed Common Assessment Tasks (CATs) were introduced for Year 12. An Investigative Project on a centrally determined theme and a Challenging Problem selected from centrally set problems were undertaken over 4 weeks and 2 weeks, respectively. The latter task was part of a problem solving and modelling work requirement. Both tasks were initially school assessed using centrally developed criteria contributing 50% of the final result. The other two tasks were end of year examinations. At the end of 1992 the Challenging Problem was eliminated as a common assessment task due to "concerns over authentication of student work" and the tendency of such concerns "to erode the credibility of the assessment of problem solving" (VCAB, 1993, p. i).

Throughout the decade following its introduction the investigative aspects of the mathematics curriculum, and consequently the scope for modelling, became more restricted (McCrae, Dowsey, & Stephens, 1998), contributing less and less to final assessment. These changes were needed "refinements to optimise economies of implementation and the demands on student and teacher time" as well as "to increase confidence in the authorship of student work" (Barnes, Clarke, & Stephens, 2000, p. 647). Changes to the Victorian Certificate of Education (VCE) (BOS, 1999) implemented in 2000, where teachers set their own school assessed coursework with only suggested themes and data sets being provided by the central authority, led all but to the demise of modelling in many schools. According to Stacey (2001), these changes signalled the end of "a bold experiment of assessment driven change and a real focus of problem solving and modelling" (p. 48) in the senior curriculum. The revised Mathematics Study Design (VCAA, 2005) to take effect in 2006 is not reversing this trend although it remains "an underlying principle of the Mathematics study that all students will engage" in mathematical activities that include "situations which require investigative, modelling or problem solving approaches" (p. 7).

3. THE STUDY

Following an examination of extant curriculum documents and materials from the 1980s onwards, purposeful samples of 6 key curriculum figures [KCG] (e.g., members of expert advisory committees or curriculum managers for the statutory board), 6 teachers in key implementation roles [KTG] (e.g., as seconded project officers for the statutory board, state or regional chairs of verification panels) and 6 classroom teachers [CTG] were selected and participants interviewed using tailored semi-structured interview schedules. All were asked about their experiences during the change and subsequently and about their beliefs regarding conditions that promoted or hindered the introduction and ongoing use of mathematical applications and modelling in upper secondary classrooms in the state. Practising teachers were also asked about the impact of the changing role of applications and modelling in the various mathematics Study Designs on their teaching and assessment practices. Classroom artefacts (e.g., tasks) typifying their current practice in the area of mathematical applications, and modelling at the upper secondary level were also collected.

A grounded theory approach to data analysis was taken using the reformulated method of Strauss and Corbin (1990). Data were analysed using NUD.IST software (QSR, 1997) to identify central ideas and represent these as fully defined categories. The identification of influences on change and conditions impeding and sustaining curriculum change in this area was the goal of this analysis.

4. SUSTAINING IMPLEMENTATION

A number of possible influences on whether the change in the Victorian curriculum would be able to be sustained were investigated. These included (a) how teachers were positioned as professionals by the implementing authority and implementation model used (e.g., externally induced change with participant involvement); (b) pressures from conservative political groups (e.g., general public, parents, teachers, academics); (c) considerations of discipline oriented versus whole curriculum approach to curriculum change (e.g., system-wide assessment requirements); and (d) changes in society (e.g., desire for equity with respect to socio-economic status). Two of the major conditions that threatened the ability of the Victorian system to sustain the change were (a) the extent of the change and (b) the pace of the change.

The extent of this change was "massive" on several levels. Initially, only three subject disciplines including mathematics were involved in the new curriculum design trials in a few schools but when the new VCE was implemented state-wide all subject disciplines came on board at the same time firstly in Year 11 then Year 12. Thus the change was system-wide and it was to be "a curriculum structure and framework that would apply in all areas but would be common" [KCG2; i.e, person no. 2 of the researcher selected Key Curriculum Group]. This had implications for in-servicing of teachers about the practicalities of the change and the rationale for it, the amount and cost of resources and support needed, and the accessibility of those resources or supporting mechanisms for teachers once the change began. For example, a university course on mathematics and modelling designed to update and renew the mathematical knowledge of teachers of the new VCE mathematics was instigated at the request of the Victorian government. Unfortunately, its major impact was limited to the two years before the VCE was fully implemented as there were "very big groups until the VCE came in.... The people who prepared early were well prepared but the people who didn't realise what was going to hit them were actually too busy.... they couldn't come because they were too busy doing the VCE" [KCG6]. The system wide change also meant solutions to issues arising (e.g., authenticity of student work or comparability of teacher assessed work) could not be solved only within a particular discipline. All discipline areas had to address issues or the whole system remained threatened. "We were basically told that maths had got it right and it was working well. They weren't upset about us but they were a bit worried about the cheating in the other areas...Maths was okay, but the others were having to look at what they were doing. And suddenly it was all gone, no more of this extended stuff, everything had to be done in class in a few periods and all this extended stuff went" [KTG5; i.e., person no. 5 of the researcher selected Key Teacher Group].

Within the mathematics curriculum itself the extent of change expected was enormous with various interviewees describing it as "a radical change" [CTG1; i.e., person no. 1 of the researcher selected Classroom Teacher Group; CTG6], "a revolution" [KTG5], "a huge shock" [KTG4], a need to "shift ground'' [CTG2], "a big challenge" [CTG5], "the biggest change" [KTG6] and "in many ways it was so very progressive change" [KCG3]. In a sense there was a need for teachers to spend some time "re-constructing what they think mathematics is" [KCG3] as this was the type of debate that informed the changes. "There was the relevance question, but that was addressed not just through problem solving, but also through investigations that other people were proposing as being a valid way of doing mathematics but there was also a sense that we were trying to have students be mathematicians and that was the message that the modellers were bringing to the discussion. Let's engage students in being mathematicians and in doing mathematics, not just regurgitating other people's maths" [KCG1]. And this change was to be central to the curriculum and "all of the kids are going to do problem solving and modelling" [KCG1] not just some. There was also "a real re-conceptualisation of the curriculum" [KCG3] as initially there was a quite different curriculum structure of subjects based on one particular area of mathematics similar to curricula in the United States. It was "very broken down into specific areas of study as opposed to what we have always had which is a more integrated approach in our curriculum" [KTG6]. In addition, teachers' judgements now were to play a vital role in high stakes assessment. "A state wide verification model meant that every teacher was involved in a discussion with other teachers and a chair. So there was clearly a big payoff in time to reflect upon and talk about what was a good CAT" [KCG5]. However, for teachers who focussed on "delivering a course to get the results for their students" it had to be seen that "the time spent doing the mathematics [in these tasks was] paying off in covering and deepening understanding of key content that the students will later be examined on" [KCG5].

The pace of the change to the curriculum in Victoria also threatened its sustainability. Radical change needed time to evolve and allow expectations for changed practice to develop so there was a lasting change in the culture of the teaching profession and a genuine renewal of practice. However, "there was never sufficient allowance made for the fact that the teachers really did need a couple of years to get used to it" [KTG3]. The implementation of the mathematics curriculum was into review almost as soon as it began. The chair of the Review of the Mathematics Study Design in 1991 is reported as having said, "These sorts of seismic changes take ten years. Victoria is moving faster than any other state has thought of moving and probably faster than it should safely be" [KCG5]. Thus, "the normal process of trying something, finding out what is not so crash hot and then refining it progressively never got applied to the version 1 of the VCE. And I do think if we'd had time to go through that process, not responding to the political elements discussion, people not getting so absolutely nervous about it, some of those things would have been developed in a much more sustainable way without the need to make radical interventions" [KCG3].

REFERENCES

Barnes, M., Clarke, D., & Stephens, M. (2000). Assessment: The engine of systemic curricular reform? *Journal of Curriculum Studies 32*(5), 623-650.

Blackburn, J. (Chairperson) (1985). *Ministerial Review of Post Compulsory schooling*. State of Victoria: Melbourne, Australia.

Board of Studies (BOS) (1999). *Study Design Mathematics*. Author, Melbourne: Australia.

McBeath, C. (1995). Barriers to effective curriculum change: A case study in dissemination practice. In *Post-compulsory Education and Training Vol. 2*, Centre for Skill Formation Research and Development, Griffith University, Brisbane. Available at http://education.curtin.edu.au/pubs/mcbeath/bris95.html.

McCrae, B., Dowsey, J., & Stephens, M. (1998). School-based assessment in VCE mathematics: Ten years on. In C. Kanes, M. Goos, & E. Warren (Eds.), *Teaching Mathematics in New Times, Vol. 1* (pp. 350-357). Gold Coast, Australia: MERGA.

Niss, M. (2001). Issues and problems of research on the teaching and learning of applications and modeling. In J. F. Matos, W. Blum, S. K. Houston, & S. P. Carreira (Eds.), *Modelling and Mathematics Education* (pp. 72-88). Chichester: Ellis Horwood.

QSR. (1997). *Non-numerical Unstructured Data – Indexing, Searching and Theorizing* (NUD-IST) v.4 [Computer Software]. Melbourne: Qualitative Solutions & Research.

Stacey, K. (2001). Trends in researching and teaching problem solving in school mathematics in Australia: 1997-2000. In E. Pehkonen (Ed.), *Problem Solving around the World* (pp. 45-53). University of Turku, Finland.

Stillman, G. (2004). Sustained curriculum change: The example of the implementation of applications and modelling curricula in two Australian states. In H.-W. Henn & W. Blum (Eds.), *ICMI Study 14: Applications and Modelling in Mathematics Education*, (pp. 261-266). Dortmund: University of Dortmund.

Strauss, A. & Corbin, J. (1990). *Basics of Qualitative Research: Grounded Theory Procedures and Techniques*. Newbury Park, California: Sage.

Victorian Curriculum and Assessment Authority (VCAA) (2005). *Mathematics: Victorian Certificate of Education Study Design*. Author: Melbourne, Australia.

Victorian Curriculum and Assessment Board (VCAB) (1988). *VCE Mathematics Study Design – provisionally accredited*. Author: Melbourne, Australia.

Victorian Curriculum and Assessment Board (VCAB) (1993). *VCE Mathematics Resources*. Author: Melbourne, Australia.

Victorian Institute of Secondary Education (VISE) (1985). *Higher School Certificate Course Description: Mathematics Group 1*. Author: Melbourne, Australia

Chapter 5.4

MATHEMATICAL MODELLING OF SOCIAL ISSUES IN SCHOOL MATHEMATICS IN SOUTH AFRICA

Cyril Julie and Vimolan Mudaly
School of Science and Mathematics Education, University of the Western Cape, Bellville, South Africa, Email: cjulie@uwc.ac.za / School of Education, Development and Training, University of KwaZulu-Natal, Natal, South Africa, Email: mudalyv@ukzn.ac.za

Abstract: Learners' and teachers' engagement with the mathematical modelling of social issues is the focus of this chapter. It is contended that the mathematical modelling behaviour of both learners and teachers is dominated by mathematical modelling as a vehicle for "entry into mathematics". This, it is suggested mitigates against the development of a "mathematical temper". It is recommended that more emphasis be placed on mathematical modelling "as content" which would open windows of opportunity to deal with social issues in school mathematics.

1. INTRODUCTION

It is widely accepted that school mathematics curricula across the world do not vary much and have more similarities than differences. With differing emphases and purposes Mathematical Modelling is a feature of the school mathematics curriculum of most countries. In South Africa the intended incorporation of mathematical modelling in school mathematics is captured in both the Mathematics and Mathematical Literacy syllabi for the last three years of schooling. For Mathematics it is stated that "An important purpose of Mathematics in the Further Education and Training band is the establishment of proper connections between Mathematics as a discipline and the application of Mathematics in real-world contexts." (Department of Education, 2003(a), p. 10). Similar intentions are expressed in the curriculum for

Mathematical Literacy as evidenced in the following statement "Mathematical Literacy will develop the use of basic mathematical skills in critically analysing situations and creatively solving everyday problems." (Department of Education, 2003 (b), p. 9). Explicitly mathematical modelling must be utilised as a mechanism to contribute towards the goal of general education stated as the development of "a participating citizen in a developing democracy [who has] a critical stance with regard to mathematical arguments presented in the media and other platforms." (Department of Education, 2003 (b), p. 9). This, it can be argued, requires that in their school mathematical experiences learners engage with the modelling of issues of social import. Concommitant with this requirement is an imperative, amongst others, that prospective and practising teachers be prepared to incorporate the intricasies of the mathematical modelling of social phenomena in their teaching and that classroom teaching experiments be conducted to provide empirical evidence of the possibilities on how instruction can proceed to make the intended expression of the curriculum desire a reality. It is within this imperative that this chapter is situated.

2. THEORETICAL COMMENT

Since at least the late 1960's and early 1970's there has been debate about how mathematical modelling should be handled in teaching situations. Problem presentation, the purpose of modelling teaching and contextual authenticity are the most important issues around which the debate revolves. Central to the debate is whether mathematical modelling should be used as a vehicle for the development of mathematics or treated as content in and of itself. A common notion associated with mathematical modelling as a vehicle is that mathematics should be represented in some context. The purpose for embedding mathematics in context is not the construction of mathematical models per sé but rather the use of contexts and mathematical models as a mechanism for the learning of mathematical concepts, procedures, conjecturing and, at times, developing context-driven justifications for obtained conjectures. Mathematical modelling as content entails the construction of mathematical models of natural and social phenomena without the prescription that certain mathematical concepts, procedures or the like should be the outcome of the model-building process. It also entails the scrutiny, dissection, critique, extension and adaptation of existing models with the view to come to grips with the underlying mechanisms of mathematical model construction and the assessment and evaluation of constructed mathematical models. Regarding problem presentation in the modelling-as-vehicle view the problem is stated in the language of some reality situation and the mathematics to be applied is obtained from the cues and clues. In the modelling-as-content view the reality situation is the starting point and the mathematical problem has to be constructed. This characterization of the mathematical modelling debate is, of course, an idealization and it is best to regard the two views as the extremities of a continuum. The authenticity, in the sense of how professional and adept applied mathematicians and modellers first encounter a problem situation to be mathematically treated, of problems decreases as there is movement from the modelling-as-content pole. The classroom teaching experiment related to learners engagement with mathematical modelling and the engagement of practising teachers with it were more towards modelling-as-content pole.

3. A TEACHING EXPERIMENT IN A SOUTH AFRICAN SCHOOL

For the most part of schooling in South Africa, modelling was limited to very basic word problems. De Villiers (1994, p. 34) describes the traditional status quo in South Africa as follows:

> The secondary school curriculum has traditionally focused almost exclusively on developing pupils' manipulative skills (e.g. simplifying, factorizing, solving equations, differentiation, etc.). This focus was in part due to the pervasive belief among teachers (and curriculum developers?) that such technical skills were essential **prerequisites** for problem solving and mathematical modelling, and therefore had to be mastered.

In order to gauge the South African learners' use of modelling strategies a teaching experiment was conducted with Grade10 learners. They were given a situation dealing with a social issue that was highlighted in various media in South Africa. The situation they were confronted with was:

> In a developing country like South Africa, there are many remote villages where people do not have access to safe, clean water and are dependent on nearby streams or rivers for their water supply. With the recent outbreak of cholera in these areas, untreated water from these streams and rivers has become dangerous for human consumption. Suppose you were asked to determine the site for a water reservoir and purification plant so that it would be the same distance away from four remote villages. Where would you recommend the building of this plant?

The learners had experiences with *Geometer's Sketchpad* and were supplied with a pre-constructed diagram, both on paper and on the computer. Rather than immediately starting with *Sketchpad* (a dynamic geometry software), the learners were asked to attempt to find a solution on their own, using any previous knowledge.

All of the learners "guessed" a solution somewhere in the "middle" of the quadrilateral, but none could find an adequate solution. More importantly, very few of them tried to test their suspicion of "somewhere in the middle". This extract below from an interview with one learner (see below) was typical of several interviews.

> R: Where do you think that we should build the reservoir?
>
> L: ...don't know ... all we are only given is this diagram ...
>
> R: Do you think that you will be able to find the most suitable point? You can use any method you know, to do so.
>
> L: I don't know sir ... this is too difficult ... please don't interview me (pleading)
>
> R: Are you saying that you cannot find any way of solving this problem?
>
> L: I can't ... I'm not so good in maths ... maybe at the centre here (pointing to the middle of the quadrilateral).

The learners seemed to feel that this type of question was not within their ability to solve and in some cases the learners explicitly said that this was because this type of question had never been taught or asked of them and thus displayed a "*learned helplessness*" that is, a hesitancy to attempt unfamiliar problems. The hesitancy of the learners to deal with seemingly unfamiliar problems is aptly summarized by a learner's comment that: "*We didn't do this in class before ... I can't do it!*"

The verification strategy a few learners resorted to was construction and measurement. As one learner asserted after offering a guess that the reservoir be placed in the middle *"It's easy to understand*" and "*Ya ... can I measure with my ruler*?" However, after finding that the measurement strategy contradicted the conjecture the solution path was discarded without searching for an alternative strategy.

Another finding that stands out was the learners' lack of discussion and reference to the situation at hand. Options such as the cost of building four smaller reservoirs closer to each village, that villages may have been of different sizes or the topography of the landscape did not feature in their deliberations and that therefore finding the 'ideal' position, using the concurrency of perpendicular bisectors, would have not been an entirely appropriate solution. (Mudaly, 2004, p. 178).

4. TEACHERS' MATHEMATICAL MODELLING WORK

A broad research project on teacher behaviour when engaged in mathematical modelling is the subject of discussion in this section. Data were systematically collected on teachers' work in mathematical modelling teacher inservice courses. In these courses the situations for which teachers were required to develop models for over the years were: A salary system to bring about equity based on the principle of "equal pay for equal work" taking into account years of service, promotion criteria and qualifications; the Human Development Index and other social indexes such as a community development index; school enrolment projections and garbage accumulation. The normal qualitative data analysis techniques were followed (Julie, 2002(a); Julie, 2002(b); Julie, 2003). The data comprise of observations and video-recordings of teachers at work; the rough work that was produced during the model construction process; the final reports on the models, formal and informal interview conversations and post-activity questionnaires. The analysis rendered three major findings related to teacher mathematical modelling behaviour. Firstly, the model-as-vehicle paradigm dominated model-construction activity. This is seen as the search for a formula to describe the situation under investigation. For example, one teacher described her experiences with the salary scale activity in the open-ended questionnaire as follows:

> It was a struggle to understand the problem. The many principles, variables had your head spinning. We started by trying to get a formula from the table – excited! Oh...the equation/formula does not satisfy all the conditions... Decide first to work with one post level only to simplify the problem – point of departure gets a ceiling – highest position and highest number of years in post level. Build other formulae around the norm.

This notion of the existence of a formula that can be found from the data dominated most of both the initial collective and individual deliberations. The formula-seeking behaviour dominated over the situation-analysis. Essentially there is nothing wrong with this behaviour. The division of approaches to modelling into empirical modelling – fitting formulae to data – and axiomatic – developing a model from a set of assumptions – requires knowledgeability of formula-seeking. However, this fixatedness on formula-seeking mitigates against in-depth discussion and consideration of the nature of the social issues involved.

Secondly, the teachers' work was driven by an immediate perceived usability. When constructing models practising teachers seem to express preference for models which were relevant to their immediate work circumstances. Consider the excerpt of fieldnotes made during teachers' engagement with the school enrolment model.

> The teachers had to particularise a model for planning the supply for mathematics teachers to their schools based on the number of pupils at their school and a school enrolment model provided by Gould (1993). They presented their particularisations to the class. At the end of the presentations we engaged in a conversation around their work and the experiences with the activity. Mr K started the discussion and he said: "This was one of the first pieces of work I did where I can see how I can use it in my situation. We know that the number of teachers for a year is determined by using the enrolment of the year before. Now I can actually use this model at school and we can determine the number of teachers a few years in advance."

This contrasts with the data on teachers' reaction to the model building activity on a garbage collection activity where they had to construct a model related to the accumulation of plastic bags against school fences. A similar discussion on their experiences with this activity produced nothing about the usability of the models that the teachers developed. What engages teachers and what not is a complex issue. Immediacy in terms of what I can use in my situation as it is currently is emerging as a facet of teacher behaviour in mathematical model construction.

Lastly, teachers in their modelling work seem to settle for elementary mathematical work and premature closure. One of the rationales for the lobbying for the inclusion of modelling and applications in school mathematics is that it will play an activating role. Modelling and applications will, in addition to its usability features, be a catalyst for thinking about mathematics that learners (and teachers) did not think about before. In dealing with the Human Development Index (HDI) teachers were requested to extend the HDI by adding a fourth factor, satisfaction with the government of the day, to the HDI. They came up with

$$HDI = \frac{L.Expec. + 2Educ.index + GDP\,index}{4}$$

The teachers were contend to simply work with the categories involved in the HDI. Although their discussion included references to fair taxation; domestic production and increase in domestic production, they only doubled the education component of the index and view this as equivalent to a fourth factor. The teachers remained as near as possible to the model that was studied and their extension of the model was confined to same categories contained in the HDI. Furthermore, the teachers only used elementary mathematics. Although the HDI appears on the surface to be a simple weighted additive model, there is much deeper mathematics underlying the eventual model. Premature closure also occurred with the garbage accumulation activity. This activity can lead to mathematics in stocks and flows. The seeds of such a development are discernable in a group of teachers' description of their model. They wrote:

> We can say the Accumulation (A_{pb}) is the proportion of learners that litter, times the proportion of the community that, multiplied by the amount of days and times the amount of plastic bags used per day gives the total accumulation

Although the above description can be faulted, the seeds for moving this description and the resultant model to the mathematics involved in stocks and flows as formulated by Bartholomew (1976, p. 162) is clearly observable. Nowhere in the teachers' deliberations, even during informal discussions, were anything said or done which would point that this kind of mathematics was activated. Teachers seem to be fixated on what they perceive the task at hand to be and hence resolving this perceived task is for them the point of closure.

5. DISCUSSION AND CONCLUSION

The mathematical modelling work of learners and teachers were related to above. What come through are similarities between the learners' and teachers' modelling work with social issues. Firstly, both learners and teachers pay scant attention to in-depth qualitative discussions regarding the issues at hand. There appears to be an urgency to get to the mathematics. This contrasts sharply with mathematical modelling practice in, for example industry, where "successful industrial mathematicians ... require a high degree of communication skills in several forms – speaking, writing, and listening – and ... business interactions often continue over a long period of time, so that clear exchanges of information and ideas are crucial." (SIAM, 2001). Secondly, there is the issue of premature closure after finding a first supposed solution. In terms of the traditional modelling cycle, there is no return to the reality situation for consideration and reflection of the aptness of the obtained mathematical solution nor were there attempts to extend the mathematics underpinning the appropriate seeds obtained for such extension.

We argue that this behaviour is structured by current school mathematics practice characterized by the dominance of mathematical modelling as a vehicle. The search for a formula in the teachers' situation and the offering of measurement as the preferred justification strategy for the conjectured solution by the learners attest to this dominance of modelling as a vehicle. Uncritically it is assumed that this modelling as a vehicle will satisfy the realisation of the expressed ideal of sensibly dealing with social issues in the mathematical curriculum. This is not necessarily the case particularly if the incorporation of social issues in school mathematics is ostensibly the only way which to address the development of a mathematical temper – a spirit of dealing rationally with the desirable and undesirable effects mathematical installations in society. There is no doubt that this realisation can only be effected through mathematics programmes which aim at developing mathematical modelling as content. After all, it is during the engagement with mathematical modelling as content that windows of opportunities are opened for dealing with social issues.

REFERENCES

Bartolomew, D. J. (1976). The control of the grade structure in a university. In Andrews, J. G. & McLone, R. R. (Eds.), *Mathematical Modelling*. London: The Butterworth Group.

Department of Education (2003a). *The National Curriculum Statement: Mathematics.* Pretoria: Government Printers.

Department of Education (2003b). *The National Curriculum Statement: Mathematical Literacy.* Pretoria: Government Printers.

De Villiers, M. (1994). The role of technology in mathematical modelling. *Pythagoras, 35*(12).

Gould, E. (1993). Enrolments. *International Review of Education, 39*(4), 319-332.

Julie, C. (2003). Work moments in Mathematical Modelling by Practising Mathematics Teachers With No Prior Experience of Mathematical Modelling and Applications. *New Zealand Journal of Mathematics, 32* (Supplementary Issue), 117-124.

Julie, C. (2002a). The Activity System of School-Teaching Mathematics and Mathematical Modelling. *For the Learning of Mathematics, 22*(3) (November), 29-37

Julie, C. (2002b). Making relevance relevant in mathematics teacher education. In Vakalis, I., Hughes-Hallett, D., Kourouniotis, C., Quinney, D., & Tzanakis, C. (Eds), *Proceedings of the 2nd International Conference on the Teaching of Mathematics at the undergraduate level, July 1-6.* University of Crete, Hersonissos, Crete, Greece: CD-ROM published by Wiley Publishers.

Mudaly, V. (2004). *The Role of Sketchpad as a Modelling tool in Secondary Schools.* Unpublished doctoral dissertation, University of KwaZulu-Natal, Natal, South Africa.

SIAM (2001). *The SIAM Report on Mathematics in Industry.* From http://www.siam.org/mii/node5.html#SECTION00041000000000000000

Part 6

BIBLIOGRAPHY

Part 6

BIBLIOGRAPHY

The following bibliography contains some basic references (in English language) published on the topic *Modelling and Applications in Mathematics Education* in recent years, most of them generated by ICMI and ICTMA activities.

1. Proceedings of the ICTMA Conference Series

Berry, J. et al. (Eds.). (1984). *Teaching and Applying Mathematical Modelling.* Chichester: Ellis Horwood.

Berry, J. et al., (Eds.). (1986). *Mathematical Modelling Methodology, Models and Micros*. Chichester: Ellis Horwood.

Berry, J. et al. (Eds.). (1987). *Mathematical Modelling Courses*. Chichester: Ellis Horwood.

Blum, W. et al. (Eds.). (1989). *Applications and Modelling in Learning and Teaching Mathematics*. Chichester: Ellis Horwood.

deLange, J. et al. (Eds.). (1993). *Innovation in Maths Education by Modelling and Applications.* Chichester: Ellis Horwood.

Galbraith, P. et al. (Eds.). (1998). *Mathematical Modelling – Teaching and Assessing in a Technology-Rich World.* Chichester: Ellis Horwood.

Haines, Ch. et. al. (Eds.). (2006). *Mathematical Modelling (ICTMA 12): Engineering and Economics.* Chichester: Ellis Horwood.

Houston, S.K. et al. (Eds.). (1997). *Teaching and Learning Mathematical Modelling.* Chichester: Albion Pub.

Lamon, S. J., Parker, W. A., & Houston, S. K. (Eds.). (2003), *Mathematical Modelling: A Way of Life*. Chichester: Ellis Horwood.

Matos, J. F. et al. (Eds.). (2001). *Modelling and Mathematics Education: ICTMA-9.* Chichester: Ellis Horwood.

Niss, M., Blum, W., & Huntley, I. (Eds.). (1991). *Teaching of Mathematical Modelling and Applications*. Chichester: Ellis Horwood.

Sloyer, C., Blum, W., & Huntley, I. (Eds.). (1995). *Advances and* Perspectives in the Teaching of Mathematical Modelling an Applications. Yorklyn: Water Street Mathematics.

Ye, Q., Blum, W., Houston, S., K., Jiang, Q. (2003). *Mathematical Modelling in Education and Culture: ICTMA 10*. Chichester: Ellis Horwood.

2. Proceedings and survey articles of the ICME Conference Series

Bell, M. (1983). Materials Available Worldwide for Teaching Applications of Mathematics at the School Level. In Zweng, M. et al. (Eds.), *Proceedings of the Fourth International Congress on Mathematical Education*, (pp. 252-267). Boston: Birkhäuser.

Blum, W., & Niss, M. (1991). Applied mathematical problem solving, modelling, applications, and links to other subjects – state, trends and issues in mathematics instruction. *Educational Studies in Mathematics 22*(1), 37-68.

Blum, W., Galbraith, P., & Christiansen, I. (Eds.). (2001). Mathematical Modelling – Papers from ICME-9, Part 1. *Teaching Mathematics and its Applications, 20*(3).

Blum, W., Niss, M., & Huntley, I. (Eds.). (1989). *Modelling, Applications and Applied Problem Solving – Teaching Mathematics in a Real Context.* Chichester: Ellis Horwood.

Breiteig, T., Huntley, I., & Kaiser-Meßmer, G. (Eds.). (1993). *Teaching and Learning Mathematics in Context.* Chichester: Ellis Horwood.

Burkhardt, H. (Ed.). (1983). *An International Review of Applications in School Mathematics*. Ohio: ERIC.

Galbraith, P., Blum, W., & Christiansen, I. (Eds.). (2002). Mathematical Modelling – Papers from ICME-9, Part 2. *Teaching Mathematics and its Applications, 21*(2).

Houston, K. (Ed.). (2005). Applications and Modelling in the Teaching and Learning of Mathematics – Papers from ICME-10. *Teaching Mathematics and its Applications, 24*(2/3).

Pollak, H. O. (1979). The Interaction between Mathematics and Other School Subjects. In UNESCO (Ed.), *New Trends in Mathematics Teaching IV*, (pp. 232-248). Paris.

3. ICMI Study 14 Publications

Blum, W. et. al., (2002). ICMI Study 14: Applications and Modelling in Mathematics Education – Discussion Document. *Educational Studies in Mathematics, 51*(1/2), 149-171.

Henn, H.-W. & Blum, W. (Eds.). (2002). *ICMI Study 14: Applications and Modelling in Mathematics Education.* Dortmund: University of Dortmund.

4. Some further publications on modelling & applications

Blomhøj, M., & Jensen, T.H. (2003). Developing mathematical modelling competence: Conceptual clarification and educational planning. *Teaching Mathematics and its Applications 22*(3), 123-139.

Boaler, J. (2002). *Experiencing School Mathematics. Traditional and reform approaches to teaching and their impact on student learning.* Mahwah, NJ: Erlbaum.

Boyce, W. E. (Ed.). (1981). *Case Studies in Mathematical Modeling.* Boston: Pitman Ad. Pub.

Burghes, D., Galbraith, P., Price, N., & Sherlock, A. (1996). *Mathematical Modelling.* Upper Saddle River, NJ: Prentice-Hall.

Burghes, D., Huntley, I. , & McDonald, J. (1982). *Applying Mathematics – A Course in Mathematical Modelling.* Chichester: Ellis Horwood.

Burkhardt, H. (1981). *The Real World and Mathematics.* Glasgow: Blackie and Son.

Bushaw, D. et al. (Eds.). (1980). *A Sourcebook of Applications of School Mathematics.* Reston: NCTM.

Clements, R. et al. (Eds.). (1988). *Selected Papers on the Teaching of Mathematics as a Service Subject.* Berlin/Heidelberg/New York: Springer.

COMAP (1997-1998). *Mathematics: Modeling Our World.* Cincinnati: South-Western Publishing Company.

De Lange, J. (1996). Using and Applying Mathematics in Education. In Bishop, A. et al. (Eds.), *International Handbook of Mathematics Education V.1*, (pp. 49-97). Dordrecht: Kluwer Acad. Pub.

Edwards, D., & Hamson, M. (1996). *Mathematical Modelling Skills.* Basingstoke, Hampshire, UK: MacMillan Press Ltd.

Galbraith, P., & Clathworthy, N. (1990). Beyond Standard Models – Meeting the Challenge of Modelling. *Educational Studies in Mathematics 21*(2), 137-163.

Giordano, F. P., Weir, M. D., & Fox, W. P. (1997). *A First Course in Mathematical Modelling.* Pacific Grove: Brooks.

Howson, G. et al. (Eds.). (1988). *Mathematics as a Service Subject*. Cambridge: Cambridge University Press.

Huntley, I., & James, G. (Eds.). (1990). *Mathematical Modelling – A Source Book of Case Studies*. Oxford: Oxford University Press.

Kaiser, G.: Mathematical Modelling in School – Examples and Experiences. In Henn, H.-W., & Kaiser, G. (Eds.). (2005), *Mathematikunterricht im Spannungsfeld von Evolution und Evaluation. Festband für Werner Blum*, (pp. 99-108). Hildesheim: Franzbecker.

Kaiser, G., Blomhøj, M., & Sriraman, B. (Eds.). (2006). Mathematical Modelling and Applications: Empirical and Theoretical Perspectives. *Zentralblatt für Didaktik der Mathematik, 38*(2).

Kaiser, G., & Willander, T. (2005). Development of Mathematical Literacy – results of an empirical study. *Teaching Mathematics and its Applications, 24*(2-3), 48-60.

Klamkin, M. S. (Ed.). (1987). *Mathematical Modelling: Classroom Notes in Applied Mathematics*. Philadelphia: SIAM.

Klaoudatos, N., & Papastavridis, S. (2004). Context Orientated Teaching. *Teaching Mathematics and its Applications, 23*(4), 155-164.

Lesh, R. A., & Doerr, H. (Eds.). (2003). *Beyond Constructivism: Models and Modelling Perspectives on Mathematics Problem Solving, Learning, and Teaching*. Mahwah: Lawrence Erlbaum.

Lesh, R., & Harel, G. (2003). Problem solving, modelling, and local conceptual development. *Mathematical Thinking and Learning, 5*, 157-190.

Lingefjard, T. (2002). Mathematical modeling for preservice teachers: A problem from anesthesiology. *International Journal of Computers for Mathematical Learning, 7*, 117-143.

MAA, & NCTM (Eds.) (1980). *A Sourcebook of Applications of School Mathematics*. Reston: NCTM.

Niss, M. (1987). Applications and Modelling in the Mathematics Curriculum – State and Trends. *Int. J. for Math. Ed. in Science and Technology 18*, 487-505.

Niss, M. (1992). Applications and Modelling in School Mathematics – Directions for Future Development. In Wirszup, I., & Streit, R. (Eds), *Development in School Mathematics Education Around the World V.3*, (pp. 346-361). Reston: NCTM.

Niss, M. (2001). Issues and Problems of Research on the Teaching and Learning of Applications and Modelling. In Matos, J. F. et al. (Eds.), *Modelling and Mathematics Education: ICTMA-9*, (pp. 72-88). Chichester: Ellis Horwood.

Nunes, T., Schliemann, A. D., & Carraher, D. W. (1993). *Street mathematics and school mathematic*. Cambridge – New York – Oakleigh: Cambridge University Press.

OECD (Ed.) (1999). *Measuring Student Knowledge and Skills – A New Framework for Assessment,* Paris: OECD.

Pollak, H. O. (1997). Solving Problems in the Real World. In Steen, L. A. (Ed.), *Why Nymbers Count: Quantitative Literacy for Tomorrow's America*, (pp. 91-105). New York: The College Board.

Pozzi, S., Noss, R., & Hoyles, C. (1998). Tools in Practice, Mathematics in Use. *Educational Studies in Mathematics 36*(2) 105-122.

Sharron, S. (Ed.). (1979). *Applications in School Mathematics, NCTM Yearbook.* Reston: NCTM.

Skovmose, O. (1994). *Towards a Philosophy of Critical Mathematics Education.* Dordrecht: Kluwer.

Sriraman, B., Kaiser, G., & Blomhøj, M. (Eds.). (2006). Modelling Perspectives from around the World. *Zentralblatt für Didaktik der Mathematik, 38*(3).

Stillman, G., & Galbraith, P. (1998). Applying Mathematics With Real World Connections: Metacognitive Characteristic of Secondary Students. *Educational Studies in Mathematics 36(*2), 157-195.

Swetz, F., & Hartzler, J. (Eds.). (1991). *Mathematical Modelling in the Secondary School Curriculum.* Reston: NCTM.

Verschaffel, L., Greer, B., & De Corte, E. (2000). *Making Sense of Word Problems.* Lisse: Swets&Zeitlinger.

INDEX

NEW ICMI STUDY SERIES

1.	M. Niss (ed.), Cases of Assessment in Mathematics Education. An ICMI Study. 1992 (The 6th ICMI Study)	0-7923-2089-1 HB
2.	M. Niss (ed.), Investigations into Assessment in Mathematics Education. An ICMI Study. 1992 (The 6th ICMI Study)	0-7923-2095-6 HB
3.	G. Hanna (ed.), Towards Gender Equity in Mathematics Education. An ICMI Study. 1996 (The 7th ICMI Study)	0-7923-3921-5 HB 0-7923-3922-3 PB
4.	A. Sierpinska and J. Kilpatrick (eds.), Mathematics Education as a Research Domain: A Search for Identity. An ICMI Study. 1998 (The 8th ICMI Study)	0-7923-4599-1 HB 0-7923-4600-9 PB
5.	C. Mammana and V. Villani (eds.), Perspectives on the Teaching of Geometry for the 21st Century. An ICMI Study. 1998 (The 9th ICMI Study)	0-7923-4990-3 HB 0-7923-4991-1 PB
6.	J. Fauvel and J. van Maanen (eds.), History in Mathematics Education. The ICMI Study. 2000 (The 10th ICMI Study)	0-7923-6399-X HB 1-4020-0942-9 PB
7.	D. Holton (ed.), The Teaching and Learning of Mathematics at University Level. An ICMI Study. 2001 (The 11th ICMI Study)	0-7923-7191-7 HB 1-4020-0072-3 PB
8.	K. Stacey, H. Chick and M. Kendal (eds.) The Future of the Teaching and Learning of Algebra: The 12th ICMI Study. 2004	1-4020-8130-8 HB
9.	F.K.S. Leung, K.-D. Graf and F.J. Lopez-Real (eds.) Mathematics Education in Different Cultural Traditions- A Comparative Study of East Asia and the West: The 13th ICMI Study. 2006	0-387-29722-7 HB
10.	W. Blum, P. L. Galbraith, H.-W. Henn, and M. Niss (eds.), Modelling and Applications in Mathematics Education: The 14th ICMI Study. 2007	0-387-29820-7 HB

Information on the ICMI Study program and on the resulting publications, including Studies 1 to 5, can be obtained at www.mathunion.org/ICMI/ or by contacting the ICMI Secretary-General, whose email address is available on the ICMI website.

Printed in the USA

MIX
Papier aus verantwortungsvollen Quellen
Paper from responsible sources
FSC® C105338

If you have any concerns about our products,
you can contact us on
ProductSafety@springernature.com

In case Publisher is established outside the EU,
the EU authorized representative is:
Springer Nature Customer Service Center GmbH
Europaplatz 3, 69115 Heidelberg, Germany

Printed by Libri Plureos GmbH
in Hamburg, Germany